Undergraduate Topics in Computer Science

Undergraduate Topics in Computer Science (UTiCS) delivers high-quality instructional content for undergraduates studying in all areas of computing and information science. From core foundational and theoretical material to final-year topics and applications, UTiCS books take a fresh, concise, and modern approach and are ideal for self-study or for a one- or two-semester course. The texts are all authored by established experts in their fields, reviewed by an international advisory board, and contain numerous examples and problems. Many include fully worked solutions.

For further volumes:
http://www.springer.com/series/7592

Edwin F. Meyer III · Nickolas Falkner ·
Raja Sooriamurthi · Zbigniew Michalewicz

Guide to Teaching Puzzle-based Learning

 Springer

Edwin F. Meyer III
Baldwin Wallace University
Berea, Ohio, USA

Nickolas Falkner
University of Adelaide
South Australia, Australia

Raja Sooriamurthi
Carnegie Mellon University
Pittsburgh, Pennsylvania, USA

Zbigniew Michalewicz
University of Adelaide
South Australia, Australia

Series Editor
Ian Mackie

Advisory Board
Samson Abramsky, University of Oxford, Oxford, UK
Karin Breitman, Pontifical Catholic University of Rio de Janeiro, Rio de Janeiro, Brazil
Chris Hankin, Imperial College London, London, UK
Dexter Kozen, Cornell University, Ithaca, USA
Andrew Pitts, University of Cambridge, Cambridge, UK
Hanne Riis Nielson, Technical University of Denmark, Kongens Lyngby, Denmark
Steven Skiena, Stony Brook University, Stony Brook, USA
Iain Stewart, University of Durham, Durham, UK

ISSN 1863-7310   ISSN 2197-1781 (electronic)
Undergraduate Topics in Computer Science
ISBN 978-1-4471-6475-3   ISBN 978-1-4471-6476-0 (eBook)
DOI 10.1007/978-1-4471-6476-0
Springer London Heidelberg New York Dordrecht

Library of Congress Control Number: 2014941623

© Springer-Verlag London 2014

This work is subject to copyright. All rights are reserved by the Publisher, whether the whole or part of the material is concerned, specifically the rights of translation, reprinting, reuse of illustrations, recitation, broadcasting, reproduction on microfilms or in any other physical way, and transmission or information storage and retrieval, electronic adaptation, computer software, or by similar or dissimilar methodology now known or hereafter developed. Exempted from this legal reservation are brief excerpts in connection with reviews or scholarly analysis or material supplied specifically for the purpose of being entered and executed on a computer system, for exclusive use by the purchaser of the work. Duplication of this publication or parts thereof is permitted only under the provisions of the Copyright Law of the Publisher's location, in its current version, and permission for use must always be obtained from Springer. Permissions for use may be obtained through RightsLink at the Copyright Clearance Center. Violations are liable to prosecution under the respective Copyright Law.

The use of general descriptive names, registered names, trademarks, service marks, etc. in this publication does not imply, even in the absence of a specific statement, that such names are exempt from the relevant protective laws and regulations and therefore free for general use.

While the advice and information in this book are believed to be true and accurate at the date of publication, neither the authors nor the editors nor the publisher can accept any legal responsibility for any errors or omissions that may be made. The publisher makes no warranty, express or implied, with respect to the material contained herein.

Printed on acid-free paper

Springer is part of Springer Science+Business Media (www.springer.com)

To my Parents: Both PhDs who raised six children, each of whom has earned a PhD. Thanks for continually holding us to such high standards.

E. F. M.

To my wife, Katrina, and my friend, Marita, both of whom believed that there was more to me.

N. F.

To my sons, Anand and Santosh, who are each other's puzzle and solution.

R. S.

To my friend, Paweł Kerntopf, who, during my high-school years, introduced me to the world of mathematical puzzles.

Z. M.

Introduction to Teaching Puzzle-based Learning: What Is It and What Is This Book About?

> *Now more than ever, an education that emphasizes general problem solving skills will be important.* – Bill Gates

What is missing in most curricula – from elementary school all the way through to university education – is coursework focused on the development of problem-solving skills. Most students never learn how to think about solving problems. Usually they are just trained to apply particular formulae to a problem and most of their effort is related to "calculating the answer."

Students are not prepared for *framing* and solving problems that are given in a descriptive form: they have serious difficulties in *extracting* relevant information, *eliminating* noise (that is always present in descriptive-type problems), *building* a model of the problem, and *reasoning* about the solution. Further, throughout their education, they are constrained to concentrate on specific questions at the back of chapters.

So, without much thinking, they apply the material from each chapter to solve a few problems given at the end of each chapter (why else would a problem be at the end of the chapter?). One of our favourite examples to illustrate this point is a puzzle on breaking a chocolate bar:

> A rectangular chocolate bar consists of m × n small rectangles and you wish to break it into its constituent parts. At each step, you can only pick up one piece and break it along any of its vertical or horizontal lines. How should you break the chocolate bar using the minimum number of steps (breaks)?

If you do not know the answer, which textbook would you search to discover the solution? Textbooks on optimization? Simulation? Strategies? Games? Other textbooks? Or it might be that someone wrote a book on chocolates where in Chapter 7 there is a full discussion on efficient breaking strategies of a chocolate bar? Very unlikely. The same applies to solving many real-world problems: which textbook should you search to find an approach that would lead to the solution? Some individuals (including the authors of this book), when interviewing job candidates, would ask them to solve problems during the interview. When a candidate responds, "*I didn't have that in school*," they would reply, "*Yeah, I know you didn't; that's why I'm asking it.*"

In the introduction of *Heard on the Street, Quantitative Questions from Wall Street Job Interviews*,[1] the author writes: "This book bridges the considerable gap between the typical education and the knowledge required to successfully answer job interview questions. The considerable gap arises because interviewers must separate the wolves from the sheep. The sheep are confined by the boundaries of their education. The wolves are not. Of course, most interviewers are wolves. Unfortunately, most interviewees are sheep. The 'butchering' that can take place in these interviews is horrific."

Clearly, it is not surprising that most students are ill prepared for framing and addressing real-world problems. When they finally enter the real world, they suddenly find that problems do not come with associated formulas, instructions, or textbooks. Although many educators are interested in teaching "thinking skills" rather than "teaching information and content," the fact remains that young people often have serious difficulties in independent thinking (or problem-solving skills) regardless of the nature of a problem. As Alex Fisher wrote in his book, *Critical Thinking*[2]: "... though many teachers would claim to teach their students 'how to think', most would say that they do this indirectly or implicitly in the course of teaching the content which belongs to their special subject. Increasingly, educators have come to doubt the effectiveness of teaching 'thinking skills' in this way, because most students simply do not pick up the thinking skills in question." The curricular approach of emphasizing "remembering" over "reasoning" has dominated the educational arena – whether in history, physics, geography, or any other subject – almost ensuring that students never learn how to think about solving problems in general.

Over the past few decades, various people and organizations have attempted to address this educational gap by teaching "thinking skills" based on some structure (e.g. critical thinking, constructive thinking, creative thinking, parallel thinking, vertical thinking, lateral thinking, confrontational and adversarial thinking). However, all these approaches are characterized by a departure from mathematics as they concentrate more on "talking about problems" rather than "solving problems." It is our view that the lack of problem-solving skills in general is the consequence of a decreasing level of mathematical sophistication in modern societies.

It seems that a different approach is needed. Many individuals and educational organizations have recognized this need some time ago. Many instructors have been introducing puzzles and various problem-solving activities in more or less formal way over the last twenty years. For example, in *How to Solve It: Modern Heuristics*[3] the authors introduced a variety of puzzles to support a course on modern heuristic methods. A. Levitin and M. Levitin have been basing their courses on puzzles.[4] Tim Bell from University of Canterbury (New Zealand) incorporated

---

[1] Crack TF (2008) Heard on the street, quantitative questions from Wall Street job interviews, 11th edn. Typeset by the author, USA.

[2] Fisher A (2001) Critical thinking: an introduction. Cambridge University Press, Cambridge.

[3] Michalewicz Z, Fogel DB (2000) How to solve it: modern heuristics. Springer, Berlin/New York.

[4] Levitin A, Levitin M (2011) Algorithmic puzzles. Oxford University Press, Oxford/New York.

puzzles into their Computer Science curriculum. His *CS Unplugged*[5] is a collection of free learning activities that teach Computer Science through engaging games and puzzles. Puzzle-based problem solving has been offered as a for-credit course since 2002 at the Baldwin Wallace University (Ohio, USA) – and now there are two such problem-solving courses that Ed Meyer teaches in the graduate business program (not to mention his teaching of problem solving during the summer at Gedanken Institute for problem solving for many years now). His *Naked Physics*[6] contains 64 problems (from physics and mathematics) to develop students' problem-solving skills. At Carnegie Mellon University, Puzzle-based Learning has been offered as a very popular freshmen seminar since 2009. Abbreviated versions of this course have been offered to range of audiences from middle school, to high-school outreach programs, to graduate school, to managers and engineers in industry workshops, and to retired professionals in continuing education programs.

Clearly, many individuals all over the world experimented with such approaches over many years; however, the term *Puzzle-based Learning* emerged just a few years ago.[7] As it was the case with all earlier attempts, it focuses on getting students to *think* about framing and solving *unstructured* problems (those that are not encountered at the end of some textbook chapter). The idea is to increase the student's mathematical awareness and problem-solving skills by solving a variety of *puzzles* and reflecting on their solution processes.

So what is Puzzle-based Learning?

Puzzle-based Learning is a foundational approach to develop thinking skills, mental stamina and perseverance at solving problems. We focus on unstructured, generally context-free (i.e., does not require domain knowledge) and almost always entertaining problems, better known as puzzles.

Over the years, researchers have developed sets of rules for solving puzzles and problems and it is left to the reader to identify one that works for his or her teaching situation. However, there are a couple of places to start looking. Gyorgy Pólya[8] presented four fundamental steps to problem solving:

1. Understanding the problem (*Recognizing what is asked for*)
2. Devising a plan (*Responding to what is asked for*)
3. Carrying out the plan (*Developing the result of the response*)
4. Looking back (*Checking what does the result tell me*)

and also provided a (large) list of different problem solving approaches (heuristics) that would give the puzzler a starting point or a way to rearrange the problem to

---

[5] See csunplugged.org.

[6] Meyer EF (2011) Naked physics. Gedanken Publishing.

[7] Michalewicz Z, Michalewicz M (2008) Puzzle-based learning: an introduction to critical thinking, mathematics, and problem solving. Hybrid Publishers, Melbourne.

[8] Gyorgy Pólya was born in Budapest on 13 December 1887. For most of his career in the United States, he was a professor of mathematics at Stanford University. He worked on a great variety of mathematical topics, including series, number theory, combinatorics, and probability. In his later days, Gyorgy Pólya spent considerable effort on trying to characterize the general methods that people use to solve problems, and to describe how problem solving should be taught and learned.

make it simpler. Michalewicz and Michalewicz presented a simplified approach in *Puzzle-based Learning*, namely:

1. Understand the problem, and all the basic terms and expressions used to define it
2. Do not rely on your intuition too much; solid calculations are far more reliable
3. Build a model of the problem by defining its variables, constraints, and objectives

Whichever approach you use, and there are many out there, ensure that you know how to explain it and that you can demonstrate it in practice. As we will see, your ability to teach in a Puzzle-based Learning course has little to do with your ability to solve puzzles but far more to do with the way that you can explain your failures to your students! In other words, we believe that the course should be based on the best traditions introduced by Gyorgy Polya and Martin Gardner[9] during the last 60 years. In one of our favorite books, *Entertaining Mathematical Puzzles*,[10] Martin Gardner wrote:

> *"Perhaps in playing with these puzzles you will discover that mathematics is more delightful than you expected. Perhaps this will make you want to study the subject in earnest, or less hesitant about taking up the study of a science for which a knowledge of advanced mathematics will eventually be required."*

Many other mathematicians have expressed similar views. For example, Peter Winkler in his book *Mathematical Puzzles: A Connoisseur's Collection* wrote: "*I have a feeling that understanding and appreciating puzzles, even those with one-of-a-kind solutions, is good for you.*"

As a matter of fact, the Puzzle-based Learning approach has a much longer tradition than just recent approaches or even the last 60 years.[11] The first mathematical puzzles were found in Sumerian texts that date back to around 2,500 BC! The earliest evidence of the Puzzle-based Learning approach can be found in the works of Alcuin, an English scholar born around AD 732 whose main work was *Problems to Sharpen the Young* – a text which included over 50 puzzles. Some twelve hundred years later, one of his puzzles is still used in countless artificial intelligence textbooks![12]

---

[9] Martin Gardner was born in Tulsa, Oklahoma, on 21 October 1914. He is one of the most beloved personalities in the areas of recreational mathematics, magic and puzzles. The influence of his work is immeasurable – he was a prolific popular recreational mathematics and science writer. He wrote the *Mathematical Games* column in *Scientific American* for 25 years. The author of more than 100 books, his favorite puzzles require a sudden insight, which he termed an "*Aha!* moment". To know more about Gardner and his seminal contributions his Wikipedia entry is a good starting point.

[10] Gardner M (1986) Entertaining mathematical puzzles. Dover Recreational Math, New York.

[11] Danesi M (2004) The puzzle instinct: the meaning of puzzles in human life. Indiana University Press, Bloomington.

[12] The puzzle is the "river crossing problem": *A man has to take a wolf, a goat, and some cabbage across a river. His rowboat has enough room for the man plus either the wolf or the goat or the cabbage. If he takes the cabbage with him, the wolf will eat the goat. If he takes the wolf, the goat will eat the cabbage. Only when the man is present, are the goat and the cabbage safe from their enemies. All the same, the man carries wolf, goat, and cabbage across the river. How has he done it?*

Introduction to Teaching Puzzle-based Learning: What Is It and What Is This Book About?

The Puzzle-based Learning approach aims to encourage students to *think* about how to frame and solve descriptive (unstructured) problems. The goal is to motivate students, and to increase their mathematical awareness and problem solving skills by discussing a variety of puzzles and their solution strategies. Besides being a lot of fun, the Puzzle-based Learning approach does a remarkable job of convincing students that (a) science is useful and interesting, (b) the basic courses they are taking are relevant, (c) mathematics is not *that* scary (there is no need to hate it!), and (d) it is worthwhile to stay in school, get a degree, and move into the real world which is loaded with interesting problems (problems perceived as real-world puzzles). These points are important, as most students are unclear about the significance of the topics covered during their studies. Oftentimes, they do not see a connection between the topics taught (e.g. linear algebra) and real-world problems, and they lose interest with predictable outcomes. We also believe that the main reasons behind most students' enthusiasm for Puzzle-based Learning are:

- Puzzles are engaging and thought-provoking.
- Puzzles are educational, but they illustrate useful (and powerful) problem-solving rules in a very *entertaining* way.
- Contrary to many textbook problems, puzzles are not attached to any chapter (as is the case with real-world problems).
- It is possible to talk about different techniques (e.g. simulation, optimization), disciplines (e.g. probability, statistics), or application areas (e.g. scheduling, finance) and illustrate their significance by discussing a few simple puzzles. At the same time, the students are aware that many conclusions are applicable to the broader context of solving real-world problems.
- Puzzles provide an opportunity to experience a *Eureka!* moment. When students attempt a puzzle, it is not unusual for them to be dumbfounded for 10+ minutes. They read it, reread it. Then they start to slowly chip away at it. Frame it, understand it. Wrap their head around it. Think hard. Think some more. Then a *Eureka!* moment is reached (Martin Gardner's *Aha!*[13]), when the correct path to solving the puzzle is recognized. The *Eureka* moment is accompanied by a sense of relief: the frustration that was felt during the process dissipates and the problem-solver may feel a sense of reward at their cleverness for solving the puzzle. Students unconsciously startle themselves out of their deep thinking with a jump and an audible outcry with eyes wide open. Most other classes do not provide opportunities for these moments. When a student experiences a success like this, it is a life-changing moment. They are proud, confident, and ready to take more challenges.[14]

Puzzles can lay a foundation to shift the curricular emphasis from remembering to reasoning. Puzzles can play a major role in engaging students and can be used in talks to high school students and during open-day events. Puzzles can also be a

---

[13] Gardner M (1978) Aha! insight. W H Freeman & Co.
[14] To our amusement, we have seen that students sometimes remember the puzzles used to illustrate a concept more than the concept itself!

factor in retaining and motivating students. Our pedagogical goal is to use puzzles as a means to an end; as a means for developing critical thinking and problem-solving skills as well as raising the profile and importance of mathematics.

Over the past several years, all the authors have taught full semester long courses with the theme of Puzzle-based Learning. In addition we have offered Puzzle-based Learning themed workshops in a number of professional and educational settings. Several of our colleagues have expressed interest in offering such a course at their own institution, but given the novelty of this approach a key question we would often get is: *How to teach Puzzle-based Learning*? Are there any general teaching strategies of Puzzle-based Learning? How to organize the class of students? Should we set up puzzle clubs? Are there any special class activities that can be incorporated into Puzzle-based Learning courses? In particular, should we use some on-line activities? How to organize the material in the most meaningful way? How to set assignments? What about effective assessment? How can we increase and maintain confidence among students? Is there any merit in peer teaching? How to present the material for this course? Which puzzles should be selected as warm-up exercises? And the list of questions goes on and on.

The main goal of this book is just to address these questions. This book is a *guide* for teaching Puzzle-based Learning. In this volume we put together our collective experiences of teaching Puzzle-based Learning over many years: in many different countries, in many different formats, and for different class sizes. The book is organized into three parts.

Part I provides additional motivation and information on the Puzzle-based Learning approach, from *Why Teach Puzzle-based Learning?* to *General Teaching Strategies*. In this part we discuss models for student engagement, setting up puzzle clubs, hosting a puzzle competition, and various warm-ups activities. This part concludes with an overview of effective teaching approaches that are commonly used in Puzzle-based Learning. We discuss a variety of class activities, assignment settings and assessment strategies, including peer teaching opportunities.

Part II concentrates on problem solving strategies. We start with some discussion related to issues of framing the problem (e.g. understanding the problem, building a model, drawing a diagram) and then we continue with a discussion on various solving strategies (e.g. reasoning backwards, simplify, iterate and increment). Each chapter that illustrates a solving strategy is kept in the same format – apart from a brief introduction, we present 5 or 6 puzzles/problems that illustrate the strategy (so we work through the puzzle). Further, each of these puzzles is discussed in a similar format – from learning objectives and puzzle statements, through teaching preparations (if needed) and teaching strategy (step by step guidance thru a classroom situation) to the final debriefing. Also, when appropriate, we included short paragraphs on teacher tips (what can go wrong and how to avoid it) and student pitfalls (e.g. typical mistakes, misinterpretations).

Part III contains a collection of puzzle sets that can (and should) be used during the Puzzle-based Learning event. The selection of some of these puzzles may depend on a variety of factors that range from the instruction level of the event (e.g. a seminar for a general audience vs. high-school students vs. engineering students) to its duration (e.g. a separate talk vs. full-semester course). These puzzle sets are organized into a few subsets – we start with a collection of puzzles that require probabilistic reasoning and continue with logic and geometry puzzles. The final chapter includes several puzzles that may provide a significant challenge even to experienced problem-solvers...

We hope this new book would be of interest and of assistance to many instructors (on all levels) who have experimented or plan to experiment with the Puzzle-based Learning approach. We would like to express our gratitude to many "generations" of students at Carnegie Mellon University, Baldwin Wallace University, and the University of Adelaide, where Puzzle-based Learning courses have been offered on a regular basis. In particular, Ed Meyer would like to thank the students of Baldwin Wallace University for being so eager to tackle challenging problems. Their valuable feedback has allowed him to fine-tune the presentation of the problems as well as the problems themselves. Raja Sooriamurthi would like to thank his friend and colleague, Randy Weinberg, for his ardent support for introducing Puzzle-based Learning in our curriculum and beyond. Finally, Nick Falkner and Zbigniew Michalewicz would like to thank Peter Dowd, the former Dean of the Faculty of Engineering, Computer and Mathematical Sciences, for introducing Puzzle-based Learning in all Schools of the Faculty at the University of Adelaide, and Paweł Nowacki, the Chancellor of the Polish-Japanese Institute of Information Technology, who introduced the course in his Institute. Further, a number of people, in different capacities, have influenced the production of this book and its content. So we thank many instructors from all over the world who shared with us their experiences, comments, suggestions, and insights. Also, our thanks go to Samantha Meyer, Meridith Witt, Edwin F. Meyer II, Thomas Weise, Lizbeth J. Phillips, Claudia Szabo, and Peer Johannsen, who provided excellent comments on an earlier draft of this text. Finally, our thanks are due to Simon Rees and Wayne Wheeler from Springer's office in London. Wayne observed the enthusiastic reception of a Puzzle-based Learning workshop we conducted at SIGCSE 2012 (Raleigh, North Carolina) and extended an invitation to write this book. We thank them both for their support and patience during the long gestation of this project.

We certainly stand on the shoulders of many giants. Unfortunately, it is not easy to give full credits to all contributors – as in most cases it is difficult to trace the origin of a puzzle and acknowledge the inventor. Many puzzles (often in slightly different form) have surfaced many times in many different places, while others were simply passed on as word of mouth. This notwithstanding, we would like to acknowledge several puzzles that were published earlier in a variety of sources. Many puzzles were found in journals (e.g. *The American Mathematical Monthly* or *Scientific American*), while others were adapted from books by Martin Gardner, *My Best Mathematical and Logic Puzzles* and *Entertaining Mathematical Puzzles*, and from other books: *How to Lie with Statistics*, by Darrell Huff; *Which Way Did the Bicycle Go?*, by Joseph D. E. Konhauser, Dan Velleman, and Stan Wagon; *536 Puzzles and curious problems* and *The Canterbury Puzzles*, by Henry Ernest Dudeney; *The Gedanken Institute Book of Puzzles*, by Edwin F. Meyer III and Joseph R. Luchsinger; *Fifty Challenging Problems in Probability with Solutions*, by Frederick Mosteller; *Mathematical Puzzles: A Connoisseur's Collection*, by Peter Winkler; *The Moscow Puzzles*, by Boris A. Kordemsky; *Puzzles for Pleasure*, by Barry R. Clarke; *Innumeracy: Mathematical Illiteracy and Its Consequences*, by John Allen Paulos; *One Hundred Problems in Elementary Mathematics*, by Hugo Steinhaus; and *The Lady or the Tiger? and Other Logic Puzzles* by Raymond Smullyan.

In its many forms (one hour talk to semester long course; middle school to senior citizens), we all very much enjoy teaching Puzzle-based Learning. We hope that this book will facilitate many instructors to experiment with the Puzzle-based Learning approach in their teaching curricula.

Enjoy!

References

1. Crack TF (2008) Heard on the street, quantitative questions from Wall Street job interviews, 11th edn. Typeset by the author, USA. http://www.amazon.com/Heard-Street-Quantitative-Questions-Interviews/dp/0970055234
2. Danesi M (2004) The puzzle instinct: the meaning of puzzles in human life. Indiana University Press, Bloomington
3. Fisher A (2001) Critical thinking: an introduction. Cambridge University Press, Cambridge
4. Gardner M (1978) Aha! insight. W H Freeman & Co, New York
5. Gardner M (1986) Entertaining mathematical puzzles. Dover Recreational Math, New York
6. Levitin A, Levitin M (2011) Algorithmic puzzles. Oxford University Press, New York/New York
7. Michalewicz Z, Fogel DB (2000) How to solve it: modern heuristics. Springer, Berlin/New York
8. Michalewicz Z, Michalewicz M (2008) Puzzle-based learning: an introduction to critical thinking, mathematics, and problem solving. Hybrid Publishers, Melbourne
9. Meyer EF (2011) Naked physics. Gedanken Publishing, Berea, Ohio

Contents

Part I Motivation and Teaching

1 Motivation 3
    References 11

2 Getting Started 13
    2.1 The Instructor 13
    2.2 Motivating Students 14
    2.3 Hosting a Puzzle Contest 17

3 Icebreakers 21
    Reference 35

4 Effective Teaching Approaches 37
    4.1 Cognitive Apprenticeship 38
    4.2 Class Activities 42
    4.3 Online Activities 49
    4.4 Measuring and Assessing Puzzle-based Learning 52
    4.5 Effective Assessment 57
    4.6 Increasing and Maintaining Confidence 62
    4.7 Peer Teaching 62
    Reference 63

Part II Tools, Tips, and Strategies

5 Understand the Problem 69
    5.1 Take Inventory 69
    5.2 Build a Model 73
    5.3 Draw a Diagram 83

6 Reasoning: Logic and Reasoning Backwards 95

7 Pattern Recognition 107
    Reference 121

8	**Enumerate and Eliminate**	123
9	**Simplify!**	137
10	**Perform a Gedanken: "What If?" and "So What?"**	149
11	**Simulation and Optimization**	159
	11.1 Simulation	160
	11.2 Optimization	170
	Reference	184

Part III Challenges

12	**Probabilistic Reasoning**	187
	Reference	232
13	**Logical Reasoning**	233
14	**Geometric Reasoning**	259
15	**Grand Challenges**	287
	Reference	334

Summary	335
List of Puzzles	337
Index	343

Part I
Motivation and Teaching

In designing a new course, instructors have three concerns: (1) What knowledge and skills should students learn? (2) How can I facilitate their learning? (3) How do I determine how well they have learned via formative and summative feedback? These three concerns are viewed under the umbrella that pedagogy is all about *learning* and not teaching. As with the role puzzles play in Puzzle-based Learning, teaching per se is just a means to the end of effective learning.

Chapter 1 sets the tone for the rest of the book building upon the introduction. It provides further motivation for teachers as to why they should consider teaching a course on puzzle-based learning. From the interview process of companies to current research on System 1 and System 2 thinking, we discuss our perspective of the continuum of Project-based, Problem-based, and Puzzle-based Learning.

At workshops on education-themed conferences, we often get the question *"How can I start teaching Puzzle-based Learning at my university?"* Chapter 2 suggests various possibilities to get started. A crucial component is communicating the joy and value of Puzzle-based Learning. We conclude with a discussion of running a puzzle contest which is a fun activity both for organizers and participants.

Perhaps more than any other course, Puzzle-based Learning makes the instructor and students feel vulnerable as they have to expose not just their final answers but more importantly their step-by-step reasoning. It is very important to establish a supportive classroom environment of constructive critique and mutual exploration.

Chapter 3 discusses various icebreakers we have used in our classes to infuse a sense of excitement of things to come and to establish an ambience of cognitive camaraderie.

From elementary school to senior citizens, from 15-people seminars to 300-people lectures, from one-hour talks to full-semester courses, we have, over the past several years, successfully taught puzzle-based learning in a range of educational settings. Chapter 4 shares our collective experience in teaching Puzzle-based Learning. Whereas each teacher will have their own pedagogical principles and style, we share what we have used in our own classes to form a basis that could be molded to suit one's own academic environment.

Motivation 1

> *If you want to build a ship, don't drum up people to collect wood and don't assign them tasks and work, but rather teach them to long for the endless immensity of the sea.*
> – Antoine de Saint-Exupery

Consider the following puzzles. Some of the solutions to these are discussed in detail in further chapters. For now, just ponder the puzzles themselves.

- Given two eggs, for a 100-story building, what would be an optimal way to determine the highest floor, above which an egg would break if dropped?
- Suppose you buy a shirt at a discount. Which is more beneficial to us: apply the discount first and then apply sales tax to the discounted amount or apply the sales tax first and then discount the taxed amount? What do stores do?
- If you have a biased coin (say, comes up heads 70 % of the time and tails 30 %), is there a way to work out a fair, 50/50 toss?
- A $10 gold coin is half the weight of a $20 gold coin. Which is worth more: a kilogram of $10 gold coins or half a kilogram of $20 gold coins?
- A farmer sells 100 kg of mushrooms for $1 per kg. The mushrooms contain 99 % moisture. A buyer makes an offer to buy these mushrooms a week later for the same price. However, a week later, the mushrooms would have dried out to 98 % of moisture content. How much will the farmer lose if he accepts the offer?
- If you heat a metal washer with a hole in the middle, what happens to the size of the hole?

What is common to all of the above? Apart from being fun to ponder, solutions to these puzzles exemplify several problem-solving heuristics.[1] What general problem-solving strategies can we learn from the way we solve these puzzles?

---

[1] A heuristic is an easy but imperfect way of answering hard questions. Emerging from experience, heuristics may provide satisfactory solutions via mental shortcuts that ease the cognitive load. The word has the same etymological root as Eureka, emerging from the Greek word *heuriskein* meaning to find or discover.

There are two main reasons to incorporate Puzzle-based Learning in schools' curricula:

1. Puzzles are *autotelic*; they are inherently *fun*. As Marcel Danesi discusses,[2] we humans are wired to solve puzzles: "*The puzzle instinct is, arguably, as intrinsic to human nature as is humor, language, art, music, and all the other creative faculties that distinguish humanity from all other species.*" It is natural for people to want to explore puzzles and experience both the tension and exhilaration of figuring things out. A class on Puzzle-based Learning is designed to help the student experience this joy.

2. As entertaining and engaging puzzles inherently are, they are just a means to our pedagogical end of fostering general domain-independent reasoning and critical thinking skills that can lay a foundation for problem-solving in future course work. Problem-solving is regularly identified not only as one of the key *skills* required in successful employees, but also it represents a general skill that will be used in all aspects of life from financial problems, through relationships to all matters of daily decisions. Puzzles can lay a foundation for acquiring and developing this skill.

Puzzle-based Learning is rapidly becoming a bigger and bigger part of the curriculum as there is no guarantee that a traditional education will provide students with enough practice and experience to develop problem-solving skills. The rapidly changing face of employment and technology means that the problems that we train people to solve today are probably not the problems that they will be solving in ten years. When our current education system tends to favor highly focused learning of rigid approaches to predictable problem sets, there is no guarantee that our students will be flexible enough and resilient enough to cope with open-ended problems with no guaranteed solution.

William Poundstone[3] in chronicling the interview process at Silicon Valley and other technology companies highlights the same: "*Why use logic puzzles, riddles, and impossible questions? The goal of Microsoft's interviews is to assess a general problem-solving ability rather than a specific competency. At Microsoft, and now at many other companies, it is believed that there are parallels between the reasoning used to solve puzzles and the thought processes involved in solving the real problems of innovation and a changing marketplace. [...] When technology is changing beneath your feet daily, there is not much point in hiring for a specific, soon-to-be-obsolete set of skills. You have to try to hire for general problem-solving capacity, however difficult that may be. [...] Both the solver of a puzzle and a technical innovator must be able to identify essential elements in a situation that is initially ill-defined. It is rarely clear what type of reasoning is required or what the*

---

[2] Danesi M (2004) The puzzle instinct: the meaning of puzzles in human life. Indiana University Press, Bloomington.

[3] Poundstone W (2004) How would you move Mount Fuji?: Microsoft's cult of the puzzle – how the world's smartest companies select the most creative thinkers. Little Brown and Company, Boston.

precise limits of the problem are. The solver must nonetheless persist until it is possible to bring the analysis to a timely and successful conclusion."

Teachers are often required to conform to the combined constraints of a fixed, or strongly prescribed, curriculum and a repeatable and predictable assessment scheme, where achievement can be clearly (if not accurately) quantified in neat percentages. How can we allow students the opportunity to explore problems that may not have solutions,[4] if we can only give them "full marks" when they find the "correct" solution? By teaching with Puzzle-based Learning, a teacher can introduce a number of opportunities for students to develop their problem-solving skills, while still conforming to the curriculum, because Puzzle-based Learning is a set of techniques and strategies that may be applied across a wide variety of courses.

Puzzle-based Learning doesn't have to be isolated to an individual course but can be integrated throughout an entire program. For example, in software development, we go through various phases such as analysis, design, development, implementation, deployment, evaluation, and maintenance. Analysis is all about understanding the problem and the requirements of the client. In courses on System Design, we have used the following puzzle as a good example of this type of reasoning – if we understand the process involved, the correct solution emerges[5]:

Three backpackers cooked rice for dinner. The first one gave 400g of rice and the second – 200g of rice. The third one did not have any rice so he gave $6 to the other two. How should they divide the $6 in a fair way (assume they equally shared the dinner)?

The ability to reason correctly from available information is vital to all professions and all disciplines – it's not restricted to STEM[6] majors only. At its heart, Puzzle-based Learning is about not only reasoning and thinking but using the mechanism of puzzles to make the approach interesting and to free the student from having to carry around too much domain knowledge. Indeed, the authors have presented seminars at various schools and conferences on how to infuse Puzzle-based Learning throughout the curriculum.

The ultimate goal of Puzzle-based Learning is to lay a foundation for students to be effective problem-solvers in the real world. At the highest level, problem-solving in the real world calls into play three categories of skills: dealing with the vagaries of uncertain and changing conditions, harnessing domain-specific knowledge and methods, critical thinking and applying general problem-solving

---

[4] In hiring interviews, they are often known as *impossible questions* or *Fermi questions* based on estimation techniques popularized by the Nobel Prize winning physicist, Enrico Fermi. Some questions don't have precise answers, e.g., "how many piano tuners are there in Chicago?" and some questions are to be approached with back-of-the-napkin style estimates, "what is the circumference of the earth?"

[5] When we have used this puzzle in class and elsewhere, we've seen the audience suggest 3 or 4 solutions before converging on the correct solution once they have analyzed and understood the actions described in this puzzle. This puzzle is discussed later in the book.

[6] STEM education is an acronym for the fields of study in the categories of science, technology, engineering, and mathematics.

strategies. These three skill categories are captured in the three forms of learning – project, problem, and puzzle based – depicted below:

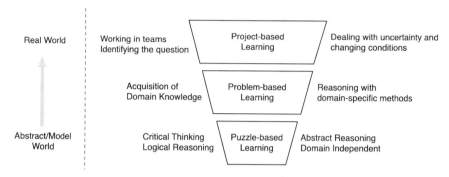

In the continuum depicted in the figure above, each layer of skills builds upon the layers below it. Both project-based learning and problem-based learning are well-established methodologies. By our description above, project-based learning deals with complex situations where usually there is no one clear unique or correct way of proceeding, for example: How can we increase the adherence of cystic-fibrosis patients to follow their treatment protocol? It can be very hard to determine the best solution. The pedagogical objectives of project-based learning include dealing with ambiguity and complexity, integration of a variety of approaches, user-testing of the value of proposed solutions, and working with a team of people with diverse backgrounds and skills. Problem-based learning on the other hand requires significant domain knowledge. This is the form of learning one typically sees emphasized in a domain-specific undergraduate course such as electromagnetism, data structures, circuit theory, etc. In both problem- and project-based learning, the problem drives the learning: students need to assess what they already know, what they need to know to address the problem, and how to bridge the knowledge/skill gap. Puzzle-based Learning focuses on domain-independent transferable skills of critical thinking and abstract reasoning. In addition, Puzzle-based Learning aims to foster introspection and reflection on the personal problem-solving process. What was I thinking? What is the solution? Why did I not see it? This leads to the question: What is the difference between a puzzle and a problem? One way of characterizing the difference is the extent to which domain-specific knowledge is needed to solve it. The general flavor of puzzles is that their solution should only require domain neutral general reasoning skills – a biologist, a musician, and an artist should all be able to solve the same puzzle. The different styles of reasoning required for problem-based learning and Puzzle-based Learning could be compared to the difference between an in-the-field investigator and an armchair detective – one only requires reason. Having said this, you will find the term "puzzle" and "problem" used relatively interchangeably throughout the book. The core of our approach is to encourage reasoning across areas that are more than just purely analytical or methodical. When we add a puzzling aspect to a problem or require domain knowledge for a puzzle, it can be hard to really differentiate. We encourage the reader to use whichever term that they find most helpful in achieving their goals and developing their own understanding.

A great deal of research has been carried out in uncovering student behavior and in an attempt to understand what motivates "good" students to perform well, while also understanding why so-called "bad" students underperform. Rather than depend upon a simple argument of innate ability, we believe that the environment that we provide for our students has significant impact upon their ability to engage and succeed with our courses. As part of this, the research has uncovered that the most effective way to get anyone to engage with a thinking task is to let them motivate themselves, rather than praising or punishing them to get them to work "correctly." When a student is working well, and achieves something for himself or herself, then there are inbuilt psychochemical rewards that will provide more than enough benefit, without us handing out gold stars.

During the evolution of humans, survival was predicated on identifying and understanding cause-and-effect relationships. Because of this, when a new level of understanding is reached, the neurotransmitter dopamine is released in the brain. It is the dopamine that is responsible for the pleasure and excitement of the *Aha!* moment. As a teacher, a key goal is to provide opportunities for students to have these *Aha!* moments by giving them challenging problems and nudging them but immediately giving them the solution and thus depriving them of the joy of finding things out for themselves.

One of the challenges we face with traditional problems is that we often overlook the *process* by which the result is obtained. If the correct result is achieved, then full marks are available. In some cases, even with a correct answer, a non-recommended process is actively *penalized*, which further restricts a student's creativity and flexibility in solving problems. While we could make a long psychological argument to demonstrate why this is bad, we choose to use a simple analogy of riding a bicycle.

If all you ever do is ride a bike on the sidewalk on a sunny day at low speed, and that is the only activity you'll be rewarded for, then how does this prepare you for a cross-country bicycle race in a storm? It's not enough to get practice at the skill of turning the pedals; you need practice at hill riding, braking, and turning in wet conditions, and, on top of this practice, you then need to develop experience so that you can combine your increased skills with your increased knowledge to put it all together. Each rider will attack the course in their own way, even though the course is clearly defined, because they will bring a different perspective to the problem. As long as they stay within certain bounds and finish the race legitimately, nobody really cares exactly where they applied the brakes, skidded, jumped, or turned the pedals. Real-world problems are unstructured and ill-defined, lacking sometimes even the comforting side barriers of that downhill bike track. Telling students that all that matter is the output can make them highly risk-averse, fearful of making mistakes because only the right answer will get them passing marks. Telling students that they must always use a fixed set of steps, or a certain process, to solve problems will restrict their creativity or, worse, make them unable to even start working on a problem unless they have seen it before. The world is full of problems that have not yet been solved, may not be solved by existing techniques, and may not even have a solution. A well-trained cyclist with lots of experience will be able to make the best of an unknown track. We owe at least this much to our students when we try to develop them as problem-solvers, and this is one of the strengths of teaching Puzzle-based Learning: we focus on exploring ways to solve problems to produce students who are willing to take risks and can be creative because they are not locked into fixed patterns.

The advent of the modern personal computer has reduced the need for the human brain to memorize facts and to perform repetitive calculations. The Internet has lots of information stored along with a great deal of computational power. The skills that an education should develop are the skills that compliment rather than compete with those of a computer. Our education system often "teaches" students by giving them answers to remember rather than problems to solve. In this era, we have a wonderful device that has a great ability to remember and retrieve facts – a computer. But, computers are not (yet) creative problem-solvers when compared to the potential of humans. Computers don't have original ideas. Computers are not brilliant. Humans, however, have great potential for brilliance, for creativity, and for having new thoughts and new ideas. Solving new, challenging problems like the ones in this book will develop these skills.

One of the other advantages of practice is that it helps the development of mental stamina and this is as true for problem-solving as it is for any other area. One of the greatest issues many educators face is that students either won't start an activity, because they don't see the point of it, or won't persevere at an activity, because they lack the mental stamina and insight to see that continuing to work at a problem may ultimately reap rewards. However, this often reflects a lack of resilience in our students when faced with threatening activities or the possibility of failure. In a traditional assessment environment, when the wrong answer looms over a final grade or an unrecognized process can make a right answer incorrect, there is no capacity to develop risk-taking behavior because we are actively punishing students when they step outside of very fixed guidelines. Similarly, if we are not practicing risk-taking and giving students an opportunity to fail gracefully and without huge penalty, it is hardly surprising that they don't get enough opportunity to develop resilience to failure. In traditional assessment, repeated failure will naturally exclude a student for further progress – which then robs them of any benefit of resilience. And yet, the ability to experiment, the capacity to take risks, and the resilience to fail and then get back up again are absolutely vital in employment and life. This is the world for which we have to prepare our students.

A student's first reaction to a problem is often instinctive. Daniel Kahneman terms this as *System 1* thinking process.[7] System 1 thinking is subconscious reasoning, intuitive, automatic, associative, metaphorical, and low-effort. Questions that are answered by System 1 thinking are "What is $2+2$?" and "What is the capital of France?" *System 2* thinking is conscious reasoning, explicit,

---

[7] Daniel Kahneman gives a fascinating account of his Nobel Prize winning work in his book, *Thinking, fast and slow*, Farrar, Straus and Giroux, 2013.

1 Motivation

slow, rule-based, and high-effort. A simple example of System 2 reasoning is determining 19×37. Evolutionarily System 2 thinking is recent and specific to humans. Unfortunately, by the way we humans are wired, System 2 thinking tends to be slothful and tires easily. Our first reaction to a problem is System 1 reasoning. Much of the time, it is correct and has contributed towards the survival of our species. But the more deliberative System 2 reasoning is what has contributed towards the successful achievements of our species.

By the nature of these two systems of thinking and their interplay when attempting to solve a problem, intuitive reasoning of System 1 comes into play. As when ones skill and experience in *some* domains increases, System 1 gets better at giving a right answer (the 10,000 hour training in predictable, rapid-feedback, environment, e.g., chess, piloting, music).[8] But, the literature is replete with examples of how an intuition can lead us astray – numerous reasoning fallacies and cognitive biases (e.g., anchoring effect, framing effect, availability bias, etc.). Kahneman's book discusses these in more detail with astonishing real-world examples.

Intuition is a double-edged sword. A rule for problem-solving from Michalewicz and Michalewicz which we espouse is: Do not rely on your intuition *too much*; solid calculations are far more reliable. As pioneer and teacher of problem-solving extraordinaire Gyorgy Polya eloquently states *"Finished mathematics consists of proofs; but mathematics in the making consists of guesses."*[9] Our summary view is: first guess (System 1) and then prove (System 2).

When System 1 reasoning is not sufficient to come up with an immediate answer, or even when System 1 reasoning does come up with an answer, given how prone it is to be wrong, we need to resort to System 2 reasoning. Unfortunately, by its slothful nature, System 2 reasoning does not get utilized to the level that it could.[10] When this fails, or is perceived as too complicated to undertake immediately, many students stop. This is when we would like students to pause, reflect, and reorganize their thinking (reactivate their System 2 thinking process) and then try again. But, too often, this is the point at which many students decide that the problem is impossible and they wait for the teacher to cue them or to move on to another activity. What is important here is to realize that teachers are also set up to

---

[8] For an entertaining account of this skill please see Malcolm Gladwell's book *Blink: The Power of Thinking without Thinking*, Back Bay Books, 2007.

[9] In 1966, the Mathematical Association of America (MAA) filmed a university class of Polya where he led a discussion on solving the five-plane problem (how many parts are created when a three-dimensional space is cut by five planes). The video recording, *Let us teach guessing: A demonstration with George Polya*, is highly recommended for both its content and to see the maestro at work. A local library may have a copy of this recording. At the time of this writing, it is also available online at http://vimeo.com/48768091

[10] An interesting phenomenon termed the *Einstellung Effect* (related to the confirmation bias) is partly due to the slothful nature of System 2 thinking. This effect is about our thinking tendency to stick with a familiar solution to a problem – the one that first comes to mind (via System 1) – and to ignore alternatives (which could arise from System 2 thinking). The interested reader is referred to the article *Why Good Thoughts Block Better Ones*, Scientific American (March 2014), 310.

deal with students in this way, when we think that a student isn't understanding what is going on or is not going to "get it," we either move on to something else or simply give the student the answer. We want the student *and* the teacher to persevere, to keep trying and to approach every problem as an opportunity to learn *even when the problem isn't solved*. As Einstein said, "It's *not* that *I'm* so *smart*, it's just that I stay with problems longer," and this is advice that everyone can learn from. The puzzle-based approach gives us a framework to work on problems that reward exploration and persistence, but in a way that is not as threatening or high stakes as many larger, domain-specific problems. Sometimes finding the right solution to a problem is to go through all of the incorrect ones first, but this is impossible without mental stamina, a willingness to take risks, and an overall resilience in the face of failure.

One of the other advantages of teaching with Puzzle-based Learning is that it provides a good basis for exposing fallacies of System 1 thinking in how we reason about problems. The best way to illustrate such an error in thinking is to catch someone in the middle of making the mistake and make them aware of what they're doing. Given the instinctive way that many students first approach a problem, there are many opportunities to catch these errors. Common cognitive fallacies include the gambler's fallacy, where too much weight is given to previous events in statistically independent events (e.g., suppose you toss a fair coin nine times and each time it comes up heads. What is the probability that the 10th toss will also come up heads?), or the framing effect, where the way that the question is asked changes the way we think about the data (a procedure with 90 % survival rate is preferred to one with a 10 % mortality rate).[11] Such examples, which read as rather dry and mathematical (and have the "won't happen to me" characteristic), become much more relevant when we can identify someone making the mistake as they make it. This is also a cognitive bias in itself, where students do not think about the need to seek help until they are having a problem, which means that start-of-term lectures on the campus help line and psychologists fall on relatively deaf ears and are not remembered when problems actually occur some 6–10 weeks down the track.

Most of our assessment mechanisms are designed to provide an indication of the degree to which a student can perform a task or demonstrate knowledge. For a number of (very good) pedagogical and practical reasons, the behaviors and processes that are rewarded are those that are defined as productive, saving the highest reward for the production of a known right answer. The problem that many teachers have in assessing reasoning and problem-solving behavior is that the point

---

[11] Perhaps the most famous of these errors is the conjunction fallacy as represented in "The Linda problem" originated by Kahneman and his collaborator Tversky. *Linda is 31 years old, single, outspoken, and very bright. She majored in philosophy. As a student, she was deeply concerned with issues of discrimination and social justice, and also participated in anti-nuclear demonstrations.* Which is more probable? (1) Linda is a bank teller or (2) Linda is a bank teller and is active in the feminist movement. Majority of those asked, including students from a top business school, incorrectly choose option 2, even though every feminist bank teller is a bank teller; adding a detail can only lower the probability.

at which the correct solution is found may take place after a large amount of previously fruitless activity. If we interrupt this process before the outcome is produced, how do we assess the effort involved in thinking about a problem? Another advantage of the Puzzle-based approach is that the work required to correctly frame the problem and to select an appropriate solution strategy allows a student to demonstrate to both the instructor and themselves that some activity has taken place. It is very hard to determine, based on what is produced, exactly how much effort is being applied to solve a problem, but the techniques discussed in Puzzle-based Learning allow us to show progress towards a state where the puzzle or problem is ready to be solved. As noted, the process of puzzle and problem-solving can be very sensitive to the amount of time available, and this is at odds with the usual method of providing a very fixed (and often quite limited) time to undertake an assignment.

In summarizing our motivation for teaching Puzzle-based Learning, the ultimate goal is to lay a foundation for students to be effective problem-solvers in the real world. More specifically, our pedagogical objectives for Puzzle-based Learning are to introduce students to

- A range of general problem-solving strategies that transcend disciplines
- Introspection and the value of meta-level reasoning of one's problem-solving process
- Transference and the ability to reapply a prior result or method in a new context

We aim to provide a didactic framework that engages, educates, and motivates students to become better problem-solvers.[12]

Puzzle-based Learning shares many of the pedagogical goals of the emerging foundational paradigm of Computational Thinking (CT).[13] Puzzle-based Learning resonates with the Computational Thinking emphasis on abstraction and analytical thinking. With reference to the figure presented earlier, Computational Thinking straddles the whole problem skill spectrum but places more emphasis on problem-based and project-based learning. With its emphasis on domain-independent, rigorous, and transferable reasoning, we believe that Puzzle-based Learning lays a basis for CT in the curriculum.

References

1. Bilalić M, McLeod P (2014) Why good thoughts block better ones. Sci Am 310:74–79
2. Danesi M (2004) The puzzle instinct: the meaning of puzzles in human life. Indiana University Press, Bloomington

---

[12] In addition to the pedagogical value discussed in this chapter and illustrated in the remainder of this book, solving puzzles also provides mental health benefits. See the book *The Playful Brain: The Surprising Science of How Puzzles Improve Your Mind*, Riverhead Trade, 2010. Neurologist and neuropsychiatrist Richard Restak and master puzzle creator Scott Kim discuss how puzzle-solving improves three aspects of our mind: memory, perception, and cognition.

[13] Wing J (2006) Computational thinking. Commun ACM 49(3):33–35.

3. Gladwell M (2007) Blink: the power of thinking without thinking. Back Bay Books, Boston
4. Kahneman D (2013) Thinking, fast and slow. Farrar, Straus and Giroux, New York
5. Poundstone W (2004) How would you move Mount Fuji?: Microsoft's cult of the puzzle – how the world's smartest companies select the most creative thinkers. Little Brown and Company, Boston
6. Restak RM, Kim S (2010) The playful brain: the surprising science of how puzzles improve your mind. Riverhead Trade, New York
7. Wing J (2006) Computational thinking. Commun ACM 49(3):33–35

Getting Started 2

> *Well begun is half done.*
> – Aristotle

When we discuss with colleagues our motivation and experience in teaching Puzzle-based Learning, a question that quickly follows from those interested in exploring this paradigm further is: *How can I do this in my university?* Given our engagement with teaching Puzzle-based Learning in a range of settings and to a range of audiences, in this chapter we discuss how an instructor could start teaching Puzzle-based Learning and also how to initiate students to such course.

2.1 The Instructor

At our respective institutions, a course on Puzzle-based Learning was initially targeted at freshman level. The objective was to lay a foundation in domain-independent reasoning that could be used in future courses. Since then, Puzzle-based Learning has worked its way up, down, and outside our curriculum. We have included Puzzle-based Learning themes in other undergraduate courses (e.g., System Development, Intelligent Decision Support Systems) and graduate courses (e.g., Heuristic Problem-Solving, Big Data Analytics). We have also used Puzzle-based Learning in outreach programs in high school, middle school, and even elementary (3rd grade) school. In addition we have offered industry workshops and continuing education courses. As our experience with teaching Puzzle-based Learning improved, we realized that there was also a broad demand for such pedagogy and the theme could be molded to fit the needs of a wide range of audiences. Based on our experience we feel that a new instructor has a number of choices to explore the teaching of Puzzle-based Learning.

Having said this, we would recommend initiating a teaching experience in Puzzle-based Learning in a limited setting such as an outreach effort or as a teaching tactic in another course. Next, we would recommend that a new instructor offer a course on Puzzle-based Learning as an elective so that those in the course are self-selected and are truly interested in the material. (In one of our institutions, Puzzle-based Learning is a required course for all engineers. This has led to challenges in motivation and assessment. We discuss this further in Chap. 4.)

In addition to courses that exclusively focus on Puzzle-based Learning, puzzles can be added to other courses for a number of reasons. They can be added to provide a diversion that has an educational aspect related to the current course. They can be placed in a difficult course to provide respite. Conversely, very challenging puzzles can be put into a simpler course in order to keep students from becoming bored once they have met all of the existing challenges! However, as we discussed in the motivation, it is the linkage between the puzzle-solving process and its position as a stepping stone from domain-free problem-solving to highly contextualized Project-based Learning activities. In future chapters we discuss puzzles that can be used to emphasize a domain-specific concept, e.g., the value of iteration in software development.

In addition to the obvious requirement of an innate curiosity in puzzles, we recommend an additional characteristic for any instructor of Puzzle-based Learning: *resilience*. In just about all other courses that we authors (and other instructors) teach, we are the domain expert. The skill and knowledge gap between instructor and student is tangible. But, as discussed in Chap. 1, solving puzzles does not require any specific domain knowledge (which is how we differentiate a puzzle from a problem), only reasoning skills. Hence it is not uncommon for students of a Puzzle-based Learning class to solve a novel puzzle before the instructor.[1] Pedagogically this is fine as the goal of any puzzle-solving effort is the reasoning and not the final solution. Hence, as an instructor of a Puzzle-based Learning class, one needs to be comfortable being stumped and using such instances as an opportunity to examine in more detail an incomplete reasoning process.

2.2 Motivating Students

One of the challenges in any course is encouraging, developing, and maintaining student interest and engagement. A bored student is unlikely to take part in activities and is also more demanding in terms of what a given course or class can do for them. This can be a particular problem in a Puzzle-based Learning course as it can be more difficult for students to see how solving puzzles is going to be of help to them in future studies. While students may undertake dull and repetitive activities, such as memorizing mathematical tables or complicated formulas, because it will "be on the test," they may not be willing to take the steps required to get the most from a Puzzle-based Learning course.

As we will discuss in more detail, students need to think that what they are doing is useful now and be motivated to try and must also see some value in the future, if they are going to have the highest motivation to take part. Given that Puzzle-based

---

[1] We discuss one common instance of this in Sect. 4.2 under the theme of "puzzle of the day."

2.2 Motivating Students

Learning sometimes requires students to step outside of their comfort zone and risk being incorrect in public, we need as much motivation as possible for the students to take part. Given that Puzzle-based Learning is a highly thoughtful exercise, it's not enough to force students to participate or to try to control their behavior with marks; we have to provide the right environment and approach to help students to realize that this is worth doing.

However, communicating this to students requires us to know who the students are and what matters to them, and this is specific to the class, the educational level, the teacher, the country, etc. – so many variables that we can't list them all here. It is fair to say that many students, through years of training, often regard courses of study in terms of what can be achieved and, all too often, in terms of how easy it will be to achieve a passing or excellent grade. Thus, any argument that depends upon "you should solve puzzles because it's good for you" is unlikely to make much progress. A number of companies have employed puzzles as part of their job assessment strategy, so any industry speaker you can find to support the utility of this in job interviews will be valuable.

An approach we have used to motivate Puzzle-based Learning to our students is discussing the "big picture" in their education. Consider the below diagram depicting paths through four years of undergraduate education:

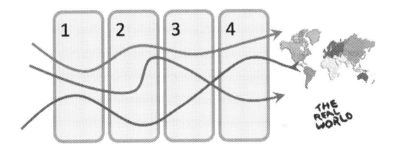

Undergraduate education is expected to be a transformational experience. At the end of their UG experience (hopefully, all), students emerge into the so-called real world. Some continue onto graduate school, many enter into the work force, few start their own businesses, etc.

As an instructor, an intriguing question to propose to your class is: *What are the characteristics of the real world?* How would you describe the real world? What adjectives might you use? We've done this as an in-class exercise in a number of formats (poll each member of the class, group exercise on the board, etc.) ultimately resulting in a pooled collection of thoughts. Typical responses have ranged from scary and unforgiving to colorful, fun, and exciting.

After discussing these responses, a follow-up question is: *What skills do you need to succeed in the real world you've just now described?* Responses often include problem-solving, critical thinking, perseverance, resilience, etc. Typically one will be able to cluster these skills into two groups – soft and hard skills.

Over the years, a number of researchers have investigated similar questions. Under the direction of Cynthia Atman at the Center for the Advancement of Engineering Education (CAEE),[2] a longitudinal study of important design activities was conducted among undergraduate engineering students. For the following 23 design activities, which do you view as the six most important?

Abstracting	Generating alternatives	Making trade-offs	Synthesizing
Brainstorming	Goal setting	Modeling	Testing
Building	Identifying constraints	Planning	Understanding the problem
Communicating	Imagining	Prototyping	Using creativity
Decomposing	Iterating	Seeking information	Visualizing
Evaluating	Making decisions	Sketching	

While there are variations from 1st year to 4th year (e.g., the value of *iterating* was perceived to be higher in 4th year than 1st), four of the top five skills are consistent across the years and also match top five skills suggested by practicing experts: *understanding the problem, communicating, identifying constraints, brainstorming*. Experts also had *seeking information* in their top five, while seniors had *making decisions*. As discussed in Chap. 1, we believe Puzzle-based Learning provides an opportunity to explore and practice some of these skills.

One thing should be established early on. The course is about getting into good mental shape – not getting the answer. Having the answer does not build the brain; thinking builds the brain. To explain this to the student, you can make an analogy between a physical workout and a mental workout. Spending two solid hours thinking about a problem and not getting the answer is OK. It's analogous to leaving your house and jogging a three-mile loop. In both cases you didn't get anywhere, but that's not the point. The point is to develop strength and stamina. Puzzles can develop mental strength and stamina, while jogging develops physical strength and stamina.

The first week of the course should be devoted to convincing the students that Puzzle-based Learning will help make them successful as adults. A Puzzle-based Learning course provides the opportunity for the student to increase the number of neurons in their brain and the connections between them as well. A Puzzle-based Learning course is a workout for the brain. The way to become a good problem-solver is to solve problems – hard ones.

The specific answer to *Why Solve Puzzles?* will vary by student, and by assessment scheme to an extent, but some useful guidelines are as follows:
- Draw on any industrial or practical applications of puzzle-solving, in terms of concrete skills.
- Conduct exercises to show students how simple puzzling can help them to think.
- Reinforce that it's a way of looking things from another angle.

---

[2] http://www.engr.washington.edu/caee/

Puzzles are, in the main, domain-free challenges that require very little formal progress in other areas of study to be enjoyable and exacting while still approachable. Puzzles help to get students thinking about the kind of problems that they will face outside of the educational experience: problems where no one gave them a chapter to read and a set of questions that derive from that chapter.

A well-constructed puzzle course will help students to realize that they are capable of much more than they think, help them to frame problems in a useful way, and remind them of ways to deal with the kind of situations they will encounter later on in life. In later chapters on effective teaching approaches and problem-solving strategies, we go into a lot of detail about how to make an environment suitable for the kind of student community that will enjoy solving puzzles, but this assumes that you've got them in the door in the first place!

Some of the most effective Puzzle-based Learning environments, regardless of whether at school or college, have a strong element of play and participation to them and starting on a playful note will set the tone for the rest of your time with students. Can you invite students with a puzzle? Can you put up puzzles outside of class or around your school or college so that you start community formation before the first student has entered your classroom? We already know that students will invest large amounts of effort into certain types of games, often for little real reward, if there is enough motivation, so try to tap into that "game" effort early as locating a few students who are keen will help you to form a more solid environment.

A Puzzle-based Learning course can be enjoyable, but very few of the puzzles will remain enjoyable if they immediately segue into arduous or complex mathematical proofs, especially for younger students. Rather than consider Puzzle-based Learning a gateway to a particular concept, it's better to plan for it as a parallel development of thinking skills, so that traditional content and Puzzle-based Learning content are linked thematically rather than sequentially. Many of the probability puzzles will help students think about probability, but there is no great benefit to setting 10 puzzles and then moving into a detailed discussion of the Z test.[3]

When explaining to the students why the course is valuable, be genuine. Tap into your personal experiences. There is bound to be a reason that you have chosen, or been asked, to conduct a Puzzle-based Learning course, based on what is believed to best for your students. Use that to communicate to the students why they should be interested.

2.3 Hosting a Puzzle Contest

Puzzle-based Learning is a course that increases in effectiveness as more people get involved – a course that relies upon one teacher to keep moving students forward will quickly become arduous for the teacher and unrewarding for the students.

---

[3] http://en.wikipedia.org/wiki/Z_test

When a puzzle-based approach is working, students will actively seek out new challenges and look for like-minded people to work with, and, in many cases, they will look for new people to stump with interesting and challenging puzzles!

One way to create an awareness of and to develop interest in Puzzle-based Learning is to host a puzzle contest. The simple reason is that puzzle contests are fun! Both the participants enjoy solving puzzles under contest conditions and the organizers enjoy assembling puzzles to stump their peers. We have organized puzzle contests in a variety of settings: as part of the activities of a puzzle club, a component of an outreach effort, and a capstone experience for students in a class on Puzzle-based Learning. The following are some factors and suggestions we have for conducting a successful contest. Naturally the format and content of any contest will have to be tailored based on the target audience, objective of the contest, duration, and background of the organizers.

As with any public event, many issues need to be considered:

Publicity One needs to consider three parts: (1) before, (2) during, (3) and after the event. Unless the contest is being held as part of an outreach event where it is known beforehand (approximately) how many people will attend, promoting the contest is crucial for a good turnout. Social media, the school newspaper, eye-catching creative posters across campus, and a general e-mail to the faculty to announce in their classes are all great ways of getting the word out. The "buddy system" is also effective – if each organizer were to bring 3 friends, one can augment the audience. Media coverage of the event itself (audience solving puzzles, winners, organizers) is critical for sustaining the effort. Be sure to contact the local news station, the school photographer, and other faculty members.

Sponsors and Prizes Our experience has been that students like to compete for the fun of it and also for bragging rights. Prizes certainly help. Local companies, campus recruiters, and bookstores are often supportive of such student-organized events. As with all campus events, food is almost a must and is a great way to retain and engage the audience while the winners are being determined. Local restaurants can be explored to support this component of the contest.

Logistics A primary decision to be made early on is indoor vs. outdoor, as the puzzles that can be used will depend on the venue. The main criterion is to have sufficient space for the participants to work. As with all campus events, the day and time of the event can determine attendance. Whereas a puzzle contest as part of an outreach event could be longer, given the tight schedule of college students, an event of at most 2 hours is effective. This duration would support 5–7 puzzle sets to be solved in 10–15 minutes.

Contest Structure An effective way to run a contest and to keep *all* participants engaged till the end is *not* to have a knock-out style contest. We have found the following to work well:
1. To add a social component, teams of two people participate. The ability to mutually discuss a puzzle increases the engagement factor.

2.3 Hosting a Puzzle Contest

2. The contest is conducted with multiple rounds of puzzle sets. Each puzzle consists of a pair of puzzles. Puzzles vary in difficulty from easy, medium, and hard and have different point values (e.g., 3, 5, 8).
3. While all teams work on the same puzzle set at the same time, the constituency of the puzzle sets will vary (e.g., easy to medium, medium to hard, etc.). Contestants are made aware of the point value of each puzzle.
4. Each puzzle set has a fixed time limit (e.g., 10 minutes). Once a publically viewable countdown timer starts (many are available on the web), the teams can start working on the puzzles. When a team is ready to submit their answers (written on the puzzle sheet itself), the time taken (or left) is also noted to determine tiebreakers. In order to ensure that all teams proceed in lockstep, if a team finishes before the allotted time, they will need to wait till the time for that puzzle set runs out.
5. After the predetermined number of puzzle sets, the winner is determined by a combination of their points and time taken. Given the nature of the contest, assessment of team's answers will need to be binary – full points or 0. Unlike in classroom assignments, puzzle contests do not support the ability to give partial credit.

Running the Contest Depending on the size of the event, you will need many assistants. Some of the tasks involved for which you will need 2–3 people each are (a) registration of the participants and teams, (b) entering scores and times for each puzzle set, (c) handing out and collecting puzzle sets, and (d) grading the contestant answers.

Pre and Post Puzzle Set As people register for the contest and await the start, it is fun to handout a sheet of sample puzzles for the contestants to ponder. Some pre-contest puzzles we have used are given below (answers are left to the reader). Once the contest is over and the winners are being determined, as the contestants mingle over food, it is fun to display a visual puzzle (say, projected on a screen) for them to consider.

Puzzle 1 The proprietor of a rural farmer's market would like to be able to weigh out any integer amount of grain from 1 to 40 pounds in only one weighing using a two-pan balance. What is the minimum number of weights that will accomplish this and what are their weights?

Puzzle 2 You have the misfortune to own an unreliable clock. This one gains exactly 12 minutes every hour. It is now showing 10 pm and you know that is was correct at midnight, when you set it. The clock stopped four hours ago, what is the correct time now?

Puzzle 3 A pie was stolen from Bakery Square by one of five suspects. Each suspect gave a statement:

Dave: It wasn't Jen. It was Eric.
Eric: It wasn't John. It wasn't Jen.
John: It was Jen. It wasn't Dave.
Meghan: It was John. It was Eric.
Jen: It was Meghan. It wasn't Dave.

The police identified each suspect told exactly one lie. Who stole the pie?

Puzzle 4 Below, 10 countries have been broken into chunks of letters. These chunks have been mixed up, no chunk is used twice, and all chunks are used. Can you determine what the 10 countries are?

```
EZU  ITZ  ZIL  ELA  BRA  GI   IA   FI
PAN  MBA  BEL  AND  BER  ZI   NL   BO
CAM  VEN  DIA  AND  UM   SW
BWE  MEX  ERL  ICO  JA   LI
```

Icebreakers 3

> *Individual commitment to a group effort – that is what makes a team work, a company work, a society work, a civilization work.*
>
> – Vince Lombardi

The amount and type of teacher–student interaction in a Puzzle-based Learning course can vary widely based on the number of students in the course. In our experience, we have taught this course to as many as 300 and as few as 10. Irrespective of the size of the class, it is important that students are active rather than passive. The goal of the teacher should be to get students to use their System 2 thought process as much as possible. Students should treat the course as if it is a workout for their minds, as a mental gymnasium.

Puzzles can be classified in many ways, from simple to complex, linguistic to mathematical, general to specific, and there are probably as many interpretations of the classifications as there are classifications. A simple breakdown of puzzles can put them into the following four groups[1]:

(A) *Icebreakers*: These are special, challenging puzzles that are designed for maximum participation and to get your class taking to each other and to you.

(B) *Warm-ups*: Warm-ups are puzzles that rehearse students in a particular technique or way of thinking. While not as blatant as *"Turn to Chapter X and do the problems there,"* a good warm-up will put students into the appropriate mindset to look at more complicated puzzles in the same vein.

(C) *General puzzles*: While these vary in difficulty, there is no assumption of advanced learning in any discipline, although specific problem-solving techniques may help with their solution.

(D) *Discipline-specific puzzles*: These include mathematical puzzles, which assume high-school or college mathematics, and any puzzle where you have to state prerequisite knowledge from a course above middle school.

---

[1] Later on we also categorize puzzles as in-class, exam-oriented, and assignment puzzles.

We apply a range of different puzzle-solving techniques, but we recognize that a good course is made up of a combination of the four types of puzzles, where the exact composition will vary depending upon the skill and experience level of the students and the teacher.

In this chapter we present some mental workouts that are designed to be used over the first few class meetings that will set the tone for the course. The goal of the icebreakers is to get the students to know one another and to make them comfortable volunteering their thoughts and opinions – both in small groups and in a formal lecture setting.

There are probably more icebreakers presented here than you will need to use. Invariably, as you teach this course, you will select your favorites. Do go through all of the below and decide which one would be best to introduce your students to the style of Puzzle-based Learning as well as their classmates.

Icebreaker 1 We have successfully used this puzzle socially among friends for 30-plus years. It also serves as an excellent warm-up puzzle or icebreaker for a talk or workshop on Puzzle-based Learning. Audiences from a wide range of backgrounds – from 7-year-old elementary school kids to 70-year-old adults as part of a continuing education program – resonate with this puzzle.

Apart from serving as an icebreaker, this puzzle also conveys (a) the potential of working as a group in solving a problem and (b) how easy it is to miss the obvious.

Start by telling the students that you will be asking them to individually write down multiple responses to the puzzle on a card. You will ask a question to one person of the audience and that you want only that one person to answer. Pick a random person at this point in time. To alleviate any social pressure/anxiety on the person, it helps to continue to have a dialogue with the class including the chosen person, e.g., *"Now I'm going to ask Robert a question and I would like Robert alone to answer. We all will listen to Robert's answer and proceed from there. Robert – this is a really simple question with multiple possible answers so no need to feel concerned. I'm not going to ask in which country is Timbuktu* ☺*."* Then, you ask: *"Name one and only one body part that has exactly three letters in it. No naughty words, abbreviations, or slang."* These are all common anatomical terms that any 3rd grader would know.

Some of the more common responses to this question are eye, ear, leg, etc. Recall that only one response is to be given. Reaffirm with the class the response. "Robert has said leg." Now, including "leg," write down 10 body parts that have exactly three letters in them. Then let the audience work. While they are working, mention that items like "lid" and "cap" don't count as they are parts of hyphenated words (e.g., eye-lid, knee-cap); "abs" doesn't count as it is an abbreviation; "gut" is colloquial (though sometimes this is contested). Saying that "fat" is not a body part often brings out chuckles among the audience. (Writing "toe" ten times also doesn't count!)

After the participants have had a chance to work on this puzzle for a couple of minutes, poll the class to see how many they have written down. We usually start with 5 and work our way up. Only once or twice in the numerous times we have tried this puzzle have we seen a person write down 9 three-letter body parts. The typical pattern is for people to get 5 quickly, and then it starts to taper off at 7 or 8.

Whenever you feel that their progress has plateaued, we sometimes offer hints. For example, of the 10 three-letter body parts, 5 of them are above the neck. While intriguing, as this narrows down the search space, typically this produces an improvement of one additional body part or so. Rarely do individuals reach 10 even at this stage.[2] If the logistics allow, you can ask the students to compare answers with their neighbor.

To finish off the icebreaker, poll the class as a whole and ask them to call out names of those parts that are above the neck. The ones that are normally harder are *gum* and *jaw*. If the participants get stuck on these, additional hints can be given (e.g., a boxer often aims for this). Often, one gets exclamation of *Oooh*! from the participants as they identify their missing items.

Of the body parts below the neck, the harder one is usually *hip* followed by *rib*. Here again, a hint can be given. Amusingly, the same hint as earlier can be given: a boxer often aims for this!

In the spirit that all our puzzles are educationally motivated, ask the class *"What may be the learning objective of this puzzle?"* An answer often given is *"It helps to work in teams to solve problems."* An additional answer (which may require some prompting) is *"How easy it is to miss the obvious."*

Both of these learning objectives can be related to other themes students may be studying. For example, when teaching a class on programming or software development, requirements need to be elicited from business clients. During this process it is not unusual to miss obvious requirements (e.g., a save button to aid the filling in of a long online form).

A charming aspect of this puzzle is the incremental solution and the sense of satisfaction in moving towards the solution. It is also a universally applicable puzzle (from elementary school to senior citizens) and in a variety of situations. We have used this puzzle in social situations, e.g., when among like-minded friends waiting in a restaurant for food to arrive, when waiting in an airport, etc. This puzzle is also popular among kids in family gatherings with competition among siblings.

Icebreaker 2 Here is a quiz that we have used on the first day of class. Consider handing it out before any introductions. As soon as the students are in their seats, your first words to them can be *"Here is the first quiz; you have five minutes."*

---

[2] We have given many hints in this discussion. We leave it as an exercise to the reader to assemble the complete list of three-letter body parts.

Here it is[3]:

> Instructions: Check one of the two boxes below.
> BOX A ☐
>
> BOX B ☐
>
> Grading: If all the students in the class check box A, all the students will receive a score of 20/20 on this quiz. If at least one student checks box B, all the students who checked box A will receive a score of 10/20 and all the students who checked box B will receive a score of 15/20.
> BONUS: Guess the number of people in the class that will check box number B. [No points for this; just for fun.]

Students will get the idea right away that this is not a normal class. Some students will be stressed. Some students will complain. One thing that they should all be doing, however, is thinking. This grading format of this quiz should give the students a feeling of connectedness because their grade was not determined by the teacher; it was determined by their classmates!

As soon as you collect the quiz, start a class discussion by asking, "*Who wants to volunteer which box they checked and why?*" In our experience, answers can be grouped into three categories: First, the people who checked box one and are incredulous that anyone would check box B. Second, the people who checked box A on principle, knowing that there probably will be someone that checked box B. Third, the people who checked by B and are incredulous that anyone would check box A. We have found that students feel very strongly about their answers and will be eager to reveal their choice and defend it passionately.

We have experienced heated discussions among the students that have lasted the entire first class regarding the "correct" box to check. This is good stuff. The students are actively thinking. They are examining their belief system, their humanity, and they are engaging in discourse. Perfect!

You can further stimulate discussion among the students by asking some of the following:
- Who based their answer on the size of the class? How would the number of people in the class have to change before you switched your answer?
- Who looked around at the other students in the class to try to gauge how they might respond?
- If I gave the same quiz tomorrow, would your answer be different?
- For all of the students that chose box A, how many points would box B have to offer before you switched to B? 17? 18?

---

[3] This is a variation of the prisoner's dilemma game-theoretic example.

3 Icebreakers

- For all of the students that chose box B to get a sure 15 points, what is the minimum number of points offered for box B that would cause you to switch to box A? 8? 10? 12?

Icebreaker 3 This puzzle serves as an excellent introductory exercise that immediately involves all students and forces them to think![4]

The following competition is announced in the class (it might be a good idea to offer a worthy price for the winner. Note also that there is *at most* one winner for this competition: no tie is possible). We ask students to take a piece of paper and write their names (so we can identify the winner) together with a natural (positive integer) number. Make it very clear that the lowest number they can write is 1 and that there is no limit on the largest number. You can also volunteer that 1,742,169 is likely to be unique in the class, but it probably will not be the lowest unique number.

Now, the winner is the student who wrote the lowest number that is *unique* among the class. So, the goal of the exercise is to think of the lowest number that no one else will write down. To further wrap the student's heads around the problem, you can even say, "*After all the numbers are collected, I may poll the class to see how many students wrote down each number, starting at one and going up. The winner will be the first person to be the only one with their hand up.*" The best "theater" for this, however, is described below.

Give students 5 minutes to think about this problem and then collect their pieces of papers. If the students want more than five minutes, you have a good class. Often, before you reveal the results, it might be good to start a general discussion on this problem. Clearly, students' intuition often points to a simple heuristic rule which says that "the larger the number of participants, the larger number should be written on the piece of paper" – and generally, this is true. You can illustrate this by displaying a curve:

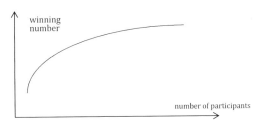

and, of course, do not display any numbers here – this is purely an illustration of the idea (and, of course, the exact shape of this increasing curve is arbitrary). Clearly, if there is one participant, any number would "win" this competition.

You can also provide your students with additional piece of information. The Swedes ran this competition as a lottery every day for several months with the

---

[4] Alternatively we have also used this puzzle to close our workshops on Puzzle-based Learning and to award a prize to the winner.

average number of participants per game of 54,000. Within these several months, the lowest winning number was 162 and the highest winning number was 4,465.

Before you reveal the results, you can also tell your students what you expect (based on your previous experience). In many cases, when we run this competition in our classes, we noticed a few interesting phenomena:

(a) There were *always* some students (more than one) who selected number 1. The reasoning usually is that no one else would select this number, as the probability of this number being unique is very low.
(b) There were *always* some students who selected *big* numbers (e.g., 1,985,359). The reasoning usually is that other students would concentrate on low numbers, and thus, they would be eliminated.
(c) Some numbers are *much more* popular than others – thus, you expect some frequency peaks for some prime numbers (3, 5, 7, and 11). For some reasons, students (like many "ordinary" people) are attached to prime numbers – the popularity of 4, 6, or 8 is much lower!
(d) Try to make a prediction – tell your students that you would not be surprised if one of these three numbers (4, 6, or 8) would be the winner (for a reasonable size of the class, say, between 20 and 100). Further, tell them that you are willing to take a bet – you believe that the winning number would be even.

Now we are ready for the show. Prepare your blackboard (whiteboard) by drawing a horizontal line where you mark all numbers from 1 to, say, 20 – the final category might be marked as "21 or more."

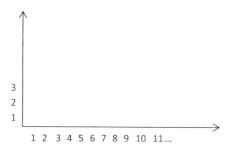

Then ask one student to read numbers from the pieces of paper, one by one, while you update the graph; after 12 entries, your graph may look like:

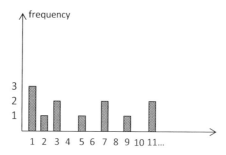

Consider also using a simple program if you can display the screen of your laptop for your class – there are a number of possibilities you may consider to build the momentum in the class. Note that at every iteration there is a "current winner" (like the student who selected number 2 in the above display) – and any new number read from the next piece of paper may eliminate that "current winner" forever.

This is great theater and great fun, with audible groans and *Ooohs*! coming from the audience as each number is read aloud. This is another icebreaker that really sets the tone for the course. There are no prerequisites. There are no protocols or flash cards. They just have to sit and *think*.

At the end of the class you may consider going a bit deeper into this problem – which was studied by a few mathematicians – see, for example, the article by Qi Zeng, Bruce R. Davis, and Derek Abbott.[5]

If conditions are right, give the same quiz the next day, grading it more quickly with a poll rather than the histogram. Many times when we have done this, the winner on the second day was someone who wrote *one* on the previous day (unless no one wrote down *one*)!

This icebreaker can have a lot of amusing results. One time we did this exercise for family night at a high school. There were thirty people in attendance and the winner was being determined by a poll. There were about five people that wrote down *one*; when *two* was announced, two people raised their hand. They were a father and his teenaged daughter sitting next to each other. She was indignant and demonstrated her displeasure with an audible "*DAD!*" and by smacking him on the arm. Dad responded, "*How was I supposed to know you were going to write 'two'?*" She didn't get the prize, but she, as well as the audience, and the authors got a great story.

The next two icebreakers are physical puzzles that can lead to a high-level discussion of topology.

Icebreaker 4 Here is an icebreaker that has been used every summer at the Gedanken Institute for problem-solving. Attendees are 11–17 years old and there is always a lot of giggling and even some success at this topological twister.

You will need about one oversized sweatshirt (also called a "jumper" in some countries) and a pair of "handcuffs" for each group or 3–4 students to perform this one. The handcuffs are made from clothesline (or other string) and cable ties. Do not let the students put the cable ties around each other's wrists because they can pull them too tight.

Here's the challenge. One student in the group (usually someone that is not too worried about messing up his/her hair) volunteers to put the sweatshirt on inside out and frontwards. So, the tag is in the back on the outside. The volunteer is then placed in handcuffs. The challenge is to reverse the sweatshirt so it is on the volunteer the right way without breaking the handcuffs. The helpers will provide

---

[5] Zeng Q, Davis BR, Abbott D (2007) Reverse auction: the lowest unique positive integer game. Fluct Noise Lett 7(4):439–447.

all kinds of advice, most of which just tangles up the volunteer. It makes a great photo op!

Icebreaker 5 Here is another one that can get a bit rambunctious. Students will work in pairs; it is recommended to pair students of the same gender. You will need 3–4 feet of rope (about 0.5 cm thickness) for each student. Each student of a pair will tie a loop at both ends of their rope and slip it over their wrists. The two ropes of each pair will also be interlocked as shown below:

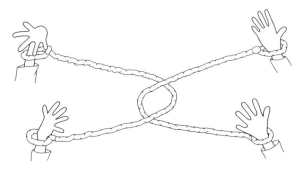

The goal is for the two students of each pair to disentangle themselves without removing the rope from their wrists.

The fact that it seems impossible to separate the two students contributes to the fascination with this puzzle. Students sometimes get completely tangled up; the younger they are, the more tangled up they will be. Hence, it is useful to have a pair of scissors ready if needed!

This puzzle has been known from around the mid-1700s. The students have to recognize that the arms, body, and handcuffs of each person do *not* form complete links as the links of a chain do. There is a gap, the gap that must be exploited to separate the two students. When successful, each student will still be wearing handcuffs, but they will not be linked to each other.

After a reasonable amount of time, if no pair has figured out how to disentangle themselves, the following hint could be provided: remove the rope from the arm of a student; slip a thick loop of rope or a rubber band (e.g., 15 cm diameter and 1 cm wide) up their arm; place the rope back on their wrist. Now ask them to remove the loop from their arm without removing the rope from their wrist. The solution to this will be obvious. With this hint, students may be able to generalize to disentangle from the original puzzle.[6] If a pair of students is able to disentangle themselves, ask them to reverse the process and see if they are able to link themselves back together.

Solution:

Step One   Step Two   Step Three

Icebreaker 6 One of the ways to get the students in the class to know each other is to have a class tournament with the tournament bracket projected on a big screen. The tournament should be based on a novel game that preferably no one in class has played before. The game should always produce a winner – no ties. We have used the game "Bridg-It"[7] successfully many times. Here is the grid we use:

---

[6] This is an example of a problem-solving strategy: *when you can't solve a problem, look for a similar but simpler problem that you can solve.*

[7] The game was popularized by Martin Gardner in the early 1960s in his Scientific American column where he called it the Game of Gale. Even before that, computing pioneer Claude Shannon created a device to automatically play a very effective game based on an ingenious analog heuristic and also demonstrating that the first person to move can force a win. For a discussion of some amazing contributions of Shannon in this area, please visit http://boardgamegeek.com/geeklist/143233/claude-shannon-the-man-the-games-and-the-machines

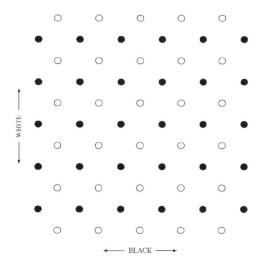

The play consists of each player drawing a line between two nearest neighbor dots of his/her color either horizontally or vertically. A line can't cross a line that their opponent's has drawn. The goal of the student playing the black dots is to make a complete path left to right and the goal of the student playing the white dots to make a complete path top to bottom. Consecutive moves do not have to be connected to one another.

This game develops the students' deductive reasoning and their ability to think both offensively and defensively. It also puts the student in a one-on-one battle of wits, and this prevents them from being in a passive mode. Because of these qualities, we have used this game in job interviews, paying close attention to how much the candidate improves from the first game to the second game. The better the quality of the players, the greater the number of moves needed to complete the game.

It is a good idea to try to finish the class tournament in one class period. Also, for students who are eliminated early, have extra playing grids and perhaps a couple of other activities to keep their minds active. For more on the subject of using competition among the students in the classroom, see the part of Sect. 4.1 on *Sociology* in this book.

Students who are waiting for their next opponent to finish a game should feel free to watch that game to perhaps pick up strategies or weaknesses. Also, they should feel free to practice when they have a bye. During outreach events where we have participants of a range of ages, we have adopted the rule that the younger player gets to go first. Finally, we have projected the championship game on the big screen so the class can follow the play. If you can offer prizes, that usually increases the fun and the motivation.

3 Icebreakers

A couple of other games that are good candidates for this icebreaker include Hex and Sprouts. Appropriate sized grids for these two games are shown below.

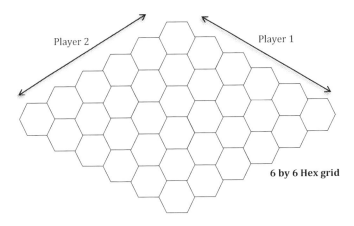

Sprouts starting grid.

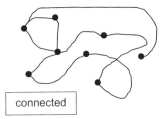

In the game of Hex, players take turns marking a hexagon as theirs. This can be done by placing an identifying colored dot or letter in claimed hexagon. The first player to make a connected path of claimed hexagons across the grid is the winner. In this way, Hex is very similar to Bridg-It.

In the game of Sprouts, players take turns drawing a line between two existing spots or from a spot to itself and adding a new spot somewhere along that line. The line may be straight or curved, but must not touch or cross itself or any other line. No spot may have more than three lines attached to it. The player who makes the last move wins the game.

Icebreaker 7 Form pairs of students: player A and player B. You will need a sheet of paper for each pair and a pen for each of the players.

Draw a number (say, n) of dots on the paper. The two players move alternatively with **A** moving first. At each move, a player can join two points with a line (not necessarily a straight line), but they cannot join points which were already joined directly or join a point to itself.

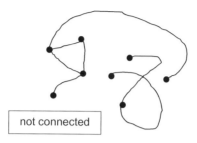

In mathematical terms, the players are building a *graph* (with predefined n *nodes*) by connecting some of the dots with *edges* (lines). The winner is the player who makes the graph *connected*, i.e., there is a sequence of lines (known as a path) between any two nodes of the graph.

The game is very challenging. A player should be careful not to make an easy connection for the other player. For example, in the "not connected" graph above, the player with the move would win the game by connecting to disjoint components of the graph.

This icebreaker provides an opportunity to discuss with students "winning strategies." In some games (like Bridg-It) we know that the player with the first move has a winning strategy – that is, a sequence of moves that would lead to a win regardless of moves of the opponent. There are games in which the player who makes the second move has a winning strategy. There are also games (e.g., chess) for which we do not know if there is a winning strategy and, if one exists, whether this winning strategy is for the first or second player.

It is a good idea to start "small" – ask your students to start with $n = 3, 4, 5$ or $n = 6$, just to "feel" the game. (We explore further the idea of investigating smaller instances of a problem in Chap. 9). Which player has a winning strategy in these cases? Clearly, for $n = 3$, the second player has a winning strategy – whatever connection is established by the first player, the second player would make the graph connected.

Also, it might be worthwhile to present a few cases of "games in progress" – in all three cases below, it is player **B**'s turn. Who is going to win?

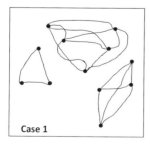
Case 1

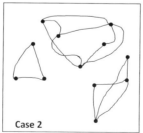
Case 2

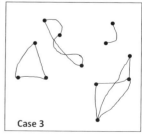
Case 3

Then we can move to larger instances of n: if you start with $n = 7$ or 8 dots, who has the winning strategy?

After an extended discussion, it might be a good moment to challenge your students by asking: *What is the winning strategy for player A if the number of dots $n = 13$ or $n = 14$?*

At some stage the students might be ready for a grand puzzle: *For what values of n does A (the player that makes the first move) have a winning strategy and for what values of n does B (the player that makes the second move) have a winning strategy?*

Here are a few thoughts related to this grand puzzle that may help in leading your students towards the solution:

- With a small number of dots, it's not too hard to just list all the possible games. Drawing a tree of all the different decisions is a useful way to organize your thinking.
- If the game comes down to three separate pieces, then if one player connects any two of these, then the other player connects the remaining one and wins. So the game can be reduced to thinking about three separate "blobs."
- If there are three separate "blobs" left, then neither player wants to connect blob to blob, so they must connect *within* the blobs. If it's player A's turn and there are an even number of within-blob connections left, then once they're all done, it will be player A's turn again and player A will have to lose. On the other hand, if there's an odd number of within-blob connections left, then player B will win.
- So somehow just before you get to three blobs, you need to arrange to get the right number of within-blob connections.
- Of course, if you play right, you can always arrange to make some of your blobs just single dots or pairs of connected dots. So it's enough to focus on just these scenarios.

Icebreaker 8 This puzzle works well in conjunction with the "three-letter body part" puzzle. An attractive aspect of the Icebreaker 1 is that one is able to make incremental progress to a solution – first identify 4, then 6, then possibly taper off at 7 or 8 identified body parts. Contrary to that the solution to the below puzzle exhibits a Martin Gardner style *Aha!* moment.

Below is a schematic of a printed circuit board:

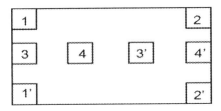

Your task is to do the following:
(A) Connect each number to its prime (1–1′, etc.) by a continuous line. The connecting lines need not be straight but (a) the lines cannot intersect, i.e.,

touch each other; (b) the lines cannot go outside the bounds of the board; and (c) a line cannot go through a contact point (e.g., a line connecting 1 and 1' cannot go through 3).

or

(B) Provide a convincing argument that it is not possible.

Among the pedagogical lessons from this puzzle is that it captures some of the nature of real-world mathematical problem-solving – we often don't know what outcome we are working towards during out initial explorations.

We've also seen that this puzzle tends to exhibit the "Roger Bannister effect" – once one person has figured out the correct answer (solution given at end) and announces that they have solved it, in quick succession, others tend to figure it out too, although sometimes students have declared that it is impossible after another student claims to have solved it.

A deceptive aspect of this puzzle is the way the contact points are numbered 1 and 1', 2 and 2', etc. Solvers often try to connect 1–1' first and then notice that either 3 is now cut off from 3' or 4 is cut off from 4' and hence deem that the puzzle is impossible to solve.

An alternative numbering sequence is given below.

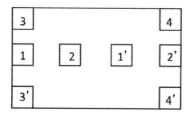

When given this version, solvers again try to connect 1–1' and 2–2' and then notice the channel through which 3–3' and 4–4' can be connected.

If the participant pool is large enough (say 30+ people or so), one could print out both versions of this puzzle (perhaps on different colored paper) and distribute one version on one-half of the room and the other version to the other half of the room to demonstrate that the numbering sequence makes a difference in terms of the number of people who solve the puzzle.

Solution Step 1:

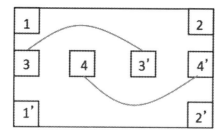

Solution Step 2:

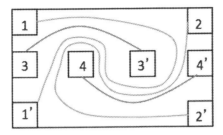

The purpose of the icebreakers in general is twofold: first to give the students a good idea of what the class will be about and second to make them feel comfortable with their surroundings and the other students. Here we have presented a variety of icebreakers that should get students thinking and having fun. These can be used anytime during the course when you feel that the students need to have some fun interaction. As mentioned in the chapter, they are also great for special functions, seminars, and meetings. We have handcuffed people together at dinner parties, company retreats, and even family reunions!

Reference

1. Zeng Q, Davis BR, Abbott D (2007) Reverse auction: the lowest unique positive integer game. Fluct Noise Lett 7(4):439–447

Effective Teaching Approaches

> *Puzzle-based Learning doesn't need a lecturer, it needs a Ringmaster.*
>
> – Nick Falkner

One of the challenges in implementing a Puzzle-based Learning approach is taking a love of puzzles, or a desire to make students think in a more open-ended fashion, and making it work in a classroom environment. Many courses reward students for sitting quietly and, when prompted, answering a set of well-defined questions with rehearsed answers built from what their teacher has said. When we describe an effective teaching approach to support Puzzle-based Learning, we are not just talking about buying a book or finding some problems, we are talking about a complete change in the way that many of us think about working with our students.

The person who knows best a group of students is always the teacher who is working with them, so all of the advice in this section must be filtered and tempered through the teacher's experience. However, the elements contained here have been tried and improved over many years of cumulative experience between the authors, across a range of disciplines and countries. These are the approaches that we have had the most success with.

Any new set of materials or resources can be challenging, even to the most seasoned teacher, so we highly recommend taking time, where possible, to gain the highest familiarity with any puzzles or lesson plans that are intended for students. A good Puzzle-based Learning class is a well-balanced mix of prepared material and spontaneous reaction to the material – teachers may be surprised by some of the directions that students take once they become excited by the material.

We don't wish to become too bogged down in pedagogical theory and details, but the two most important considerations in putting together a Puzzle-based Learning course are *cognitive apprenticeship* and *play*.

To recap, the idea behind cognitive apprenticeship is that someone who has mastered a skill then instructs a novice in order to teach them that skill. By using a genuinely supportive environment and presenting the puzzles in a way that encourages activity and interaction, peer to peer and student to teacher, we provide a *situated learning environment* and make the students active in the classroom. For the Puzzle-based Learning teacher, you don't have to be a master of puzzle-solving, but you do have to be familiar with those puzzles you are demonstrating, so you can explain them with mastery, and you should also work on developing your skills at locating puzzles, attacking puzzles, and trying to make your way through puzzles. By watching you, and learning from you, students will be able to develop their own skills, by learning from the steps you take, the way you handle obstacles, and your perseverance. You will also learn the common pitfalls and fallacies that students may encounter. In many cases, if you can find someone else to explain it to, you will be able to enhance your own learning in the area by reflecting on the problem as you articulate your own journey through it.

A *playful* environment is essential if the class is to achieve the right kind of environment that supports creativity and risk-taking. Experience has shown the authors that requiring the "right" answer, and only marking for or rewarding that, is the best way to build resentment, break engagement, and increase dissatisfaction among a group that could be enthusiasts, if they had more time and freedom to experiment. Our goal is to make our students think and this may require us to step back from our traditional assumptions and requirements over focusing on the marks students achieve, to look at the process they are following – even if the solution is wrong or the process is not what we were expecting.

4.1 Cognitive Apprenticeship

Collins, Brown, and Newman's cognitive apprenticeship[1] work discusses several important concepts that are needed to support a successful cognitive apprenticeship model. We focus here on methods, sequencing, and sociology, as a background to all of the activities and approaches listed afterwards. While this is familiar to many teachers, we summarize the approaches here for ease of reference.

Methods This set of methods helps the teacher to work out what they can do in encouraging certain student behaviors. Many of these will already be in practice but, within the Puzzle-based Learning course, it is helpful to remember that we do not very often concentrate on the solving process as a domain-free activity. Most problem-solving is heavily situated into a given domain (SOH-CAH-TOA for Trigonometry and "I before E except after C" in English are two examples), and the notion of a "free-form" problem-solving methodology is often new to both teachers and students.

---

[1] Collins A, Brown JS, Newman SE (1989) Cognitive apprenticeship: teaching the crafts of reading, writing, and mathematics. In: Resnick LB (ed) Knowing, learning, and instruction: essays in honor of Robert Glaser. Lawrence Erlbaum Associates, Hillsdale, pp 453–494.

4.1 Cognitive Apprenticeship

Our goal is to produce students who use puzzles to develop their general solving skills, and to do this well, we have to start by giving them a good example of someone who is confident in their problem-solving, can explain what they are doing, and can help the student to get up to their level.

The first method is to show the student how *you* solve the problem. This is called *modeling* and is a crucial first step towards encouraging students in developing their own skills (this is not the same as building a model, which we encounter later in this book; see Sect. 5.2). One easy way to do this is to conduct a *think-aloud* exercise so that the students can hear what you are thinking as you are solving. Most, if not all, of the teacher's cognitive activities are usually hidden from students because the thought processes that lead to the outcome are well internalized and we no longer need to articulate the steps.

The teacher should act as a coach, supporting the students as they go through the problem-solving process, providing feedback, reminders, and scaffolding, and customizing their tasks as their expertise develops in order to bring their performance closer and closer to mastery. The scaffolding provided will vary based on the level of the student and the nature of the problem, but we would expect this support to *fade* over time as confidence and skill develop. (A student who still needs scaffolding after extensive instruction may require special attention.)

One of the goals of a Puzzle-based Learning class is to get the student to articulate their thought process. This can be done one-on-one informally between the student and teacher or, if the class size is large, students can be grouped and articulate their thought processes to each other. If nothing else, this gets discussion started, even when no solutions are forthcoming.

When the work on a particular problem is complete, get the students to reflect on their strategies and their progress. They can compare their solutions with those of peers, experts, or any other model that should represent expertise. The *postmortem* is a common approach, where the teacher takes students through the steps of solution for a given puzzle, but it is essential to highlight what the differences are between different versions of a solution and why one is more or less suitable. What can be challenging, for a new Puzzle-based Learning teacher, is that reflection can uncover a previously unknown (but valid) solution to a puzzle.

Once the student becomes comfortable with the problem-solving techniques, reducing their need for scaffolding, they might start to explore the puzzle domain themselves: finding challenging puzzles, adding onto a current puzzle, or finding new solution approaches.

Sequencing A puzzle-based course is never just one lesson and this gives teachers an opportunity to build up the puzzles in their course over a period of time. While the students who are in your class will be driving the difficulty and puzzle type in more advanced puzzle courses, we have provided some simple guidelines, which are in accordance with usual teaching practice, to help a teacher build a sound progression.

There is often a temptation to start with very interesting puzzles, without realizing that puzzles with beautiful artifice or cunning tricks can be extremely frustrating to the novice puzzler. As always, building in complexity is very important but a student's background will make a great deal of difference to how a particular puzzle is perceived in terms of complexity. Students who are using a second language for their education will have great difficulty with more wordy puzzles, students who have little to no mathematics may perceive even a sniff of the arithmetic or geometric as impossible, and students from other cultures may find simple analogies impenetrable.

One way to deal with the complexity issue and keep puzzles at the same level while keeping the challenge up to the students is by presenting a wide range of puzzle types. For example, you might follow a puzzle on probability with a puzzle involving word patterns. As we add more puzzles with a wider range of possible solution strategies, we firstly allow students to learn something new, but we also implicitly increase the difficulty of selection, simply by giving them more things to choose from. A common concern with existing, simplistic approaches to teaching mathematics is that the students read, for example, Chapter X of a mathematical textbook and then undertake the problems at the end of that chapter, which reduces their requirement to think about which solution methodology they should be using. Increasing diversity allows you to mix problems and pose rhetorical questions in yet-unexplored space and starts to require your students to think rather than pattern match.

To get students interested in solving puzzles, they have to believe that solving puzzles is something that *they* can actually do. As part of our scaffolding, we can move away sections of puzzles that are not as important to focus on higher-level concepts. By doing this, we can introduce interesting puzzles, with simpler models, and then develop them over time to start to look at all of the skills that go into their construction. A common example is the Monty Hall problem (see Problem 5.5 in Chap. 5), which has both a simple argument to explain it and a more complex Bayesian analysis explanation. Ultimately, we want the students to understand the general layout of the concepts, before they focus on the detail. Then, when detail is introduced, it is within a conceptual map that is already familiar.

Sociology One cannot assume that every student in a room is ready to engage immediately with a Puzzle-based Learning course. The teacher should make a deliberate effort to make the classroom, or space, a place where Puzzle-based Learning is going to work.

There is no doubt that this approach to a course will be quite unusual for many teachers and most students. Because of this, we must pay special attention to reducing the possibility of the classroom itself from being an obstacle to adoption.

We have already discussed that the student needs to understand the purpose and use of what they are learning, and they need to practice this in an environment that prepares them for the application of the skill. What needs to be stressed here is that the student is an active participant in the process, rather than a passive receiver.

The Puzzle-based Learning approach works best when students are involved and see the point – puzzles are easy to dismiss as unrealistic and irrelevant and teachers will have to work to provide an environment where these two (potentially valid) criticisms hold no weight.

Early on in the course, the teacher should try to establish a culture of expert practice by demonstrating how an experienced problem-solver would approach the problem. The best time to do this is immediately after the students have completed their work. By turning the classroom into a place where methodologies and approaches are regularly discussed, we support the students on their path to expertise. Group activities can be crucial to this formation as we need to continue externalizing our thought processes and allow the students to constantly be exposed to how problems are solved by people who have more experience than them at solving problems.

When the students begin to develop these skills, there is a good chance that they will be intrinsically motivated to tackle harder problems. This is really necessary because extrinsic punishment-/reward-based systems do not work well at encouraging independent student activity and have a significant negative impact on creativity and innovation. While carrot-and-stick works well for simple, relatively unthinking, tasks, Puzzle-based Learning is a combination of cognitive and metacognitive skills that depend upon a willingness to persevere and a capability to think creatively, especially when faced with obstacles. Where possible, puzzles and problems should be entertaining and rewarding. As we discuss further in Sect. 4.5, Effective Assessment, some of the most successful Puzzle-based Learning courses have a very unusual approach to assessment and marks, with little emphasis on the "correct" answer.

Once the students gain confidence and skill, they will be more likely to contribute when working in small groups. This will automatically extend the learning resources beyond the teacher (because we now have more than one "teacher"). In many cases, students will provide scaffolding to each other, as it is rare that every student in the class is at the same level. At the same time, we must take care to avoid overburdening a student with the same role all the time. This can happen automatically when the types of problems are varied continually. Ideally, every student should be involved in solving, critiquing, thinking, or supporting. Early warm-up exercises can allow students to demonstrate their worth to each other and to their groups, which allows students to be seen as valuable by their peers. Research into student cooperation in small groups has shown that the perception of peer value is vital if students are to cooperate fully and participate in group exercises. However, as always, the choice of assessment mechanism will have a large impact on this. Students can often be more resistant to work in groups if they feel that this may reduce their chances of performing well, by depending upon other students. Students take different approaches to understanding and learning, with some students being more mathematically focused and others focusing on linguistic or shape-based interpretations. The benefit of students working together includes providing different views of both problem and the approach to a solution.

Let's conclude this section with a discussion of competition within the classroom. The purpose of competition is to have fun solving problems. A competition can be problematical in some situations. One significant problem is that competition can be highly inhibiting to individual students and, especially where extrinsic reward mechanisms are in operation, can lead to a focus on "win at all costs" strategies that undermine teaching value and move into antisocial or unethical areas. The way that errors are dealt with is crucial, because focusing too much on the negative can easily lead to an environment where an initially unsuccessful student does feel intimidated and resentful of their more successful peers. One way to approach this is to embed students in groups and then have groups competing with other groups. Another solution is to keep any competitive behavior relatively low risk and low value, both from a mark and reward perspective. Certainly, the latter of these two will go a long way towards the goal of having the students enjoy themselves. As always, however, the teacher knows his/her students best and should tailor any competition to their particular students. One idea for a competition is to have the students in the class design and host a competition for students who are not in the class.

4.2 Class Activities

Based on the discussion in the previous section, one will probably be able to infer that there is a considerable amount of overlap between teaching a math oriented course (e.g., discrete mathematics) using active learning techniques and teaching a course on Puzzle-based Learning. The key difference is the significant reliance on domain independent reasoning. In this section we discuss some classroom guidelines and activities we have used in our courses.

Given the range of pedagogical environments – small audience seminar style to large audience 'lecture' style different activities may have different levels of effectiveness. The key is to aim for cognitive engagement. Given a choice, we highly recommend offering a course on Puzzle-based Learning to a self-selected small group of students (e.g., as an elective). A smaller class size facilitates more engagement.

In a 16 week semester long course, we recommend issuing homework assignments at least once a week, and if possible twice a week (i.e., homework due at the start of each class meeting).[2] Flipping the classroom is an emerging practice at the undergraduate level.[3] While the jury is still out on whether flipping increases learning and retention, it certainly increases engagement and helps

---

[2] If your institution supports interdisciplinary classes we encourage you to explore that option. We have had very good experience with Puzzle-based Learning classes involving a mix of students from different majors (e.g., Information Systems, Computer Science, Psychology, Statistics, Cognitive Science, Economics, and Physics).

[3] Students do background reading/watching before coming to class and most of class time is spent working problems to strengthen and assess one's understanding of the material.

4.2 Class Activities

calibrate for different learning speeds. Given the nature of puzzles, a class on Puzzle-based Learning can be significantly flipped. In fact, in some instantiations of our courses, there is practically no traditional lecturing. Rather all content and strategies arise out of attempted puzzles. A class can start with a student led discussion of the due homework.

In a smaller sized class, all students will be able to participate and contribute. For example, after the introductory classes, each session could start with a *puzzle-of-the-day*. One student would present a puzzle of their choice. The class as a whole (including the instructor who hasn't seen the puzzle before) would try to solve the puzzle with hints and guidance provided by the puzzle poser. Students are required to submit a one page write-up of their puzzle, solution, and most importantly their reflection on the puzzle: what did they find interesting in the puzzle, variations, how does the solution tie into the general class discussions, etc. In addition to the puzzle-of-the-day, one could also conduct a *puzzlethon* where students again present a puzzle of their choice. But this time the class votes on the best puzzle (a combination of presentation plus the nature of the puzzle). During a discussion of scientific induction and mathematical induction, given the smaller size of the class, Robert Abbott's inductive game of Eleusis can be played.[4] This game which models the process of scientific method, where each class member tries to stump others, is highly recommended.

While we have broken the suggested activities down into a fairly traditional split between face-to-face, online, summative, and formative, we are approaching a point where blended learning, the mix of online and face-to-face, is becoming more widely understood and adopted. We are not suggesting that the activities listed can only take place in certain spaces: where something "looks" like a lecture theater, the class activities will work, much as anything that "looks" like an online space may in fact be a physical space with students working on tablet devices. Perhaps the best suggestion we can make is that experimentation with different approaches in different environments is almost always rewarding.

Class activities are the core of a Puzzle-based Learning course, because we can conduct warm-up exercises as a group and get students used to the idea that other people have the same difficulties while solving problems. As well as this, student groups can work on problems where different perspectives, levels of experience, or age can help a group to reach a solution, built from the contributions of everyone in the group.

In many respects, the class environment can be the most challenging, as there is often the widest range of puzzle-solving abilities and interest, and there is the implicit constraint of the time at which the class is held. Considering the impact of the time, the composition of the group and their receptiveness can greatly affect

---

[4] The reader is referred to Robert Abbott's page on Eleusis http://www.logicmazes.com/games/eleusis/index.html and to the exposition of Martin Gardner *Penrose Tiles to Trapdoor Ciphers*, Mathematical Association of America, 1997.

that degree to which students will actually be willing to step outside of their comfort zone.

Whenever it takes place, the traditional lecture or class, where a teacher stands at the front and reads out a lesson, is not an ideal way to teach Puzzle-based Learning for a number of reasons. These are as follows:

1. The traditional lecture is not collaborative. Students are waiting to be told what to do, rather than being active in their own learning.
2. Only one person is talking, which limits the amount of discussion students can have about a problem and therefore the amount of exploration.
3. Students do not have an opportunity to practice their skills, as they are listening to the lecturer.

Not all activities in class have to be collaborative, but, like many skills, puzzle-solving gets better with practice and we have to provide this opportunity if we want students to do anything other than try to memorize the solutions to existing problems. Many students find puzzle-solving challenging because we are asking them what they think – to have a new idea – rather than asking them to give us back some received wisdom or memorized fact. From the cognitive apprenticeship perspective we presented earlier, student engagement and activity are crucial if they are to develop the correct approach to puzzle-solving. The remainder of this section outlines the key considerations for building a strong class activity environment from the ground up.

We cannot stress enough the importance of selecting appropriate puzzles for the students to tackle – especially early in the semester. They have to be chosen based on what the class can currently do, or are likely to be able to do, rather than those that we pick based on our preferences and expertise. If your class is highly mathematical, then they may find mathematical puzzles both interesting and enjoyable, but in a mixed class, there are many puzzles that have a broader appeal and are more than disguised mathematical equations. Realistically, your class will be diverse and you will have to pick the path that will appeal to most of them. A good puzzle is *infectious*, in that students will want to keep working on it after class is over and pass it on to their friends and family. We have provided a selection of problem sets in Part II and Part III of this book but it's far better for a teacher to select a few of these puzzles from these problem sets that will engage and interest the class than to try and work through the puzzles sequentially. Don't feel the need to work through in a strict order or to do everything! If you think that – for any reason – a particular puzzle is inappropriate, don't use it.

The icebreakers presented in Chap. 3 also allow you to identify early potential problems with students in your group. There are often students who are happy to contribute but are not ready to accept criticism or suggestion for improvement. There are also students who are highly dismissive or critical of the approach ("*Why are we doing this?*", "*Will this be on the test?*"). Identifying students who might need special focus during a warm-up is a very handy tool for the teacher.

After the icebreakers, you will have to present the students with challenging puzzles to solve. One of the most important preparatory steps for a teacher is to be able to show the students how you think about that puzzle if they get stuck, so that they can start to model your behavior and make progress towards the solution. There are millions of puzzles available – make sure you pick one that you are comfortable with and can clearly explain.

This may limit the number of puzzles available to you if you believe that you are not overly mathematical or don't have the required skill set for a particular type of problem. There is a tremendous range of puzzles in this book, for a wide variety of student and teacher levels of preparation and experience. It is completely possible for the solo teacher to form a set of the problems that they are comfortable to present and meet their teaching requirements. However, a teacher support group, for those teachers who want to start puzzling, is as important as the construction of the student support groups in the class. Working as a pair, or a larger group, is one very good way to make preparing for class both easier and interesting. Having a peer to bounce ideas, suggestions and attempts off, without having to worry about dealing with a student, is a great way to flex your puzzle-solving muscles and get ready for a class full of challenging questions.

No matter how much preparation goes into a particular puzzle presentation, there probably will be times that a student asks a question that you can't answer or proposes a solution that is different from yours. If this happens, the teacher should not feel inadequate. When a clever student makes an interesting proposal or asks a good question, you don't always have to have the answer. Remember, the course is not about "knowing answers," it is about "thinking new thoughts." When a student makes an observation that you hadn't considered or poses a question you can't answer, it just means that you have done your job as a teacher getting him or her to think independently.

Time pressure, like any extrinsic constraint, is going to reduce creativity – which is not the point of this approach at all. Where it's possible, be generous in your allocation of time to digest and chew over problems, and also provide enough working space and writing or sketch tools for students to be able to explore their solutions. Infectious puzzles raise student interest, so letting students take puzzles out of the room to share with others is a constructive teaching tool, rather than an information leak. Setting a puzzle as something to think about and discuss, with a timeline of days or weeks, will often allow far more useful and engaging solutions than setting a deadline of ten minutes. Of course, once the puzzles are "out," there is the chance that the students in the next course offering will have them. This, of course, defeats the purpose of the course, which is developing the students' ability to independently solve problems rather than know the answers. To reduce the temptation (and negative aspects) of students accumulating a store of previous puzzles to short-circuit the learning process, the quest for new puzzles is perpetual. After three to five years, a teacher can probably assume that any puzzles not yet presented will be "new" for the majority of students. While many puzzles will be recycled to illustrate key points, it is helpful to have a large bank of puzzles that can provide "sight unseen" opportunities for challenge over a five-year period. The size of this will vary by course and ability level, of course! You can change a puzzle by altering the components or names or sometimes by simply changing the number of elements in play.

Many puzzles provide opportunities to get the students active. If a puzzle has a physical component, try and bring it into the classroom. Even one of the oldest puzzles we know (farmer, wolf, goat, cabbage; see footnote 13 in the Introduction section of this book) can become a physical exercise for younger children if we use building blocks or toys like Lego®. The rather dry "back and forth" of the fundamental logic puzzle can be animated by a performance of sketching on the board or by pushing a boat backwards and forwards and having the "wolf" attack the "goat" the moment the farmer pushes off. Multimodal learning environments, where we combine different types of learning activities, are going to make your class more interesting and memorable. Feel free to be theatrical with this. Indeed, any puzzle-solving lecture should have a bit of theater about it, especially where we are trying to bring a puzzle to life. Much as adding a physical component can bring an interactive dimension to a puzzle, being prepared to "act out" elements of the puzzle will help those students who are not really interested in the puzzle as an intellectual curiosity but are willing to get involved in something interesting. The degree to which will suit the class does vary greatly, by student age, culture, and expectation, so match your degree of performance to the class. As always, students looking away, looking down, or not responding probably means that more energy is required. Students cringing or trying not to meet your eye may mean that you've gone too far!

Match your tone and your language to the puzzle. If you're running a Pirate Puzzle (see Problem 9.5), don't ask for volunteers; say "*I need five pirates!*" Immersing the students in the language and environment of the puzzle takes them out of the traditionally didactic learning space and gives them an opportunity to experiment and enjoy this alternative experience.

Not surprisingly, props are very useful in creating a fun atmosphere, but they also help the student visualize and frame the problem.

If you are setting probability puzzles, then why not have cards or dice handy? Are you showing students the "Monty Hall" problem (Problem 5.5, Chap. 5)? There are many pieces of software and physical demonstrations of the problem that will allow students to see how the situation unfolds and help to guide them to the correct solution. However, the props should be tailored to the classroom space available, as 300 students flipping coins in a traditional lecture space quickly gets noisy and slightly dangerous. Match the activity to the space and choose your props accordingly. Some puzzles dramatically increase in interest and fun with props. The "Pirate Puzzle" (Problem 9.5) is a classic example of this, as a few eye patches, a bag with a skull, and crossbones with some plastic doubloons will provide even the most resistant student with a chance to have some fun.

While the "Pirate Puzzle" certainly is a great intellectual challenge, there can be sensitivity issues with some students. If the class is presented in a part of the world where piracy is still a problem with tragic consequences, the presence of the pirates can be eliminated without any change in the meat of the problem. Many of the problems presented in this book were restructured to avoid the game of Russian roulette, committing suicide, and other topics that might be sensitive.

4.2 Class Activities

When tailoring a puzzle for your students, make sure that they understand all the terms and be sensitive to the fact that different cultures and languages are familiar with different things. If a puzzle contains something that students aren't familiar with, the load can easily rise to a point where students give up. Much as you will happily hop over a fence but wouldn't try and scale a building, the perceived size of the obstacle is extremely important.

One example is anything involving a sporting situation. Not everyone is either interested or knowledgeable about sports, and sports are highly acculturated. While cricket metaphors and puzzles would be understood over most of Australia, baseball would not, where the reverse is true in the United States. Even within one country, rugby is the dominant Football Code in several states of Australia, where Australian rules football is dominant elsewhere.

In one class in an urban setting, we presented the following warm-up: Farmer Brown keeps cows and chickens. If the cows and chickens together have 54 legs and 20 heads, how many chickens are there? The problem seems safe enough, until we realized that a few of the students had never seen a chicken and didn't know how many legs a chicken had!

When you customize a puzzle, you should be thinking about:

- What can you expect your students to reasonably know?
- Are you making the puzzle easier for your students to understand?
- Have you substituted one idiom or metaphor for another, with the same problem?
- Have you tried to be concise, but at the expense of precision?

There is a line between adding color or context to make a puzzle more interesting and then making the color and context the dominant part of the puzzle. The final point, brevity vs. ambiguity, is of particular importance in assessment, as you always wish to communicate enough that the student can solve the puzzle, while balancing the requirement to make solving a challenge of some sort.

When selecting puzzles for use in the classroom, it is important to have in mind the purpose of the puzzle. Is it a warm-up exercise? Is it designed to help groups form? Is it to help students practice their analysis of the text of the puzzle? Are you trying to build confidence? Are you rehearsing for assessment? Very few puzzles can do all of these at the same time and a puzzle lesson is generally composed of a range of puzzles that start from warm-up, move through group and individual skill development, and may end with challenge or assessment puzzles. Much like sporting warm-ups and training, individual activities exercise individual components of the student. Picking a puzzle in the hope that it can do everything will often raise the cognitive load to an intimidating point. One way to address this is to use a step-based approach within the same puzzle to build towards complexity. However, be careful with how you break puzzles up, as students regard our guidance on how puzzles and problems can be decomposed as the "gold standard" for decomposition. If the puzzle doesn't neatly break into pieces, don't force it or you may end up showing students a bad example of decomposition.

Students are going to get stuck at some stage on the puzzles and you have to decide how much help to give them in making their way across the stumbling blocks. Your decision should be guided by the goal of the course – to get the students to become independent, skilled problem-solvers. Hints should virtually always be in the form of a question. For example, *"Did you consider simplifying the problem?"* Knowing some good nudges or hints to get students thinking along the right path, without making the solution obvious, is a very important part of preparation. Trying to come up with these "on the fly" often leads to problems, either because the hint is more obscure than the puzzle or it leads to the solution dropping out immediately. Remember the role of the puzzle and the ability of the students when trying to set the hint level. You'll see in Parts II and III that some of our problems explicitly say not to give the students hints at certain stages, to encourage the students to explore as far as they can. If you find that students are always waiting for the hint, you might want to think about not giving hints for a while!

Another part of puzzle preparation is the decision about the manner in which the students will solve the problem. Will it be individually of groups? Will it be a class discussion led by the teacher or will it be all three in sequence? If it is a class discussion, you might find that many students find it intimidating to participate in a more open environment – especially in a large classroom. Any exercise that gets students talking and raising their hands is going to prime them for more activity later. A given class may need priming every time or they may eventually come to accept that this is what happens in this class. (It's important to note that the same students who are vocal and active in your class can be silent and inert elsewhere, or vice versa. Students are highly context sensitive!)

Seeding is an attempt to get the flow going by selecting specific students who you can rely upon to contribute and encouraging them, before or at the start of the lesson, to get things going. This has to be authentic and genuine, however, as insincere contribution will quickly be picked up. If all else fails, always be prepared to talk to yourself until someone else chimes in. If the problem is sufficiently interesting, someone probably will.

When no one is volunteering to contribute and you really want an answer, you have to be patient and just wait. This can be the longest 60 seconds of your life but, if you want students to answer, they have to accept that you will wait until someone answers or, if no one does, you will move on to another problem without giving the solution. In other words, "Don't blink." You want the students to be involved but it requires you to commit to not rewarding people for inaction. It's hard and it's very uncomfortable, but once students realize that there is no escape, you will usually see participation rise. If you feel this would help, you can try this joke. Tell the students, *"I know that many times it is hard to be the first person in the class to respond, so what I'm going to do is to take the second response first. Who wants to respond second?"* Often, this is the push the students need to participate and it lightens the atmosphere as well.

Even if your preparation is impeccable, and the students are eager to contribute, there is always the possibility that the class will solve the problem much faster than you expect[5]. Perhaps a student had seen it before and wants to show off to his/her classmates or perhaps a student simply had a brilliant insight. There are a number of ways to address this, the safest of which is to have more puzzles on hand than you think you will actually need for the class. This is also useful where a student walks up to you after class and asks for more puzzles. Having a few on hand, especially written up in a format that a student can take away and work with, will make it easier for you as a teacher when the demand appears. There are plenty of good candidates for "extra" problems in Part III of this book. If you have a know-it-all, consider giving him or her a problem from Chap. 15, "Grand Challenges."

Perhaps the best solution is to hand out a set of challenging problems at the beginning of the class as "grand challenge" problems. This set does not have a due date, but every once in a while you can ask if any students have made progress on the grand challenge problems at the beginning of class. When you ask about them, the students will be reminded and motivated to be able to raise their hand and claim success in front of everyone. This problem set is also useful anytime a student is done before his/her classmates.

Eventually, you might get to the point where the students are making fantastic progress on a particular problem without your help. You are sitting quietly and the students are fully occupied and engaged with a challenging problem – totally ignoring you. Don't interrupt them. Savor that moment. Congratulate yourself. This is the goal of the Puzzle-based Learning course. If students spend the whole class in relevant discussion and only solve one problem out of four that you had prepared, this is far preferable to trying to force them through many problems, where they don't have time for discussion and reflection.

4.3 Online Activities

In many respects, online activities are more challenging as we have greatly reduced the number of possible ways we can interact with students – we cannot pass them a bag of dice or gauge the atmosphere of the room while attempting to communicate the key parts of a puzzle. While many of the aspects from classroom puzzles still apply, there are some specific considerations for online activities.

One of these is the development of an online community. Although, as we will note, communities come with increased risk of certain problematic behaviors, a moderated or supervised discussion community is an excellent partner to an online environment. Online activities can be done at any time and these times often don't match when teaching staff are available or even awake. Another student can answer a student question on a community forum before you have woken up and eaten breakfast, reducing the frustration on the part of the inquirer and developing the problem-solving mastery of the person who answers. Online environments tend to

---

[5] In a puzzle-of-the-day context, an instructor needs to be comfortable with students reasoning out an answer before the instructor!

form their own structure and timetables – take advantage of this but seed controlled environments where students can safely discuss problems and help each other.

Another restriction of the online environment is limited interactivity. While tablet PCs of various types are increasing in market share, many students will be experiencing your online activities through a traditional keyboard-and-mouse environment. This starts to limit students to click, drag, and type, when it comes to how they work with a puzzle, and this starts to place load into the kinesthetic arena, as students may wish the puzzle to do things that it can't do. A geometric block puzzle is far more interesting as a physical object than a difficult to orient and flip virtual object.

If your teaching institution has handed out iPads, then an iPad app for the activity will be easier to use than a web page. Similarly, if the student electronic environment is laptops, customizing the delivery to the most likely environment will have the highest impact. However, certain formats do not work across all platforms, most notably Flash animation, which does not currently work on any Apple iOS devices. When choosing a means of providing the online environment, it is essential to work out how many students you are excluding, if you use a format that is not universal.

Of course, not all teaching institutions have handed out iPads. Similarly, not every student has access to online activities outside of school and even asking about this can be embarrassing for students and their parents or guardians. Depending on student level, placing online activities into a computer lab setting will make these resources available to everyone. If labs are not provided or available, and there is no prescribed device at the teaching institution, it is probably unwise to make completion of online activities a compulsory component of student activity.

Another drawback of the online community is it has limited ability to provide feedback that is anywhere near the customization and richness of human feedback. Many environments only allow simple answer checking, textual pattern matching, or some low-level rubrics, with matched responses that have been customized by the teacher. The other option is allow students to undertake the online activity and provide feedback at a later date, which then separates the constructive advice for correction from the activity and reduces the benefit of correction. If possible, providing an online element that is simple enough to allow for useful and authentic automated feedback, paired with elements that require human intervention, combines an immediate piece of improvement advice with the delayed gratification of a full consideration of the work.

Finally, it is important to be on the lookout for the unpleasantness of cyberbullying whenever working in an online community. Many schools have a "no connected environment" policy, and it's worth checking to see if the online activity planned is going to meet the requirements of your institutions. Obviously, where wider access is available, or where the online activity takes place on a site where nonstudents can access it, a risk assessment of the cyber safety issues should be conducted as requiring students to access resources beyond those of the school or college could introduce problems.

4.3 Online Activities

Another important aspect of students' safety involves links to Internet sources. Any links provided should be active and able to be clicked on directly to reduce possible transcription problems from students typing it in a box. However, given that names and web addresses can change, it is good practice to check all of the links to anything outside of your online environment every time you run a course. Even simple things like advertising providers can change on a site that has been trustworthy for years, and now "adult" content is showing next to an interesting guide to solving probability problems. This is especially important when you allow students to link to other puzzles. In certain cases, it may be wise to prevent other students from seeing these links until they have been checked.

If you have directed a student to something that is online, it should work as you expect all the time, be available for the duration of the time you want students to access it, and, somewhere or somehow, there should be a copy of it so that catastrophic failure doesn't compromise your teaching and learning activity. The computing industry is full of products that have been cancelled with little or no notice, and the number of "free" or "cloud-based" services that are cancelled is increasing. The more effort a student invests into these online activities, the more it will dishearten them when the service or system becomes unavailable, especially if you have been building up a portfolio or accumulating some sort of progress metric. Backing up your problems, student solutions, and any student marks is a minimum requirement of any system. If you don't know how it's being backed up, assume it isn't and work to change that.

There is often a large divide between the expectation of what a student should be able to do and actual evidence of student experience. All online environments should be tested to make sure that they are doing what you expect them to do. If possible, get more senior students (or teaching assistants) to run through the activity so that they are familiar with all of the error conditions, possible pitfalls, and problems. Load testing, where we see how a system works with a full class, can be hard to simulate, and the first time an online activity is run is often the first time that the system is really tested under load. Testing everything else before this point reduces the number of things that can go wrong.

Even when things are working as expected, all students may still have problems, especially if the system isn't intuitive. A step-by-step usage is often a worthwhile investment, especially when accompanied by screenshots. These guides should be kept up to date for each new version of the system, and paper copies can be as valuable as online copies, as these can be put next to the computer and consulted, without the student having to switch backwards and forwards between reading the guide and working.

"Teaching" in an online environment is very different from teaching in the classroom where we would normally be able to gauge if a class was not following our explanation and their level of interest in a problem. There is no equal measurement in an online environment (in the absence of highly intrusive keystroke logging and eye tracking). To track the point at which the students become lost or disengaged in a puzzle, it can be broken down into stages of progress towards a solution.

By breaking the activity into steps, we can start to assess those points where students slow down, disengage, or give up.

By collecting data on what students are doing, monitoring their connectivity and progress, we can start to assess how they are responding to the online activities and environment. If, on the day before it is due, no one has even gone to the web site, then there is a low level of interest. If, however, lots of students have gone and completed activity 1, but then stopped before activity 2, it's probably worth going to have a look at activity 2 and asking students why they're not making progress.

Many online environments use standard activity and question formats, which mean that you can transfer all of your hard work when the system you are using is replaced or becomes obsolete. Proprietary formats, especially those that require point-and-click interaction to create, can be hard to copy across to other systems and this level of "lock in" is both undesirable and a waste of time when, inevitably, a new system is chosen.

4.4 Measuring and Assessing Puzzle-based Learning

Puzzle-based Learning can be particularly effective in an environment where there is no traditional marking or assessment, as the quantification of exactly how much a given solution is *worth* is a very challenging problem in itself. If two students come to the same (correct) answer using two different approaches, are they worth the same marks or is there one approach that is more "correct" than another?

In this section, we first look at setting assignments as a general problem, recapping many of the issues that we have previously discussed, and then we discuss the assessment problem in more detail.

All of the research that we do in this area reminds us that extrinsic motivational factors are the natural enemy of creativity and innovation, and yet the nature of assessment is often to function as a carrot-and-stick approach, which seems to be at odds with the philosophy of puzzle-solving in general. You can no more demand a student to "be creative" than you can set them a challenge to "write an excellent symphony" in five minutes.

In this area we discuss aspects of goal theory and the value of assignments, to help you to get your students participating in, and enjoying, your assignments.

Much as we did for the class activities, we have a checklist for any assignment that will allow us to select the correct one, with some additional aspects to help us manage the assignment. We'll then discuss each one as a separate topic.

1. What is this assignment supposed to achieve in terms of learning outcomes?
2. What level of knowledge do your students have in this area?
3. What level of knowledge do you wish your students to have after this assignment?
4. How much time do you have for this?
5. Have your students seen a similar or related puzzle before?
6. Have you demonstrated this in class?
7. Do the students have a reason to do this assignment?

4.4 Measuring and Assessing Puzzle-based Learning

8. Do the students understand the value of this assignment?
9. Do you understand how to do this assignment?
10. Have you tested it?
11. Does the assignment require additional equipment or resources?
12. Do you need to provide a guide?
13. Do you need to provide a rubric?
14. Do you have a list of hints?
15. What is the impact if a student cannot solve any of the problems?

What Is This Assignment Supposed to Achieve in Terms of Learning Outcomes?

If featured in class, an assignment will often be used to illustrate a point, provide an example, or lead students into a new area. If used outside of class, it might provide practice or an assessment opportunity. We want students to focus on gaining knowledge of problem-solving strategies and developing the skills needed to utilize them effectively as a pathway to being able to solve new and interesting problems. Students shouldn't see the knowledge here as an end point but a required stepping stone to *application*. Each assignment should either be introducing knowledge or providing a means for testing knowledge and skill. While traditional courses can often assume that an assignment will practice certain skills (such as numerical drills in mathematics), Puzzle-based Learning is often more subtle in terms of what an individual puzzle will actually achieve.

What Level of Knowledge Do Your Students Have in This Area?

What is the worst-case situation for knowledge of either the techniques or the context of the puzzle that you are using? If the puzzle is strongly written/verbal, do you have a large number of students who do not have the puzzle language as their primary language? Have you set a puzzle with mathematical notation?

Within the solving sphere itself, how much do students already know about this technique? Is this the first time you've shown them a probability puzzle or a pattern recognition puzzle?

What Level of Knowledge Do You Wish Your Students to Have After This Assignment?

What is the learning outcome that you wish to achieve? Does the assignment provide enough examples, practice opportunities, or assistance in conceptual mapping and step formation to take the "average" student from where they currently are to where you want them to be?

How Much Time Do You Have for This?

Some puzzles take time because there is a lot of thinking time involved or, in many cases, a lot of dead ends and false starts. A good puzzle leads the solver back towards the solution, making little steps of progress, and this can take time, especially if you are waiting for creative insight.

Creativity cannot be rushed or scheduled. If you are depending on an *Aha!* moment to solve the puzzle, you will have to allocate enough time for it to take place. If you are time poor, then the puzzles must be simpler or less reliant upon creative insight – or you must allow time and materials for solving the puzzle outside of class.

Have Your Students Seen a Similar or Related Puzzle Before?

Is this the first time they've seen anything like this? If so, then they will need more scaffolding and potentially more time. Existing theories on the development of intellectual ability indicate that students take time to move into new domains and that transferability of existing knowledge into new areas is neither guaranteed nor necessarily predictable.

Have You Demonstrated This in Class?

A large component of cognitive apprenticeship is allowing the novices to see a master in the solution process. Have you demonstrated something in sufficient detail that would provide a framework for students to apply in solving this assignment? You need not demonstrate everything, but have you provided enough that students can at least get started? Because a puzzle-based course is, by its nature, highly diverse and wide-ranging, there are far more opportunities for us to present work for which students don't even have a rudimentary solving model.

Do the Students Have a Reason to Do This Assignment?

As we discuss in Sect. 4.5 "Effective Assessment," the carrot of good marks, or the stick of bad marks, does not necessarily lead to desired performance, especially when our goal is to make our students think. Is the assignment interesting while still being sufficiently rigorous? Is it challenging, without being impossible? Does it fit into what you have been teaching and is it a natural fit for people thinking inside this space?

More subtly, is this a problem where the solution requires a human to think about it or are you asking your students to solve a problem that is neither useful in developing their knowledge for this course nor important? Students are trained from an early age to pull out their calculators and are likely to not react well if you force them to be inefficient, error-prone wetware calculators.

As part of goal theory, if the students have the freedom to attack the problem, the skill set to actually attack it, and a reason to do it, then they are far more likely to undertake it willingly and with creative capacity.

4.4 Measuring and Assessing Puzzle-based Learning

Do the Students Understand the Value of This Assignment?

Again, marks are not the only value, although research shows that the local task value does have some impact. A student will build their value of the assignment from three things: its intrinsic value, their intrinsic level of motivation, and their understanding of the *instrumentality* of the assignment. Instrumentality reflects their knowledge of how solving this assignment now will help them achieve a goal in the future.

Highly motivated students will probably attempt all tasks that we set them, regardless of any other value, but, in the absence of such motivated students, we must make the task itself valuable (by making it interesting and relevant and giving the students a good reason now) and we must also clearly indicate how completing the task now will help the student in the future. This reinforces the need to tie the task to specific learning outcomes or professional and life skills that a student will need. For some students, it may be enough that this will help them with the final exam. Others may require more far-reaching goals. However we do it, we need to keep reminding the students that these things are important – potentially at the beginning of each teaching session with the students!

Do You Understand How to Do This Assignment?

We are all profoundly busy and it is completely understandable that a teacher would grab a puzzle book and set an assignment from it, while possibly not having the time right now to go through and solve it themselves. This does, however, leave the teacher exposed if questioned on it and it also does not provide the students with the opportunity to see any of the teacher's own solving steps, whether demonstrated in class or indicated by the construction of the assignment.

The simple rule here is that if you can't do the assignment, it may be wiser to find something else until such time as you've had a chance to solve it.

Have You Tested It?

Again, any problem that comes in should be tested to make sure that there are no typographical errors in the text, transposition errors, or ambiguities. When reading the text, do you have questions? When looking at the symbols, do you have to look any up? Does the solution exist and is it correct? Is there more than one solution?

Does the Assignment Require Additional Equipment or Resources?

Does the assignment require a certain book or a set of blocks? Will you provide them or are you expecting students to provide their own? Do you own the licenses for any software? If so, where are they valid? Are all of the resources and equipment available when you need them to be?

Do You Need to Provide a Guide?
Is the first assignment of its type or so unusual that you need to provide more supporting and scaffolding material? Does it need to be online, printed, or face-to-face? How will you coordinate access and production? Who will test your guide?

Do You Need to Provide a Rubric?
Are you expecting the solution to take a really heavily specified format or to be delivered in a particular way? A rubric, where we specify all of the key aspects of the solution and provide exemplars to students, is very handy here as it also provides a basis for marking. If there are multiple solutions, but you only expect students to discuss two of them, you can mention this here and then students have no ambiguity or uncertainty about what they need to do. We return to this point in the next section.

Do You Have a List of Hints?
One of the key aspects of the cognitive apprenticeship is understanding how a puzzle-solver solves a puzzle and how hints may provide useful steps and nudges to guide novices into the correct paths. Does this assignment need hints provided? If so, how many and what form do they take? How will they be delivered? Do they come at a cost of some sort – is there a limit on the number of hints available? Remember that hints shouldn't make the problem drop out but that they shouldn't make the problem any more difficult. Good hints can be hard to develop and wide testing with peers is very helpful. When you find a good hint, record it and share it!

What Is the Impact if a Student Cannot Solve Any of the Problems?
Is this assignment a vital part of a student's development, in your course plan? What happens if a student gets stuck? What happens if every student gets stuck and gives up? Do you have contingency plans, replacement assignments, or more obvious hints? In some cases, you may have to make the decision to strike the assignment and attempt another approach. It is generally better to do this than to pretend that the fault lies with the students and sticking one's head in the sand.

The solution isn't everything! Too many people think that spending 50 minutes on a problem that they don't solve is a "waste of time." If you run for an hour on a treadmill, you go nowhere but by exercising, you are training yourself for running and increasing your fitness. Similarly, working through numerous approaches and partial solutions will rehearse students in valuable "brain exercise," as well as developing their problem-solving stamina.

4.5 Effective Assessment

We have taught Puzzle-based Learning both in academic (for grade) and non-academic (for fun or training) settings. Given that emphasis is put on the process of reasoning and not the final answer, students often express concern on how they will be evaluated. Following are some strategies we have used.

We have classified puzzles into in-class, exam, homework, and grand challenge (discussed in more detail in Chap. 15). In-class puzzles are puzzles we expect students to figure our during class time, perhaps with some hints and a nudge. These are formative exercises. Exam puzzles are in the spirit of in-class but involve novel reasoning. We anticipate students should be able to figure them out in about 15–20 min. Puzzles that require longer time to ponder are issued as homeworks. Grading exercises in a Puzzle-based Learning course is similar in spirit to grading proofs in a maths course. In both, the line of reasoning is the critical component. Hence we provide generous partial credit for solutions that are heading in the right direction.

We have also found it useful to grade on a qualitative scale (A/A−/B+, etc.) rather than on a numeric scale to avoid students (and graders) to have to differentiate between 8/10 and 7/10. The larger the class size the more the need for graders and hence more the need for a uniform rubric[6] and guidelines.

Any assessment of thinking processes is hard because we are not merely concerned with the answer, which is usually easy to mark, but we are interested in the development of skills and the demonstration of skill and knowledge that was required to provide the answer. However, we must balance the work involved with the degree of correctness. There is little point in providing a good grade for meandering and confused solution approaches, but there is also little point in awarding a good grade for a single answer written on a page, with no evidence of what transpired to achieve it (in this context).

Puzzle-based Learning has been used in a variety of settings and is often easiest to run when students are provided feedback on their work, but without a marking scheme that will then quantity their efforts into some scale of suitability for solution. Quantifying effort, attitude, and perseverance is very hard, yet this is what we should be looking at here, assuming that the effort is being usefully applied. Puzzle-based Learning has been highly successful as additional material in traditional courses (without it being directly assessed) and within the Freshmen Seminar environment (where grading is not relevant). Puzzle-based Learning is also useful in a club-like atmosphere, as an additional activity to traditional teaching.

At its core is the requirement to provide constructive feedback that will develop a student's skills and knowledge, while (where assessment is required) combining this with a grade that does not dishearten or confuse the student if full marks are not available. A symmetrical trap is that of awarding full marks and then having

---

[6] In the educational context, a rubric is set of performance standards for a given assignment or course, represented as a set of guidelines to illustrate what type of performance corresponds to a particular grading level.

nothing to say to the student about their future development – constructive and supportive criticism should accompany all marks, not just those that are not "perfect."

While there are general principles that apply to all ethical assessment in education, we will focus on those that have particular application to Puzzle-based Learning. These are listed here and then explored in subsequent subsections:

1. The assessment should be useful to the student.
2. The marking scheme should be fair, transparent, and equitable.
3. Valid effort and outcome should both receive recognition.
4. Assessment should be based on the student, not the class.
5. Assessment should be repeatable.
6. Assessment should always be accompanied by constructive feedback.
7. Where possible, err on the side of less formal assessment rather than more.

The Assessment Should Be Useful to the Student

Rather than running quizzes or assignments as part of a regular schedule, consider reducing the amount of formal assessment, where grades are assigned, to focus on key points where you have already spent a large amount of formative assessment. Ideally, any summative assessment in Puzzle-based Learning is the culmination of a large body of demonstrations, exemplars, collaborative activities, and feedback opportunities. This is not because the subject is "soft" but it reflects the innate difficulty in providing an assessment environment that fosters creativity. Every piece of summative assessment must provide a good window into the student's level of progress and the students must be clear that this is going to be of use to them.

The Marking Scheme Should Be Fair, Transparent, and Equitable

Assessment items should be chosen with the same care we give to all assignments but with definite attention to making sure that all students in the class can make a reasonable attempt and have an expectation of doing well, assuming that they have participated in the preceding activities.

Rubrics provide an excellent way to show students what you want and guide their developmental steps and remove any ambiguity about what is expected. Publishing a rubric prior to the assessment item being due is vital as it may also expose questions or uncertainties that you weren't aware of.

If two students hand up work that is (legitimately) similar, then their marks should be very close to each other. If similar work can attract very different marks, it may be worth looking at the weighting that is being assigned to each marking criterion. Again, a rubric is a good way to visualize your expectations for students.

4.5 Effective Assessment

For example, at the University of Adelaide, the following rubric has been used:

	Assumptions and discussion (25 %)	Modeling (50 %)	Final result (25 %)
Fail	Assumptions chosen that result in oversimplification or otherwise poor modeling of the question without any justification	No model presented or model does not match problem description and/or chosen assumptions	No result(s) included or presented result(s) bears little or no resemblance to the correct result(s)
	No discussion of general and/or special cases	Little/no working presented. Incorrect/no justifications made for each deduction. No intermediate states shown	
Pass	Assumptions well chosen but poorly justified OR poorly chosen but well justified	Model mostly matches problem description and chosen assumptions	Partially correct result(s) presented
	Limited discussion of general and/or special cases	Small amount of working presented. Very little justification provided for each deduction. Incorrect or no intermediate states shown	
Credit	Most assumptions well chosen and well justified	Model mostly matches problem description and chosen assumptions	Results presented differ only slightly from the correct result due to processing errors or correct results not clearly presented
	Small amount of discussion of general and/or special cases	Most working out and justifications presented for deductions. Limited number of correct intermediate states shown	
Distinction	All assumptions well chosen and fully justified	Model correctly matches problem description and chosen assumptions	Correct result(s) clearly presented in their entirety
	Significant discussion of the general and/or special cases of the problem	All working out and justifications presented for all deductions. Limited number of correct intermediate states shown	
High distinction	All assumptions well chosen and fully justified. Some discussion of alternative interpretations included	Model correctly matches problem description and chosen assumptions	Correct result(s) clearly presented in their entirety
	Complete discussion of the general and/or special cases of the problem	All working out and justifications presented for all deductions. All intermediate states shown	

An ideal marking scheme would not reward knowledge that was gained outside the course, unless you can guarantee that all students would have that knowledge. The marking scheme should be self-contained to the Puzzle-based Learning instance that you are running.

Valid Effort and Outcome Should Both Receive Recognition

One of the most difficult problems for any teacher is marking any partial attempt towards a solution that does not result in a correct solution or a correct solution when presented with no explanation (if it is required). Firstly, assignment items would ideally always be designed so that arriving at the right answer requires either the correct process or the most astounding luck. (Apart from the obvious reason, this reduces the possibility of students being confused as to which process they should follow if a correct and an incorrect formulation can both give the correct answer. This is particularly important for probability questions.)

Quantifying what constitutes valid effort is difficult, unless clear guidance is given to the student. This is, again, where supplying a rubric or marking guide to the students can be extremely useful. If you are looking for a thoughtful solution to a problem and a student, instead, solves it analytically by an exhaustive search with a computer – is this worth marks? It is probably not enough to, *post hoc*, tell students that you didn't want them to do it a certain way. It is far more preferable to clearly outline the parameters at the outset.

Courses that the authors have run have recognized problem reformulation, identifying the key aspects of the problem, then the process, and, finally, the solution. By reducing the value of the solution to 25 % of the final mark, this naturally encourages students to show their work and emphasizes that it is their process we are interested in. However, we now have to weigh the effort and nature of their solution process and, at the same time as we are saying "any solution path is valid," mark students down for not using a particular approach.

In the early stages of a Puzzle-based Learning course, students will need clear guidance and well-established scaffolding, which we can fade as they learn when to apply their skills and in which context. However, we must always be ready for a student to solve something through a clever application of technique that we had not expected – and then give them all of the marks.

Assessment Should Be Based on the Student, Not the Class

While curve grading is heavily used, the application of curve grading takes the intrinsic pitfall of grading students against a rigid marking scheme and makes it even worse. Now a student can fail because they were not only idiosyncratic, but they were individual – both facets that we are trying to encourage in this type of class. How can we maintain a student's engagement and enthusiasm if all of their achievements and improvements are being overshadowed because they are slightly below an increasing class average?

Assessment Should Be Repeatable

The marking and assessment scheme should be clear enough that a marker will arrive at a similar mark. To be more precise, there should be no implicit assumptions that are needed to be made in the head of the marker, nor should we have to depend upon the knowledge of the class itself or an individual student to arrive at a grade for a piece of work.

Assessment Should Always Be Accompanied by Constructive Feedback

A mark of 7 out of 10 tells a student nothing about how to find those remaining 3 marks. Where assessment is carried out, it should always be accompanied by the information that students need to start bridging the gap to full marks. If the entire assessment plan has been carried out, then there should be clear indicators as to what the student has to do. However, there is almost always scope for encouragement in this type of feedback but it must be honest. The absence of any work results in the award of zero marks. However, if a student has done nothing, then 0 out of 10, *"You haven't done anything here. Would you like to talk about it? Have you thought about approach X,"* is a much better start than just 0 out of 10.

Where Possible, Err on the Side of Less Formal Assessment Rather than More

Questions on the techniques of puzzle-solving are easier to assess fairly, under traditional mechanisms, than questions that involve puzzle-solving. But, even where we are assessing Puzzle-based Learning, we must be aware that the assessed items have the highest risk of not actually meeting the intention of the course. A course that is rich in examples and collaborative and formative activities, with a carefully contained set of assessment items, is more likely to generate creative puzzle-solvers than a course built on weekly high-value quizzes and lots of exams.

Constructing the assessment for a course is always a challenge, but we can reduce the risk of students freezing up or doing irreparable damage to their grade by devising a set of smaller activities, which are worth smaller amounts individually. Students will find certain puzzle areas more or less easy to work with, so allowing a choice of questions can also be very useful, especially in a final examination situation.

With smaller value assignments, it is easier to be generous in marking, as the individual weights of each point are lower. If the argument is between 7.5 or 8 out of 10, it is easier to award the 8, unless there is a very clear and rubric-based reason to award 7.5. If the difference between 7 and 8 is only 0.5 of an actual final mark percentile, then this is not worth worrying about. The benefit of that extra half mark to the student, in terms of confidence and encouragement, is probably worth the 0.25 benefit that they have received in their overall grade.

Tying all of the problems into one coherent framework can help students to assign value and link together the required tasks, based on the necessity of achieving the identified "grand challenge." If you can take a project-based approach to assessment, where students are trying to solve a larger problem made up on many small puzzles, marking sensitivity can be decreased as you focus on how students are progressing on their large project.

4.6 Increasing and Maintaining Confidence

The currency of Puzzle-based Learning is confidence. It is earned by solving puzzles and it is spent on hunting for solutions, draining away very quickly if a student feels that they are achieving nothing. Students start a Puzzle-based Learning course in three modes: no confidence, confidence based on previous experience, and false and overinflated confidence. In our experience, warm-up exercises are essential to starting to build confidence in group 1 and to address the problems implicit in group 3.

From a learning design perspective, teachers should consider regularly inserting some simpler problems or "quick gets" to remind everyone how far they have come. If the class is heading towards a known obstacle, in whichever area, then a few shorter exercises or highly rewarding group activities, especially those with physical interaction and some theater, will boost confidence to bolster the students.

Language use is essential and, as always, it's important to minimize negative language while still being honest about what is being said. If any part of as a student's solution can be used or praised, do it. This can assists all of the students in forming a better mental model of the solution. However, if something is not relevant here, then say so, without editorializing, and move on.

Puzzle-based Learning is challenging and is, once again, unlike most of what our students do. They are already in a potentially uncomfortable and unfamiliar space, which requires more scaffolding than usual.

Students who do not try cannot find a solution. Students who are too scared of failing won't try. Students who have no confidence in themselves or the course will be worried about getting a bad grade or looking stupid. The course needs to be built up in a methodical and efficient way so that students have confidence in what is being taught. Then, by showing students how to solve things and training them, we can build up their personal confidence and increase their willingness to try, because we are reducing their fear of the risk of failure.

4.7 Peer Teaching

Peer teaching is one of the most effective ways to get students involved, engaged, and confident, because we move away from the power structure between teacher and student and move to a peer-to-peer model. Students are more likely to be able to successfully explain solution steps to other students because they are much closer to each other in age and context. There will always be a wide variety of students in a class and similar acculturation is not guaranteed but a large class can be scaled up quite effectively by getting students used to working in small groups and trusting each other's feedback.

Students must value each other's feedback if they are to invest time in sharing solutions. Group formation is one thing but group collaboration is another. Once groups have formed, the teacher moves into a facilitation role to move around the groups and provide feedback, inject hints, identify good steps or solutions, and disseminate information to all of the groups.

There are many roles a student can play in a group: leader, supporter (who provides positive support of suggested ideas), critic (who questions the solution), scribe, and so on. Rotating these roles will require all students to get involved and will reduce the chances of one or two students doing all of the work while the rest sit back. A teacher/facilitator should be watching for group members who are active or passive and make sure that they inject enough questions to make every group member active at some point. When a solution is being presented, a successful group will be full of students trying to contribute.

Students within a group should be explaining to each other why a given solution or solution step is required. When consensus is reached, it is then time for that group to seek out another group to try and convince them.

Consensus may be hard to reach and sometimes a group will want to present two solutions. While you can hear both, it is useful to then settle on a key point of distinction between the solutions and focus the group on this point, as this will often break the deadlock. It is essential to avoid having group members "cut out" of the group as they will disengage and not participate in further activity.

Warm-ups are an excellent way to get groups working together and the groups formed will continue to operate outside of the classroom, extending the peer-teaching model to informal external study groups and even social activities. A good peer group gives students a measure of expected performance and a reason to come along to activities, with the benefit of providing peer teaching as well.

To summarize Part I, three questions that we all have as instructors of a course are: (1) What knowledge and skills should students learn? (2) How can I facilitate their learning? (3) How do I determine how well they have learned via formative and summative feedback? Part I of this book has discussed our experience in addressing these questions in the context of Puzzle-based Learning. In Parts II and III, we contextualize these questions using a wide variety of puzzles.

Reference

1. Collins A, Brown JS, Newman SE (1989) Cognitive apprenticeship: teaching the crafts of reading, writing, and mathematics. In: Resnick LB (ed) Knowing, learning, and instruction: essays in honor of Robert Glaser. Lawrence Erlbaum Associates, Hillsdale, pp 453–494

Part II
Tools, Tips, and Strategies

In this part of the book, we explore puzzle-solving in a way that will help teachers to understand important aspects of structuring puzzles, how each technique works, and how it can be applied in the classroom. It is easy to say "Apply this technique" or "Get the students to understand the problem," but it can be far harder to achieve that outcome. Our experience in the classroom clearly indicates that good preparation depends upon a sound understanding of *how* to apply a technique. As teachers work with the puzzles in this section, we will share our thinking and our experience to assist Puzzle-based Learning teachers in using these techniques successfully to bring new and innovative puzzles into the classroom.

We begin (Chap. 5) by looking at determining what the problem is that we are trying to solve, a process that we refer to as *taking inventory*. Once we have listed all the important aspects of the problem, we can then find out if we have described everything that is *actually* important. Many students get stuck because they have overlooked an element that is crucial to achieving a solution.

As well as taking inventory, we will also discuss how the skill of *modeling* a problem can take a seemingly impenetrable problem and lay it bare for the solving. Modeling, as a mechanism for abstraction, also allows us to draw on the strength of analogy, where we can draw upon a student's existing knowledge in a related area and show them that, yes, they already know how to solve this, if only they modeled it correctly.

The ability to draw a useful *diagram* is a very important skill and we tackle this next, focusing on only drawing what is useful and avoiding the pitfall of drawing an intricate and beautiful picture that draws us no closer to finding a solution. Diagramming, as a skill, is very rewarding when done correctly, but a useless diagram may often consume more effort and time than a useful one, because a lack of direction leads to a focus on trivial or nonessential elements.

Having built a firm foundation for understanding *what* we are trying to solve, we then focus (Chap. 6) on one of the great solving skills, that of *reasoning*. Reasoning, forwards or backwards, and an understanding of simple logic, can propel us towards a solution with a surprisingly small set of facts. The quote opening this chapter is from that great exemplar of deductive reasoning, the immortal Sherlock Holmes, emphasizing the great possibilities of a keen mind, sound reason, and a good knowledge of what is, and what is not, a fact.

Human beings see patterns in everything and Chap. 7 addresses *pattern recognition*, to reflect how we can take advantage of our existing proclivity to find ways towards a solution. We address the problems of seeing patterns where there are none, as well as helping teachers to understand how crucial context, and even cultural familiarity, can be with seeing given patterns.[1] Patterns go well beyond recognizing shapes, as many strategies depend upon seeing the pattern of the game, and we spend some time discussing how simple strategy games can be discussed and solved, once the pattern is seen.

Chapter 8 presents *enumerate and eliminate* technique and it deals with problems where we attempt to identify all of the possible solutions (enumeration) and then reduce this (potentially vast) number down to only that set which will meet our requirements. As well as providing excellent examples of problems that can be taken from mathematics, biology, or resource usage in general, we also provide a basis for the discussion of problems that have solution spaces so vast that we cannot do a full enumeration.[2] There are many subtle aspects of enumeration as sometimes the way in which we organize or label the contents of our solution space is important and sometimes it is not. This is introduced with a number of examples that should clarify uncertainty about the importance of order or designation, with straightforward notation. The examples often lend themselves to a classroom example, where the use of puzzle props allows students to experiment and interact with the problem. We use the term *manipulatives* throughout this section to refer to physical artifacts that students can handle as part of a problem work-through.

Simplification is a very powerful technique, but oversimplification allows the solving of a different, and less powerful, problem. In Chap. 9, we discuss a technique for solving a simpler version of a problem, changing the way that we represent a problem, and transforming the problem from one that we can't solve to one that we can. Some of the authors' favorite puzzles are in this section, as they clearly illustrate the benefit of *thinking about a puzzle*, and many of these puzzles contain powerful *Aha!* moments that give students a great deal of satisfaction.

Chapter 10 is on the use of *gedanken*,[3] constructing a solution through the use of thought experiments where we say "*What if this were so?*" and "*This fact is true. So what? What does this tell us?*" There are many places where the brute force application of equations will eventually yield a solution, but reasoning and, in particular, experimental reasoning where we contemplate what *could be* and what would occur if it *were true* provides us with a thoughtful and elegant way to

---

[1] As a simple example, the observation of religious icons or messages in random patterns is referred to as pareidolia, but what is seen generally depends upon the religion of the observer and their known languages.

[2] This naturally leaves open the door to the fundamental aspects of optimization that, while the detail is beyond the scope of the book, we introduce shortly to provide additional resources for PBL teachers.

[3] We adopt the term gedanken as convenient shorthand for the German word *gedankenexperiment*, which literally means "thought experiment." As the word is no longer truly "in German," we do not capitalize it.

approach a large number of potential puzzles. The correct application of the gedanken technique allows us to solve puzzles that appear to need specialist knowledge of physics and mathematics, because we can reason our way to solution by posing questions and testing the new puzzle universe that is thus constructed.

The final chapter in this part (Chap. 11) provides a high-level tour of some key aspects of simulation and optimization (as opposed to earlier chapters of this part that concentrate on particular problem-solving techniques). As noted, a detailed explanation of either of these areas is best left to other texts, but many problems presented in this book have either simulation or optimization flavor. So we believe that it is important that teachers of Puzzle-based Learning have enough knowledge to be able to answer student questions on these areas or to extend their own knowledge and vocabulary. This chapter is kept at a higher level and can safely be left until other techniques and approaches are more familiar to the reader. However, if this area is of interest, the authors strongly recommend reading more detailed texts on simulation, *algorithmic techniques*, *heuristic methods*, and *optimization*.

5 Understand the Problem

> *The mere formulation of a problem is far more essential than its solution, which may be merely a matter of mathematical or experimental skills. To raise new questions, new possibilities, to regard old problems from a new angle requires creative imagination and marks real advances in science.*
>
> – Albert Einstein

In this chapter we look at one of the biggest stumbling blocks for students and teachers alike: working out what the problem actually *is* so that we can solve the right problem. When approaching puzzles, some people feel overwhelmed because they can't even start on a path to a solution. We show you in this chapter that with some preparation and practice, most people will be able to make a good start on even the most (initially) overwhelming puzzles!

5.1 Take Inventory

When we first look at a problem, it is very important to work out what we know, what we can determine, and which solution we are looking for. In determining the facts, the unknowns, and the goal of a given puzzle, we take puzzles from being mysterious and unsolvable to better known and solvable. Students can find this process challenging because they do not understand how to define the problem and, therefore, whether they have an appropriate approach that will allow them to solve it. The act of assessing the problem, what it means and how it could be approached, we refer to as *taking inventory*. Some puzzles are deliberately constructed to make this process harder, whether they obscure a simple pathway to a solution or are deliberately misleading in order to offer a simple path that leads the puzzler away from the correct approach. Some puzzles have a simple pathway to the solution that is naturally obscured. More precisely, to take the inventory of a problem is to carry out a systematic and thorough recording of all of the things that we can know or should be able to derive from the problem description. Students need to develop their skill in working out what the puzzle is about if they are going to succeed.

Let's start with the following puzzle.

Problem 5.1

"As I was going to St. Ives,
I met a man with seven wives,
Each wife had seven sacks,
Each sack had seven cats,
Each cat had seven kits:
Kits, cats, sacks, and wives,
How many were there going to St. Ives?"

Discussion 5.1 This is a classic problem, but the inventory of the problem is incomplete, despite us having a very large number of cats, sacks, and kittens to count, because we have an ambiguous aspect as to who is going where and, in a very pedantic reading, what it even means to be *going* somewhere. (The cats are carried and have no agency in the matter: Are they going or being taken?) Most problems are simpler than this. Let's refer back to a river-crossing puzzle (footnote 13 of the Introduction):

> *"A farmer must transport a wolf, a goat and a cabbage across a river but has a boat that will only hold him and one other object. If left alone, the wolf will eat the goat. Similarly, without the farmer present, and not having already been eaten, the goat will eat the cabbage. How can the farmer safely transport everyone to the other side? The farmer must be in the boat to move it from one side to the other – cabbages can't row."*

> **Teacher Tip**
> Don't forget the value of humor in getting important points across.

There are many versions of this puzzle, with the original variant using the fox, the goose, and the bag of beans, but the inventory of the problem is the same. We have a set of objects (farmer, boat, wolf, goat, and cabbage), and that is where many people stop because their inventory of the problem does not include the other facts. Let's list them:
1. The farmer must be in the boat for transit.
2. The boat only holds two objects, one of which must be the farmer.
3. The wolf and the goat cannot be left together without the farmer.
4. The goat and the cabbage cannot be left together without the farmer.

5.1 Take Inventory

These are the explicit statements, and, like many logic puzzles, you may find that students start to ask other questions, some of which will be easier to answer than others. (The fact that a cabbage takes up the same space as a goat or that someone is carrying a wolf around for any reason, given the issues it causes, can be an interesting and amusing diversion for older classes. Injecting humor is a good way to get engagement underway!)

There are other pieces of information that we can derive from the puzzle statement, namely, the following:

1. The farmer, wolf, goat, or cabbage must either be on one bank or in the boat.
2. The wolf doesn't like cabbage (although this is actually an assumption as the problem doesn't work otherwise).
3. The farmer can safely be left with any item. (Again, otherwise the problem doesn't have a solution.)
4. There is no other way across the river except for the boat. (Sometimes this is explicitly stated as students will often ask if they can throw things across the river or similar.)

In terms of implied inventory, we may appear that we are "making things up," and this can be a source of confusion and frustration to students who may feel that we are arbitrarily allowing some fabrication while not allowing them to state that "the farmer has a balloon" or "I brought a wolf muzzle with me." While short, concise puzzles are always desirable, your class may require more detail to guide them away from trying to cross the river on the back of steam-powered dinosaurs. (Again, play can be very useful, but you do need to solve the puzzle at some stage.)

The benefit of the inventory, for this problem, is that we can now clearly talk about every aspect of the problem in terms of where things are and then limit the number of possible movements down to only those that will obey the rules as we've listed them.

In more precise terminology, for more advanced students, we have captured the *legal states* and the *legal state transitions* for the problem – solving this problem is now a matter of applying transitions until we arrive at a solution or prove that no solution is available. This also provides a good way to get students to explore the inventory of a problem – if they make an assumption that cannot result in a solution, then they'll have to step back and look at their inventory.

> **Teacher Tip**
> If students want to know more about states and transitions, there is an entire area of work called "finite state automata" that may be very interesting for more advanced students.

A good inventory can be summarized as follows:
Understand the problem and all the basic terms and expressions used to define it.

In the puzzle above, students often add extra rules and block themselves. When looking at why a student has become stuck, it often helps to ask them to go through their inventory. Puzzles that are over-constrained are usually insoluble, whereas puzzles that are under-constrained may appear trivial, because the students who create a deus ex machina[1] or a *get out of jail free* equivalent, due to the lack of constraint, have not actually solved the problem.

Looking at everything that makes up the puzzle can also help students to realize that they do have enough information to be able to produce a solution – even if they think that they do not.

Another example problem, which often requires some localization, is the "jar puzzle":

> *"You have three jars. One of them is full of chocolate-coated peanuts, one is full of solid chocolates, and one jar is full of a mixture of the two. You cannot tell, by looking or feeling, which chocolate is in which jar. All of the jars have labels but also all of them are mislabelled. If you have to taste to find out what a given chocolate is, how many will you have to taste in order to work out which jar is which?"*

The inventory of this problem can be deceiving. Students will often try to work out if there is some method where, without tasting, they can determine the contents. Weighing the jars or chocolates is sometimes suggested so an additional constraint on the puzzle, which you add as students ask, can be that a given chocolate weighs the same as any other. (Not overly realistic, but who carries a wolf, a goat, and a cabbage?)

What is the problem? We want to work out which jar is which, by tasting *some number* of chocolates. What do we know? We have three sets of contents and three jars. The contents are, in simpler form, PEANUT, CHOC, and MIX. At this point, students often try to explore the inventory and ask questions such as:

- What is the mixture? Is it 50/50 or can it be as small as everything in the jar is the same except for one that is different?
- Can I pick which jar I look at first? (Some students think that they have to randomly select a jar, which is a greater level of misdirection than is required.)

When asked the "mixture" questions, it is probably easiest to answer "50/50," as this will then encourage students to attempt a solution. Solutions often include "I need to sample 50 % of the first jar I pick, plus one, as that will then tell me if it is a jar of one thing or a mix." Other students, expecting a trick, will also suggest that you have to try no chocolates – asking them why they think this can highlight inventory problems but also reminding them that a solution must have a justification!

---

[1] *Deus ex machina* is a Latin phrase that means "god from the machine" and was used to indicate where, in a story or play, the author had solved an apparently insoluble problem by bringing in a previously unseen character, fact, or (in some cases) machine to suddenly solve the problem.

> **Student Pitfall**
> When in doubt, many students seek more concrete information on the problem, and, when this is forthcoming, they may concentrate on numerical solutions instead of thinking about the general problem.

The answer is that we can solve this problem by tasting a single chocolate.

The key part of the inventory that many students miss is that while we have three sets of contents (PEANUT, CHOC, and MIX) and three labeled jars, where we assume that they are labeled PEANUT, CHOC, and MIX, for ease, we know that *every jar is mislabeled.* Therefore, we are guaranteed that the MIX label *cannot* be on the MIX contents. By tasting what is in the MIX-labeled jar, one chocolate, we then know what is in there. We can then add this fact to our inventory as we explore the problem. Say it is a PEANUT. We know that we only have MIX and CHOC contents left, and given that CHOC can't be in the CHOC-labeled jar, the MIX must be in the CHOC-labeled jar and the CHOC contents must be in the PEANUT jar.

In this case, we aren't looking at increased difficulty stemming from a complex or misleading description, but at students potentially not paying attention to a crucial part of the inventory. Getting stuck is a common indication that students have not completed their inventory process and reminding them of the need to:

- Understand the problem – what are we being asked to do?
- Extract the basic terms and concepts – what are the nuts and bolts of the puzzle?
- Look carefully to see if there are any additional facts that we can or should derive – what else can we see that is *reasonable* and *sensible* to add?
- Be able to start the problem – if there's no way forward, keep looking!

5.2 Build a Model

Solving a complicated problem usually involves two distinct steps: the first is a preparation step and the second is the solving step. The initial preparation step involves understanding the problem, framing the problem, drawing a diagram, and building a model. The second step is solving the actual problem or solving a model of the problem.

In traditional education, the students are usually just presented with a model to solve that is very often not connected with any particular problem. For example, "take the derivative of the function: $e^x \cos(x)$." In Puzzle-based Learning, the student is challenged to solve problems, and to do this the student often has to construct a representative model of the problem. This is the hard part. This is the part that requires skill, expertise, and experience. This is the skill for which employers are looking.

Models range from simplistic to complicated and come in many different dimensions and types. They also can be mathematical equations or computer models. The one thing all models have in common is they provide a representation of the problem or part of the problem.

Models have been used to provide insights into problems for thousands of years. It's hard to imagine the construction of the first wheel without a model preceding it. For hundreds of years, humans were trying to construct a model of the solar system in order to better understand it. Chemists and physicists constructed a mental model of the atom based on the way that they observed elements combining and the behavior of matter.

Today, as technology has advanced, modeling has become an even more important component of arriving at solutions to challenging problems. Indeed, many universities now offer advanced degrees in modeling and simulation – as without a model, it is hard to design a solution.

For many problems, building a model with *manipulatives*, which are simply any objects that the students can move around, can increase the efficiency of the problem-solving process manyfold. This is especially true if the problem involves a multiple-step procedure or has many components. The model does not have to be sophisticated. Indeed, scraps of paper – perhaps with an identifying letter or two scrawled on them – can be used as manipulatives.

On one exam in a problem-solving course, a question is asked for the number of different ways that a 4-by-1 strip of postage stamps can be folded into a 1-by-1 stack along the perforations (see Problem 5.2). After a few minutes of quiet thought, one enterprising student tore off the bottom inch of one of the test pages and began folding it into fourths. The sound of the tearing was audible, and pretty soon all the students had "built" a model of the stamp-folding problem.

> **Student Pitfall**
> Building a model is work, and students are often reluctant to invest the time needed to do so. Also, some students believe that actually building a model of the problem might be considered "cheating." It is the teacher's responsibility to disabuse the students of this notion.

This chapter contains a collection of problems that were specifically designed to demonstrate to the student that constructing a model is a powerful problem-solving tool.

Let's start with the problem presented just above.

Problem 5.2 In how many different ways can a 4-by-1 block of postage stamps be folded into a single pile along the perforations? Of course, the adhesive side and the printed side of the stamp are distinguishable as is the type of folding.

Discussion 5.2 Having an actual 4-by-1 block of stamps will make this problem much easier, but any sheet of paper will make a fine substitute. There are five different ways of *folding* the stamps, shown from the side in the figure:

5.2 Build a Model

Spiral   Accordion   Double tuck   Half and half again   One end tucked in

Each of these five folds has two orientations. The first four (spiral, accordion, double tuck, and half and half again) all have either portraits on the top and bottom or adhesive on the top and bottom. The "one end tucked in" fold is the only one in which the stack of four stamps has a portrait on one side and adhesive on the other. To see this, it is best to build a model. The visible part of the stack that is next to the stamp that is tucked can be either the printed side or the adhesive side. Students who don't recognize this – perhaps because they did not identify the adhesive side and printed side on their model – will get a number that is too low. It turns out that there are ten different ways to fold the stamps, given in the table below.

Fold type	Top shows	Bottom shows
Spiral fold	Adhesive side	Adhesive side
Spiral fold	Printed side	Printed side
Accordion fold	Adhesive side	Adhesive side
Accordion fold	Printed side	Printed side
Double-tuck fold	Adhesive side	Adhesive side
Double-tuck fold	Printed side	Printed side
Half-and-half again fold	Adhesive side	Adhesive side
Half-and-half again fold	Printed side	Printed side
Bottom stamp tucked in	Printed side	Adhesive side
Top stamp tucked in	Printed side	Adhesive side

> **Teacher Tip**
> Have the students clear their desks and give them a one-sheet quiz with this problem at the top. It will be interesting to note the manner in which the students solve the problem. Do they try to do it all in their head? Do they draw figures? Do they rip a strip off the bottom of the sheet and then start folding?

This problem is challenging enough even with a model that can be manipulated. Not surprisingly, the best way to see all ten different folds is to build a model. If the students don't "see" all ten ways, have them build a model and make them all. In fact, we suggest building ten models and having all ten types prepared simultaneously so they can be compared directly with each other. To ensure that the students understand the meaning of the word "different," you can have them explain why one type of folding is different from another.

Problem 5.3 Factory workers are allowed a ten-minute coffee break that must be started any time from 9:00 to 10:00 in the morning. If Fred and Ed start their breaks randomly during this hour, what is the probability that their breaks will overlap during the break hour?

> **Teacher Tip**
> This is a good opportunity to develop the students' intuition by asking them to estimate the answer. It will also teach them that making generalizations without careful thought can lead to errors.

Discussion 5.3 This problem can be solved in multiple steps by treating the three time intervals from 9:00 to 9:10, from 9:10 to 9:50, and from 9:50 to 10:00 separately. If Fred starts his break at 9:00 am, Ed must start anytime from 9:00 to 9:10 – a ten-minute window. If Fred starts his break at 9:05, Ed will overlap with Fred if he starts his break anytime from 9:00 to 9:15 – a fifteen-minute window. If Fred starts his break at 9:30, Ed will overlap with Fred if he starts his break anytime from 9:20 to 9:40 – a twenty-minute window. Finally, if Fred starts his break at 9:55, Ed must start his break from 9:45 to 10:00 for them to overlap.

Discussion 5.3 (cont) This problem can be modeled with a two-dimensional graph, as shown in the figure.

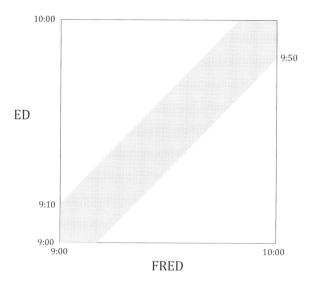

5.2 Build a Model

The start of Fred's break is equally likely to be anywhere on the horizontal axis, and the start of Ed's break time is equally likely to be anywhere on the vertical axis. If the point representing the two break times is plotted on the two-dimensional graph, the breaks will overlap if the point is within the shaded area. With this model, all that remains is to calculate the fraction of the total area that is covered by the shaded area. The entire square is 60-by-60 minutes, which represents an area of 3,600 square minutes, and the two large unshaded triangles at the upper left and lower right can be placed together to form a square that is 50 minutes on a side, making its area 2,500 square minutes. The shaded area must be the difference between these, which is 1,100 square minutes. The desired probability is then

$$P = \frac{1,100 \text{ square minutes}}{3,600 \text{ square minutes}} = \frac{11}{36} \approx 30\%$$

Student Pitfall
There are a couple of mistakes students will make here, and both involve generalizing the solution to a small part of the problem to the entire problem. The first would be, *"Assume that Fred starts his break at 9:00. For their break times to overlap, Ed must start his break anywhere from 9:00 to 9:10. Since this is one-sixth of the total time, the probability is one-sixth."* The second mistake would be, *"Assume that Fred starts his break at 9:30. For their break times to overlap, Ed must start his break anywhere from 9:20 to 9:40. Since this is one-third of the total time, the probability is one-third."*

Teacher Tip
Interpreting graphic models of data is an important skill. It is relatively easy for students to nod their heads in understanding when shown the graph. However, test their understanding by asking them specific questions about specific points. You can ask, for example, *"Where is the point on the graph that represents Fred starting his break at 9:10 and Ed starting his break at 9:50?"* Further, you can ask, *"Where is the point on the graph that represents Fred starting his break at 9:20 and Ed starting his break at 9:30?"* Finally, *"If Fred starts his break at 9:20, what time range can Ed start his break and overlap with Fred? Find this range on the graph."*

Problem 5.4 Here is a neat problem that should challenge the student's fundamental notion of the conservation of area. Consider two trapezoids and two triangles arranged in an 8-by-8 square as shown on the left. The same shapes are rearranged to make the 5-by-13 rectangle on the right. The problem is that the area of the square is 64 square units and the area of the rectangle is 65 square units. How can the total area of the four pieces possibly change when the pieces are rearranged?

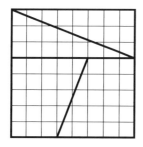
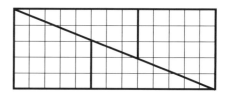

> **Teacher Tip**
> This is a mind-blowing problem, and we recommend that you let the students tackle this without any guidance whatsoever for at least ten minutes. You can have paper, pencils, and scissors available but don't rob them of the opportunity to engage their System 2 thinking and get a rewarding squirt of dopamine. It really helps to have the students working the physical objects as it makes the mystery of the disappearing area even more "magical."

Discussion 5.4 Clearly the total area of the four shapes can't be changing. If it was, you could purchase 8-by-8 sheets of gold, cut them into the four shapes, and rearrange them to get 65 square units of gold to make a nice profit. The relevant question is "*Where is the missing area?*" It must be somewhere inside the perimeters of the two shapes. The best way to find the missing area is to first carefully cut out the four shapes from an 8-by-8 square, perhaps using cardstock. Next, accurately draw a 5-by-13 rectangle on a piece of (preferably dark) paper. Now rearrange the four pieces within the 5-by-13 rectangle.

When this is accomplished, it should reveal the location of the missing area. That is, the four shapes should not completely cover the 5-by-13 rectangle. While geometry is not needed to uncover the location of the missing area, knowledge of geometry can determine the exact shape of the missing area.

The slope of the hypotenuse of the triangle is 3/8, which is 0.375. The slope of the slanted side of the trapezoid is 2/5, which is 0.40. So, the slopes do not match in the rectangle, despite the fact that they appear to form the diagonal of the rectangle in the figure. So, the missing unit of area is a long, thin rhombus along the diagonal of the rectangle.

5.2 Build a Model

> **Teacher Tip**
> There are numerous other, more devious, missing square puzzles available online. Pick one for the appropriate level of your students.

Problem 5.5 Here is the classic "Monty Hall problem" named for the host of the TV game show, "Let's Make a Deal." On the show, the host offers the contestant the opportunity to win the prize behind one of three doors. Behind one of the doors is the grand prize, and behind the other two are booby prizes. Once the contestant selects a door, the host, knowing which door contains the grand prize, opens one of the two doors that have a booby prize behind it. The host then gives this contestant the option of staying with their originally chosen door or switching to the other door, which still remains closed. So the question is: should the contestant switch or stay?

> **Teacher Tip**
> This is an excellent opportunity to develop the students' ability to express themselves logically by having a class discussion or by breaking up the class into smaller groups for more intimate discussions.

Discussion 5.5 Many students will not believe that the correct strategy is to switch doors and that the probability that the prize is behind the remaining door is 2/3. Often, logical argument simply isn't enough. When this is the case, a model of the game should do the trick. You don't need any fancy equipment, just three Styrofoam cups and anything to represent the prize, for example, a paper clip. You can do this demonstration in front of the class with one student or pair up the students and have them do, say, 30 trials, or you can do both.

> **Student Pitfall**
> There are obviously only two doors left, and the prize must be behind one of them. Students are very likely to conclude that the chance is the same that the prize is behind either door, perhaps arguing, "Well, it's either this one or that one, so it must be 50-50."

Discussion 5.5 (cont) Secretly hide the "prize" under one of the three cups, putting nothing under the two other cups. Then have the student select a cup. Then you lift up a cup that does not contain the prize and ask the student whether he/she wants to switch. Record two columns of data: whether the contestant switched and whether the student got the prize. When using a model like this, it should gradually dawn on the student that revealing a cup that the prize is not under does nothing to the probability that the prize is under the cup that the student selected – it is still one-third.

> **Teacher Tip**
> This problem also provides the opportunity to discuss statistics. You can ask, how many trials have to be performed to be able to state with confidence that the probability of getting the prize when switching is two-thirds rather than one-half? You can even talk about confidence level.

Discussion 5.5 (cont) We have found that two factors help the student have the *Aha!* moment more quickly. First, have the student place his/her finger on the cup and keep it there during the decision-making process. Second, *immediately* after the student makes a choice, lift up a cup with nothing underneath it and then pose the question.

Another way to attempt to get the student to understand that the probability of getting the grand prize when switching is not 50 % is to take the problem to extremes by considering what would happen if there were 100 doors, and after you picked one, Monty Hall opened 98 of them that he *knew* did not contain the grand prize. So, there are two unopened doors remaining: the one you chose originally and the only other door that Monty did not open. The chance that you guessed correctly is still 1 %. The probability that the prize is behind the other unopened door is 99 %.

Problem 5.6 Three humans and three zombies need to cross a river in a boat. The boat will only hold two at a time. With six humans, it would take nine trips (five going there, four going back) because one has to bring the boat back. However, the presence of the three zombies causes a problem. If the humans are outnumbered by zombies on either bank of the river at any time during the crossing, the zombies will attack the humans. Note that if there is a single human on one side of the river, two zombies can't cross to that side to drop off a zombie. Even if the "extra" zombie is in the boat, an excess of zombies on one side will lead to dead humans. What river-crossing procedure will prevent any zombie attack?

> **Teacher Tip**
> The goal of a problem-solving class is to develop the students' problem-solving skills. To get the students to both appreciate and utilize problem-solving tools, it is important that they independently come to the conclusion that the tools can be useful. To demonstrate this, you can do something like separate the class into two groups: one with six manipulatives to represent the humans and the zombies (they can be actual plastic figures or even red and black checkers) while the other group must solve the problem without manipulatives. Another possibility, which the students seem to like, is to act out the problem in front of the class with six students.

5.2 Build a Model

Discussion 5.6 The problem can be solved much faster if there are actual objects that can be moved back and forth. The difficulties in the problem will become apparent more quickly when shuttling objects rather than trying to solve the problem by drawing arrows on a sheet of paper. In fact, if the only thing that is available is the paper on which the problem is written, good problem-solvers will tear out six small pieces of paper and label three of them with a Z for zombies and the other three with an H for humans. The solution can be accomplished in eleven steps, as follows:

Move	Start	Finish
	ZZZHHH*	
1. Z&H go across	ZZHH	ZH*
2. H comes back	ZZHHH*	Z
3. Z&Z go across	HHH	ZZZ*
4. Z comes back	HHHZ*	ZZ
5. H&H go across	HZ	HHZZ*
6. H&Z come back	HHZZ*	ZH
7. H&H go across	ZZ	HHHZ*
8. Z comes back	ZZZ*	HHH
9. Z&Z go across	Z	ZZHHH*
10. Z comes back	ZZ*	ZHHH
11. Z&Z go across		ZZZHHH*

The asterisks in the table represent the position of the boat.

Problem 5.7 A red car traveling at a constant speed of 20 m/sec passes a blue car that is initially at rest. When the blue car is passed, it accelerates at a constant rate of 4 m/sec every second. How much time elapses between the red car passing the blue car and the blue car passing the red car?

Discussion 5.7 While this is a "physics" problem, there is no need for any equations other than the fact that the distance traveled is the average speed during a time interval multiplied by that time interval. The position of the two cars can be modeled by mathematical equations, with variables representing each parameter. Traditionally, the letter x is used to represent distance, and v is used for speed and t for time.

We'll define the "position" of a car as the position of its front bumper. We'll define $t = 0$ at the point where the red car passes the blue car and this starting position as $x = 0$. So at $t = 0$, the position of the red car is $x_{Red} = 0$, and the position of the blue car is $x_{Blue} = 0$. The position of the red car can then be modeled by the equation

$$x_{Red} = \frac{20\,m}{1\,sec} \times t$$

because the average speed of the red car is always 20 m/sec. In this equation, x_{Red} represents distance the red car is from the point where it passed the blue car at any time t.

The position of the blue car as a function of time is not so trivial to calculate because its speed is changing all the time. Nonetheless, the same equation applies. To get the average speed of the blue car, we first need an expression for the speed of the blue car as a function of time. The speed of the blue car is simply

$$v_{Blue} = \frac{4\,m/sec}{1\,sec} \times t$$

Since the blue car starts from rest and speeds up at a constant rate, the average speed of the blue car from $t=0$ to any time t is simply one-half its current speed. For example, if a car starts from rest and accelerates uniformly to 50 mph, its average speed while going from rest to 50 mph is 25 mph. Similarly, a car that decelerates from 70 to 50 mph has an average speed of 60 mph over the interval. So, the average speed of the blue car is

$$\overline{v_{Blue}} = \frac{1}{2}\left[\frac{4\,m/sec}{1\,sec} \times t\right]$$

where the bar above the v_{Blue} indicates average. Now we can use the average speed of the blue car to calculate the distance the blue car has traveled as follows:

$$x_{Blue} = \frac{1}{2}\left[\frac{4\,m/sec}{1\,sec} \times t\right] \times t = \frac{2\,m}{sec^2}t^2$$

This equation models the position of the blue car as a function of time. To find the time it takes for the blue car to pass the red car, we set the positions equal:

$$x_{Red} = \frac{20\,m}{1\,sec}t = x_{Blue} = \frac{2\,m}{sec^2}t^2$$

This reduces to

$$10t = \frac{t^2}{sec}$$

This equation has two solutions, $t=0$ and $t=10$ seconds. At $t=0$, the red car is passing the blue car, and at $t=10$ seconds, the blue car is passing the red car.

This can be modeled graphically by plotting the positions of the cars on a graph. The position of the red car is a straight line with a slope of 20 meters/second, and the position of the blue car is a parabola whose slope starts at zero and increases by 4 meters per second every second. The lines meet at $t=0$ and $t=10$ seconds.

Debriefing There are many types of models that can be used to facilitate the problem-solving process. Here we presented a few examples. While not always useful, the problem-solving technique of building a model should be a weapon in the arsenal of any good problem-solver. The best way to develop the students' ability to effectively utilize the technique of modeling is to present a wide variety of problems and puzzles.

5.3 Draw a Diagram

Visualization is widely used in business, professional sports, and science. Businesses use organizational charts to visualize the hierarchy of the company, spaghetti charts to depict the connectivity of steps in a process, and flowcharts to illustrate the logic of a plan. Many elite athletes use visualization techniques to enhance preparation, focus, and confidence as well as reduce their fear, nervousness, and apprehension. Scientists use graphs to represent the relationship between variables, and there are hundreds of named diagrams in the sciences.

There are also Venn diagrams, control charts, Pareto charts, fishbone diagrams, stem-and-leaf diagrams, and, perhaps the most common diagram of them all, a map. Humans have been producing maps for thousands of years. When trying to envision a path from one point to another, it is much easier to look at a map rather than a list of directions. Wherever we have a space that can contain objects, regions, or themes, we can draw a diagram to symbolically depict the relationships between them. However, as any visualization often requires sacrifices of accuracy to be made, either because of limitations in a projection or because of constraints on precision, any visualization often tells us a great deal about the person who constructed it!

For experienced problem-solvers, sketching a diagram is a fundamental problem-solving tool as it allows the solver to both identify and connect all of the key aspects of the problem in a way that is easy to understand and work with. In fact, drawing a diagram is often the first thing an experienced problem-solver does – often as a way to understand the problem before attempting to solve it.

It is easy for experienced problem-solvers to assume that students will naturally draw a diagram to help them "wrap their head around" the problem. However, our experience tells us that it is not natural for many students to grab a pencil and some paper to help them with a problem. In fact, many students will try to solve a challenging problem entirely in their heads – sometimes giving up before drawing a diagram. We often see confusion as to what we mean when we ask students to visualize. If we ask them to produce a design, then the final artifact (the design itself) is often seen as being indicative that the design process has been carried out. However, this is far from true, and many student "designs" are nothing more than a sketched symbolic representation of text, lending nothing to the solution process.

> **Student Pitfall**
> Many students seem to have a natural resistance to draw a diagram. Some even perceive it as a sign of weakness or even "cheating." You can encourage students to draw a diagram simply by ensuring that they always have a pencil and some paper handy. Perhaps the best way to convince them of the utility of drawing a diagram is to present problems whose solutions are accessed in a straightforward fashion with the help of a simple diagram. We have to emphasize to students that the production of the diagram will start them thinking about the problem and that any advantages gained from this change of representation should be seen as valid and useful!

If the student is instructed to draw a diagram in the statement of the problem or if the actual solution to the problem involves a diagram, the students will not need any further prodding to use a pencil and paper. An example of such a problem is, "*Jim has a small collection of apple saplings that he would like to plant in a formation that has five rows of trees with four trees in each row. However, he only has ten apple trees. Draw a diagram that shows how this can be done.*"

As a teacher, your goal should be to give the students ample opportunities to discover for themselves that drawing a diagram is often a tremendous help in solving and should be one of their first methods of attacking a challenging problem.

It is helpful to introduce students to the notion of the impact of a visual representation, and cartography provides one of the most important bodies of work in this area. Over time, the representations of countries have had wide-ranging political and economic impacts, all because of how people think of places, based on where they are on a map. Many students have no idea that the terms "First World, Second World, and Third World" are geographically defined, rather than economically defined, or that some countries, such as Turkey, reject simple classifications as being European or Asian.

Particularly good examples of representations that reduce the problem to a solvable scale can be seen in the maps of subway and underground systems, such as the New York Subway, the Hong Kong MTR, and the London Underground. The London Underground was the first transit map[2] where geographical detail was removed to make the map easier to read in terms of train connections – which was what the map was designed to do. Up until that time, maps tried to conform to geographical reality, which could get in the way of communicating information such as which train should be taken to get to a given stop. Most modern transit maps remove "difficult" geography to provide an abstract representation of the underlying network that is excellent for taking the train – and, at the same time, potentially misleading for navigating the local geography!

This gives us some simple guidelines for drawing a diagram as a problem-solving technique:

---

[2] http://www.tfl.gov.uk/corporate/projectsandschemes/2443.aspx

5.3 Draw a Diagram

1. *Try to model the problem in simple objects and connections.*

    Can you take the problem that you have and think of it in much simpler terms? Your visualization will be built on this mental model, so it's important that you have a mental model to work from. Be careful on introducing biases and assumptions at this point, because they will be stuck in the visualization until the end and may be difficult to remove.

2. *Find a representative visualization that makes it easier to see the problem.*

    Students should think like Harry Beck, the designer of the London Underground map, and concentrate on the problem at hand, rather than previous solutions that appear to mandate a given approach. If you want to count possible configurations, then your visualization must show the configurations in a countable way. All visualizations change the emphasis on the information that we have been presented with, and the choice of visualization can make the difference between solving the problem and not solving the problem.

3. *Only draw in the components that are important.*

    Adding unnecessary detail takes time, increases the difficulty of drawing the diagram (and hence updating it if you make changes), and distracts you from the core of the problem. Which components are important will vary by problem and by the perspective of the solver, but practicing transferring a problem into an efficient diagram is important. Reducing the complexity of the diagram reduces the amount of time wasted sketching in details that aren't required. This also stops students from thinking that they are making progress when, in reality, all that they are doing is drawing a picture.

    Cutting down on the complexity also reduces resistance from students who feel that they cannot draw. When a student is equipped with simple shapes, lines, and shading, most problems can be represented successfully.

4. *Take into account symmetry, if appropriate.*

    Many problems contain an implicit or explicit statement of symmetry – what appears to be a large set of unique configurations are actually a smaller set with lots of repetition caused by symmetry. However, it is arguable whether a student failing to see the symmetry of an item like a bracelet (which is set in a circle) is at fault as they have allocated a unique positional value in the linear representation that is not present in the ring formation (see bracelet puzzle – Problem 5.8). If there is scope for confusion, it is always more helpful to remind students about what a physical representation would look like. Not all students know what bracelets are, much as describing something in strictly mathematical terms, without a visual example, will stump many students.

5. *Think about the real world as much as you need to.*

    Very few visualizations need to take into account all of the realities of the situation, but they do intrude. We can happily talk about the pigeonhole principle[3] (called also Dirichlet's box principle) without capturing a pigeon

---

[3] The pigeonhole principle says that if you have a number, p, of pigeonholes and you have $p+1$ things to put in there, at least one pigeonhole has more than one item in it.

(or building a box), but we cannot sketch the shortest path for an aircraft across the Earth on a flat map without drawing a curve.

6. *Draw as many diagrams as you need to understand the problem.*

If the diagram that you have drawn doesn't help, then draw another diagram. Every diagram, much like every solution attempt, is a valuable exercise in thinking, and both your drawing ability and your problem-solving ability will gradually improve with practice. There is no point thinking about a diagram that you reject before drawing it, unless you are convinced that there is a flaw in the model that you are using for the diagram.

The next few problems were specifically chosen to help the students appreciate the usefulness of drawing a diagram.

Problem 5.8 Alice is producing bracelets that consist of six beads and six uniform bead connectors. She has three boxes of components: a large box of black beads, a large box of white beads, and a large box of connectors that are used to snap the beads together. How many different six-bead bracelets does she make? Bracelets are different only if they can be distinguished by the arrangement of the black and white beads, as the connectors are indistinguishable.

Discussion 5.8 Note that this problem also gives the students practice with the subject of one of the earlier sections, understanding the problem. They may ask clarifying questions about the meaning of the word "different." This is a good sign, as it means that they have grasped the importance of understanding the problem. When students ask about the meaning of "different," we have found that a good response is, "Two bracelets are different if you can tell them apart. That is, bracelets are different if you can describe a feature of one bracelet that the other bracelet does not have."

Student Pitfall

Students who are not experienced problem-solvers may rely on mathematical formulas to arrive at an answer. For example, $2^6 = 64$ is a popular wrong answer. Students reason, "*Six beads, two possible colors, the answer is 2^6.*" Also, we have received the answer of $6! = 720$ on more than one occasion.

Also, some students will draw bracelets but will draw them in a linear fashion rather than in a circle – perhaps because they are using lined paper. This leads them to double-count arrangements such as the two shown below. These two configurations will actually form the same bracelet when arranged in a circle – three black beads next to each other and three white beads next to each other.

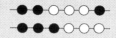

5.3 Draw a Diagram

Discussion 5.8 (cont) A good problem-solver will start in an organized fashion and draw actual bracelets as shown in the figure below.

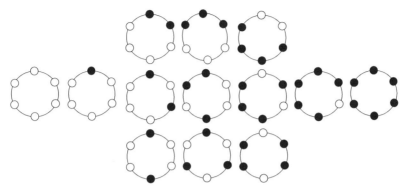

Starting from left to right, the number of black beads in the bracelet goes from zero to six. So, the answer is that there are 13 possible different bracelets.

As is sometimes the case, students with substantial mathematical training can struggle with this problem because they invest their time trying to plug numbers into formulas to get the answer rather than by drawing an appropriate diagram.

Many eighth graders, for example, often have no trouble with this problem because they adopt the straightforward method of simply drawing circular bracelets.

Teacher Tip

If you give this problem with the instructions, *"Draw all the possible bracelets,"* the students will be able to solve the problem much faster. To really help the students to solve the problem, you can provide sketches of thirteen "blank" bracelets so they can define a different bracelet simply by shading in beads. However, that is not the point of the exercise. The goal is not to help the students solve as many problems as possible; the goal is to develop the students' ability to solve problems independently. When the students struggle to get the answer without drawing the bracelets, they will learn from experience that drawing a good representative diagram is a useful problem-solving strategy.

Here is a second example for which drawing a diagram is a virtual necessity.

Problem 5.9 The surface of a traditional soccer ball is a tiling of pentagons and hexagons. Each pentagon is surrounded by five hexagons, and each hexagon is surrounded by three pentagons and three other hexagons. What is the ratio of number of pentagons to number of hexagons on any soccer ball?

Discussion 5.9 This is a nice problem for many reasons, one of which is that it offers a few different paths to the solution. Most of them start with a simple sketch of the surface of a soccer ball.

One path to the solution is to reason thusly. Each pentagon has five hexagons around it. However, each of these five hexagons is shared by two other pentagons. So, each pentagon has 5/3 of a hexagon. This is the solution to the problem. If there is 5/3 of a hexagon for every one pentagon, there are three pentagons for every five hexagons.

Another way that a diagram can be helpful is to draw dashed lines that split the hexagons in thirds – one-third to each of its three adjacent pentagons. With this drawing, it can be seen that each pentagon "owns" five hexagon-thirds, making it clear that there are 5/3 of a hexagon for each pentagon and hence five hexagons for every three pentagons.

On an actual soccer ball, there are 20 hexagons and 12 pentagons.

Yet another way this problem has been solved is to focus on the seams defining the shapes rather than the shapes themselves. With a diagram, it is apparent that half of the seams are between two hexagons and half are between a hexagon and a pentagon. Of course, there are no seams between two pentagons because the pentagons do not share a border.

Since it takes five hexagon–pentagon seams to make a pentagon and only three hexagon–pentagon seams to make a hexagon, there are 5/3 as many hexagons as pentagons. On a traditional soccer ball, there are 120 seams, 60 of which are between two hexagons and 60 of which are between a hexagon and a pentagon.

5.3 Draw a Diagram

The 60 pentagon–hexagon seams define the 12 pentagons and half of the 20 hexagons, and the 60 hexagon–hexagon seams make up the other half of the seams on the 20 hexagons.

Another problem that will challenge the student is to ask how many points are there at which three seams meet on a traditional soccer ball. This question is virtually the same as the question, "*How many carbon atoms are in the molecule buckminsterfullerene?*" which has the surface configuration of a soccer ball. Images of this amazing molecule can be found on the Internet.

Note, finally, that those students familiar with atomic arrangements in crystal structures (as an advanced chemist or physicist might be) may get the solution to this problem without drawing a figure because they can recognize the connection between the bonds that different atoms form in a crystal and the connectivity of the different shapes on a soccer ball. This problem-solving skill is discussed also in Chap. 7.

Here is another example that will clearly demonstrate to the students that a diagram can be a very helpful problem-solving tool.[4]

Problem 5.10 Two cars are on the highway traveling at constant speed. The blue car is going 120 km/h and is 200 meters behind the red car. That is, the front of the blue car is 200 m behind the front of the red car. It takes one minute for the blue car to catch up to the red car (the front of both cars are aligned). How fast is the red car going?

Discussion 5.10 Despite the fact that the only "fact" you need to know to solve this problem is

$$d = v \times t,$$

many students will struggle with this problem simply because they lack the experience to draw a clear, well-labeled figure. A picture and a logical thought process will lead directly to the answer.

Now   One minute later

200 m

---

[4] Sometimes it helps to have a soccer ball in the classroom if students are really stuck because then they can manipulate the object in space as they think about it.

At a speed of 120 km/h, the blue car travels 2,000 meters in a minute. From the figure, we can see that the red car travels only 1,800 meters in that time. Therefore, the speed of the red car must be 1,800 meters/min, which is 108 km/h.

So far we have demonstrated diagrams that involve a visualization of event or a structure. Next, we reveal another type of diagram – a chart that organizes the possibilities.

Problem 5.11 Suppose it is known that 1 % of a large population has a certain type of cancer. It is also known that a test for this type of cancer is positive in 99 % of the people who have it but it is also positive in 2 % of the people who do not have it. What is the probability that a person who tests positive has cancer of this type?

> **Student Pitfall**
> This is another problem that some students will stare at for a long time and never draw a diagram. Perhaps the reason is that it appears like a math problem that can only be solved with a set of simultaneous equations. While a set of algebraic equations can be used to solve this problem, a simple diagram will provide a better conceptual understanding of the problem.

Discussion 5.11 A simple 2×2 table, like the one shown here, will allow virtually all students to begin the problem-solving process.

		TEST Result	
		Positive	Negative
Cancer?	Has Cancer		
Cancer?	No Cancer		

In fact, if the class dynamics are appropriate, perhaps you can perform an experiment on the students by giving half the class a simple statement of the problem and the other half the same statement of the problem along with the diagram and the instructions to fill in the four spaces in the table with populations assuming that the total population is 100,000. You can also enumerate the possibilities for them as shown below:

There are four possibilities:

1. A person does not have cancer and tests negative. This is a *true-negative* test and is represented by the square on the lower right.
2. A person has cancer and tests positive. This is a *true-positive* test and is represented by the square on the upper left.
3. A person does not have cancer and tests positive. This is a *false-positive* test and is represented by the square on the lower left.

5.3 Draw a Diagram

4. A person has cancer and tests negative. This is a *false-negative* test and is represented by the square on the upper right.

Once the diagram is in place, the problem can be solved by entering numbers in the four boxes consistent with the facts given in the statement of the problem. This strategy is another problem-solving technique discussed in Chap. 9. With a population of 100,000, there will be 1,000 with cancer of this type because it is given that 1 % of the population has this cancer. It is also given that the test is positive in 99 % of the people that have it. So, of the 1,000 people that are affected by this cancer, 990 will test positive and 10 will test negative. Since this cancer affects 1,000 members of the population, this type of cancer does not affect 99,000 members of the population. The test will be positive in 2 % of this population, which makes 1,980 false positives. The remainder of the population will have true negatives. Now we can complete the table as follows:

		TEST Result	
		Positive	Negative
C a n c e r ?	Has Cancer	990	10
	No Cancer	1,980	97,020

The question at hand is: "What is the probability that a person who tests positive has cancer of this type?" We can see from the "Positive" column in the diagram that there were a total of 2,970 positive tests and the person that tested positive actually had cancer in 990 of them. The fraction 990/2,970 is one-third. So, if a person tests positive, the probability that they have this type of cancer is only one-third.

This is counterintuitive to most people, who would expect that a positive test would indicate much higher likelihood of cancer, given the stated high accuracy of the test. This illustrates why calculating the solution is such a useful activity to undertake.[5]

Problem 5.12 A survey was taken of incoming freshmen to determine the extent of their experiences. Two of the thirty questions were: "Have you ever been to Europe?" and "Have you ever been scuba diving?" There were six times as many freshmen that had been to Europe and had not been scuba diving than there were freshmen that had gone scuba diving but had not been to Europe. Also, the number of students that had done both was twice as much as the number that had been scuba

---

[5] For more advanced discussion of this, we need to look at statistics. In statistics, you will read discussions of Type I errors and Type II errors, which represent the false positive and false negative, respectively. Incorrectly accepting an alternative hypothesis that is not true is a false positive, where failing to reject an incorrect null hypothesis is a Type II error.

diving but had not visited Europe. Finally, the percentage that had done neither was eight times the percentage that had done both. What percentage of the freshmen has never gone scuba diving, and what percentage of the freshmen has never visited Europe?

> **Teacher Tip**
> In order for the students to appreciate the value of a diagram, it is best to present this problem without the prompt to draw a diagram. When you introduce the problem, don't even tell the students that it will demonstrate the usefulness of a diagram. When students figure something out for themselves, it is much more valuable than when they are directed to the solution with hints. Remember, the overarching goal is to produce students who can independently and efficiently attack a problem that they have never seen before.

Discussion 5.12 A Venn diagram is very handy here. It is very useful when showing logical relations among sets. Here we have four groups of students, those that have been to Europe but have never been scuba diving, those who have been scuba diving but have never visited Europe, those who have done neither, and those who have done both.

The set of freshmen that have gone scuba diving is represented by an oval, and the set of freshmen that have gone to Europe is represented by a different oval. The overlapping region between these two ovals represents the freshmen that have done both, and the region outside both ovals represents the freshmen that have done neither. The area of the four sections must sum to 100 % because all of the freshmen belong to one of the four groups.

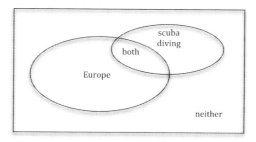

We know that there are six times as many freshmen that had been to Europe and had not been scuba diving than there were freshmen that had gone scuba diving but had not been to Europe. So the area of the lower chopped oval is six times the area of the upper chopped oval. We also know that the number of students that had done both was twice the number that had been scuba diving but had not visited Europe. Finally, we know that the number of students that had done neither is eight times as many as the students who have done both. Now we can assign relative values to the four areas in the diagram.

5.3 Draw a Diagram

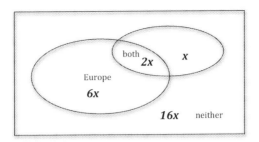

Now the only thing that remains is to add these four areas up and set the sum equal to 100 %. The equation is simply

$$25x = 100\%$$

which makes $x = 4$ %. The diagram is now

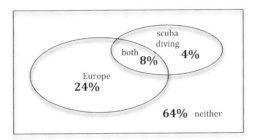

So now we can answer the question. From the diagram, we can see that 12 % of the freshmen have been scuba diving, which means that 88 % of the freshmen have never been scuba diving. We see that 32 % of the freshmen have been to Europe, which means 68 % have never been to Europe. 64 % have done neither.

Problem 5.13 A wooden cube that is 3 inches on each side is spray-painted blue on all sides. This cube is now cut into 27 smaller 1 inch cubes (ignore the kerf[6]).

(a) How many of the smaller cubes are painted blue on three sides?
(b) How many of the smaller cubes are painted blue on two sides?
(c) How many of the smaller cubes are painted blue on one side?
(d) How many of the smaller cubes are painted blue on no sides?

Discussion 5.13 This is a problem that we routinely give when introducing the Rubik's Cube. The students should immediately see the connection when trying to

---

[6] The *kerf* is the slit made by cutting the wood and normally would remove a small amount of the wood. In this case, we're ignoring it as it doesn't add anything to the puzzle. Amateur woodworkers do so in the real world at their own peril!

solve the cube. The key to the solution is simply to draw a diagram and count. To an experienced problem-solver, it is quite remarkable how many students will try to do this problem in their head without drawing a diagram.

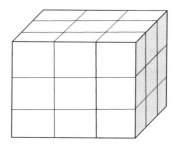

Teacher Tip
It is often the case that the students' answers to parts a, b, c, and d do not add up to 27. To help them realize that it should, you can ask, "Is there any way to check if your answer is consistent?"

Discussion 5.13 (cont) A diagram quickly reveals that a cube has eight corners and each corner piece is painted blue on three sides. A cube also has twelve edges, and each edge piece has two blue sides. A cube has six faces, and each 1 inch cube in the center of each of them is painted blue on one side. Finally, there is one cube in the very center that gets no paint at all.

Making the connection back to the Rubik's Cube, we note that there are eight smaller cubes that have three colored stickers on them, 12 smaller cubes that have two colored stickers on them, and six smaller cubes with only one sticker. The standard Rubik's Cube coloring has the white face opposite the yellow face, the red face opposite the orange face, and the blue face opposite the green face. With this knowledge, you can ask questions like:
"How many corner cubes have both a blue and a white sticker on them?"
"How many edge cubes have both a blue and a white sticker on them?"
"How many corner cubes have both a blue and a green sticker on them?"

Debriefing Over our many years of teaching problem-solving, we have seen many students nonplussed when problems are presented without a diagram. A good problem-solver will not be stymied by a complicated problem and will take steps both to better understand the problem and to eventually solve it. One of these steps is to draw a clear diagram.

Reasoning: Logic and Reasoning Backwards

6

> *In solving a problem of this sort, the grand thing is to be able to reason backwards. That is a very useful accomplishment, and a very easy one, but people do not practise it much. In the every-day affairs of life it is more useful to reason forwards, and so the other comes to be neglected. There are fifty who can reason synthetically for one who can reason analytically.*
> – Sherlock Holmes

Starting from the end of the problem and working backwards to the original condition is a problem-solving technique that should be part of any good problem-solver's arsenal. This is a problem-solving technique that is well known and is used in many disciplines. It is also known as *retrograde analysis, backward chaining*, and *backward induction*. As Holmes noted, however, familiarity with the usual approach of trying to push forwards until we reach a stumbling block, often giving up, often prevents us from trying a more successful approach of starting from the end and working backwards!

Detectives and forensic scientists often work backwards when they examine a crime scene and try to deduce what happened. Of course, they have little choice, as they must, by definition, show up at the end of the matter. Mazes in puzzle books can often be solved faster by starting from the end and trying to find a trail to the start. Why? Because we know that the place we start from *must* be part of the final path as must any single-choice options that stem from it. In business, many problem-solving sessions begin with where the company wants to position itself in the future and then move backwards to the present to discover the best way to get there.

While a formal course in logic is beyond the scope of this book, some simple logical statements can help students to understand why we say certain things about a puzzle and accept them as facts while not accepting other statements. Because of the nature of puzzles, students are often asked to extract as much information as possible (recall inventory taking, Sect. 5.1) while still not assuming so much that the problem becomes trivial. Logic provides a good basis to make the decisions as to when something assumed is correct or incorrect. Here we are concentrating on *deductive* reasoning, where we attempt to establish what follows from our assumptions.

For many of our puzzles, we usually look at all of the facts presented and then attempt to deduce any additional facts. What can confuse students is that any number of reasonable assumptions are often heavily identified with one culture, and lacking that cultural knowledge prevents inference! Here are some of the terms that you will read about associated with logic, which may be of help when explaining this to students.

Conjunction Conjunction is used to join two statements together to provide a third, and this new statement is true if both of the original statements are also true. For example, mathematically, if x is positive and y is positive, then $x + y$ must be positive as well. When students assume something, but one element is not always a fact, then we can quickly point out that we cannot assume the conjunction, because a given element is sometimes false.

True and False We often assume that we can state facts in terms of true and false. In many cases, if something is not said to be true, then, for the puzzle in question, we usually have to assume that it is false. The exception is that if through human experience or other context, we could always assume it to be true. (It's worth noting that there are schools of logic where this is not the case. If you wish to read more on this, you can look up the *open world assumption* and the *closed world assumption*.)

Revisiting the "St. Ives" puzzle (Problem 5.1), we run into problems because we do not have enough information to answer, definitively, for everything mentioned in the poem whether it is true or false that it is going to St. Ives! However, we are aware that at least one person is going, because it's clearly stated, and therefore the answer "*At least one*" is valid, because for the answer to be "at least one," we must be able to answer "True" for at least one person.

Reasoning Logically If we know that a fact is true, then if we *negate* that fact, we are saying that it is not true. This is often used in reasoning to establish whether a given conjunction is true or false. For example, let us define a dog as an animal that barks and has four legs. In logical terms, *if* it is true that this thing is an animal *and* it has four legs *and* it barks, *then* it is a dog. The *and*s in the previous sentence indicate conjunctions where we are connecting logical statements together.

How is this useful in teaching? For whatever it is we are looking at to be a dog, all of the logical requirements must be true. Therefore, for it *not* to be a dog, at least one of them must be *false*! So, if we see the negation of any of the key facts, we know it's not a dog. Thus, if it's not an animal *or* if it's not barking *or* if doesn't have four legs, it can't be a dog. This is a vital part of reasoning as it allows us to reduce the number of assumptions that we can make about a puzzle. We can reject anything where we have enough contradictory evidence: that is, if a statement depends upon a given fact being true and it is shown to be false, we can reject that statement.

This chapter contains a collection of problems that were specifically designed to demonstrate to the student the benefits of logic and that starting from the end and reasoning backwards is a powerful problem-solving tool.

Problem 6.1 Alice is looking at Bob; Bob is looking at Trudy. Alice is married; Trudy is not. Is a married person looking at an unmarried person?

Discussion 6.1 This is an interesting problem because it has three possible answers: *"Yes," "No,"* and *"We don't have enough information."* Let's look at it logically. We are asked if it is true that a married person is looking at an unmarried person. Because of the information we have, we know that there are two ways this could happen.

Since Alice is looking at Bob and Alice is married, then if Bob is unmarried, it's true (Statement 1). Secondly, since we know that Trudy is unmarried, then if Bob is married, it's true (Statement 2). Let's turn this into a simpler form. We can see that people are, somewhat simply, married or unmarried. So, in terms of logic, let's assume that people who are not married are unmarried and represent married people by their initial and unmarried people by an initial with a ~ in front of it. The facts that we have are:
A, which means "Alice is married"
~T, which means "Trudy is not"
and we don't know about Bob. However, we know that the answer to the whole thing revolves around whether either of Statement 1 or Statement 2 is true:
(S1) or (S2)
but these turn into
(A and ~B) for S1, because S1 is true if Bob is unmarried
(B and ~T) for S2, because S2 is true if Bob is married
but this means that the answer must be true because Bob can only be married or unmarried and we only need *one* of S1 or S2 to be true. If Bob is unmarried then S1 is true, but if Bob is unmarried then S2 is true. One of these is always true! Therefore, we can answer the question, and, yes, someone married is looking at someone unmarried! (This is referred to as a *tautology*, something that is always true.)

> **Student Pitfall**
> Most people assume that the answer is that it can't be solved and give up. Restating the problem logically makes it very clear that the answer exists! Students often assume that something cannot be solved as it then allows *them* to reject *the problem*, rather than potentially having to admit that they don't yet understand how to solve it. Students can seek to reclaim agency by refusing to engage in the work or rejecting a puzzle as "silly" or "unrealistic." This often masks discomfort with the teaching environment or approach.

> **Teacher Tip**
> Logical representations can be confusing especially for younger students, so you should feel free to add symbolic representations, like wedding rings if culturally appropriate, to indicate marital status on pictures of the three participants. More advanced students could logically reduce the problem down to what it really is (A and ~T) – as long as these two things are true, the problem must also be true. You can experiment with changing one of these to see what happens.

Here is an example that clearly demonstrates the importance of the ability to reason backwards.

Problem 6.2 A small colony of algae starts to multiply on the surface of a small pond on what we will call day one. The amount of the pond's surface area covered by the algae doubles every day, and it completely covers the surface of the pond on day number 10. On what day was the pond half-covered with algae?

Discussion 6.2 This is a good example of a problem that is challenging to solve in any other way but by reasoning backwards. The key question here is, "*How much of the pond was covered on day 9?*" If it took 10 days for the algae to completely cover the pond, it must have been half-covered on day 9 because the area it covers doubles every day, so the answer is 9 days.

> **Student Pitfall**
> Many students will struggle with this one because they think that there is not enough information available. It is common to get questions like, "*How big is the pond?*" and "*How much of the pond is covered by the algae on the first day?*"

6 Reasoning: Logic and Reasoning Backwards

> **Teacher Tip**
> In order to get the students to appreciate the value of the "work backwards" technique, let them struggle with the problem on their own for some time. The purpose of presenting this problem to the student is not for the student to get the answer; the purpose is to develop the students' appreciation of the strategy of reasoning backwards. This is best accomplished by allowing the students to work on the problems independently.

Here is an old problem (Russian origin) devised to have a moral as well as teach children mathematics.

Problem 6.3 Idle Ivan was lounging by a river trying to figure out a way to increase the amount of coins that he had in his pocket without doing a lot of work. The devil appeared and made him a proposition. The devil said he would double the money in Ivan's pocket every time he crossed the bridge. All he asked in return was a payment of eight coins after each bridge crossing (and after doubling the money). Ivan accepted the proposition. He crossed the bridge for the first time, and his money doubled. He paid the devil eight coins and crossed again. His money doubled again, and he again paid the devil eight more coins. He crossed for a third time, and his money doubled yet again. However, he only had eight coins left and had to give them all to the devil, thus leaving him broke. How many coins did Ivan start with?

Discussion 6.3 This problem can be solved with either of the guess-and-check or the increment-and-iterate technique. This approach allows us to explore possible solutions by guessing a solution and then checking to see if it is correct. If it isn't correct, then we modify our guess and try again. However, a more straightforward (and efficient) method is to work backwards. The last step in the problem is Ivan paying the devil eight coins after crossing the bridge for the third time. So, Ivan must have had four coins before crossing for the third time. Therefore, he must have had 12 coins before paying the devil eight for the second time and six before crossing for the second time. Finally, he must have had fourteen coins before paying the devil eight coins after the first bridge crossing, and, therefore, he must have started with seven before crossing the bridge for the first time.

The three crossings can be represented mathematically as follows:

$$(7 \times 2) - 8 = 6 \, and \, (6 \times 2) - 8 = 4 \, and \, (4 \times 2) - 8 = 0$$

When solving this one from the beginning, the solution involves a guess, whereas solving it from the end leads directly to the solution. Students can exhaust easily if too much guess and check is involved, understandably as too many failed attempts will lead to frustration! Other approaches are possible here, and this problem is also a good candidate for building a model and denoting the initial number of coins as a variable and then going on to build a set of equations.

Problem 6.4 Consider the two-player game in which the goal is to take the last pebble from a pile of twenty-one pebbles.[1] The players alternate turns, and each player has the option to remove one, two, or three pebbles at each turn. How many pebbles should the first player remove?

Student Pitfall
Many students have difficulties with games such as this because they only look at the next step, rather than any deeper into the game – and it's important to get students to think strategically. That is, they need to assume that their opponent will take the best advantage of anything that they do rather than playing a move and hoping that their opponent won't see how to defeat them.

Teacher Tip
This is a very simple game to play, and you don't need pebbles; paper clips or any other classroom commodity will do fine. If the classroom dynamics allow, pair up the students and have them play multiple games. You might want to direct the student to alternate who moves first and perhaps record the results of the games. When actually playing the game, students will usually not start thinking deeply until the end game. That is, they will use their System 1 until there is a single-digit number of paper clips left on the table whereupon they will engage their System 2. By then, however, it may be too late.

Discussion 6.4 This problem is relatively simple when it is attacked from the end game. A little thought will reveal that if your penultimate move leaves four pebbles, then you are assured a victory because the opponent must leave one, two, or three pebbles on the table and you will be able to remove them all. Stepping back further, you will see that if you leave eight pebbles after your antepenultimate[2] turn, your opponent can't prevent you from leaving four pebbles on the table after your next turn. This, as we have seen, ensures a victory. Continuing in this fashion, it is clear that the only winning move is to remove one pebble on the very first move, leaving twenty on the table. If the first player to move removes anything but one pebble, the second player can seize the advantage simply by leaving sixteen pebbles on the table after his or her turn and a multiple of four on every turn thereafter.

Problem 6.5 A high school band is having a cupcake sale at the Friday evening football game as a fund-raiser. Before the game started, 80 cupcakes were sold. During the game, they sold one-half of what remained, and after the game was over,

---

[1] This is the first example of the so-called *Nim* game – for more information, see Problem 7.4 and the following discussion.

[2] "The one before the penultimate"

they cut the price in half and sold two-thirds of what remained, leaving only ten unsold. How many cupcakes did they start with?

Teacher Tip
There is nothing wrong when starting this problem from the beginning. In fact, you can ask the students to guess how many cupcakes they started with. Let's say a guess is 200. Starting with 200, there will be 120 left when the game starts. They sold half of what remained during the game, and this would leave 60; they sold two-thirds of what remained after the game, so that would leave 20. This is too many, which means that the initial guess of 200 is too high. Have a student make another guess and then follow the same procedure. When the problem is initially solved with this guess-and-check technique, the students will appreciate the "work backwards" technique and be better prepared to use it in the future.

Discussion 6.5 Starting with the ten remaining cupcakes, we find that there must have been thirty remaining when the game was finished. Since half were sold during the game, there must have been sixty when the game started. And since eighty were sold before the game started, they must have started with 140 cupcakes. Again, producing a set of equations to model the problem could provide the solution.

Problem 6.6 Albert, Betty, and Chris leave the barn on an early winter morning from the lower right corner at positions A, B, and C, respectively. Each has a chore to accomplish in the pasture before letting the horses out. Albert has to break the ice on the water trough at A′, Betty has to repair a section of the fence at B′, and Chris has to dump a bale of hay in the feeder at C′. A fresh blanket of snow has covered the pasture, and this gives Albert, a computer science major at Montana State University, an idea. He wonders "*Can all three of us do our chores without crossing paths made in the snow?*"

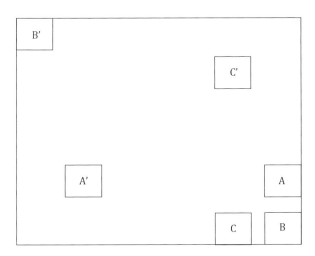

Discussion 6.6 This is actually quite a challenging problem, because the B–B' connection makes it look impossible at first glance. To use the reason-backwards technique, simply rearrange the boxes a bit and then connect them with straight lines. Now if we can move the boxes to their original positions without any paths crossing, we have the solution.

We can start by sliding the two B boxes into their positions, distorting the connecting path as necessary. Next we slide the two C boxes into position as shown, again stretching and curing the connecting path. Finally, we can slide box A', which represents the water trough, to the left to its final position, distorting the B and C paths as necessary.

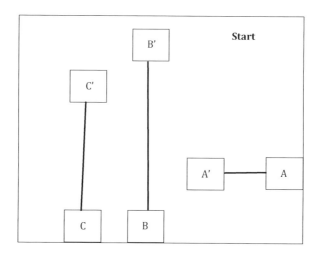

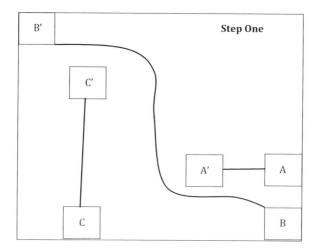

6 Reasoning: Logic and Reasoning Backwards

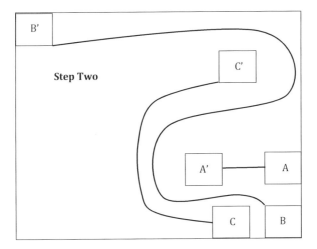

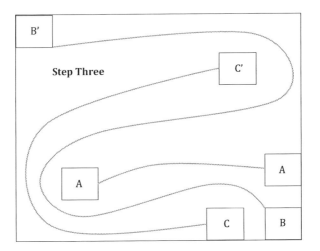

Problem 6.7 A young medical student is assigned the task of going to a remote village and returning with two 1-cc samples of blood that need to be analyzed. When he arrives at the village, he finds that his syringe broke during transport and that the only things he has to measure volume are a 5-cc vial and a 7-cc vial. The 100-cc blood sample is in a sterile IV bag that can be drained through a tube using a plastic valve. The student takes his assignment quite literally and is determined to return with two carefully measured one cc blood samples – one in each of the two vials. How can this be accomplished?

> **Student Pitfall**
> Students are used to focusing on the first step of the problem. If they invest their time trying to figure out which vial to fill first, they will probably not make much progress. In this problem, the best first step is to start from the end of the problem and work backwards.

Discussion 6.7 It is very challenging to solve this problem from the beginning and somewhat straightforward by working backwards. Since the only containers are the IV bag and the two vials, the final step must be to transfer the last cc remaining in the bag to an empty vial with the other vial already containing one cc. There is no other way. To measure one cc, we need to have three ccs in the 7-cc vial and a full 5-cc vial. By filling the 7-cc vial from the 5-cc vial, one cc will remain in the 5-cc vial. Therefore, the second-to-last step must have been pouring a full 5-cc vial into the 7-cc vial that already contained three ccs. The way to measure out three cc of blood is by filling the 5-cc vial, dumping it into the 7-cc vial, filling the 5-cc vial again, dumping it into the 7-cc vial, and leaving three cc in the 5-cc vial. Since the above procedure requires filling the 5-cc vial three times and one cc must be left in the IV bag, the position that will allow the measurement of two one-cc samples is sixteen cc remaining in the IV bag and both vials empty. These final steps are shown in the table below:

Volume in each container			
IV bag	5-cc vial	7-cc vial	Procedure
16 cc	0 cc	0 cc	Start
11 cc	5 cc	0 cc	Fill 5-cc vial from IV bag
11 cc	0 cc	5 cc	Transfer contents of 5 cc to 7 cc
6 cc	5 cc	5 cc	Fill 5-cc vial from IV bag
6 cc	3 cc	7 cc	Top off 7 cc from 5 cc
6 cc	3 cc	0 cc	Dump contents of 7 cc
6 cc	0 cc	3 cc	Transfer contents of 5 cc to 7 cc
1 cc	5 cc	3 cc	Fill 5-cc vial from IV bag
1 cc	1 cc	7 cc	Top off 7 cc from 5 cc
1 cc	1 cc	0 cc	Dump contents of 7 cc
0 cc	1 cc	1 cc	Fill 7-cc vial from IV bag

Now the only problem that remains is to get the one hundred cc in the IV bag down to sixteen cc. The most efficient way to do this is to fill the 7-cc vial twelve times, discarding the contents of the vial each time. This will result in sixteen cc remaining in the IV bag, and the procedure in the table shows the way home from there. It should be clear to the students that spending time trying to determine whether to first fill the 5-cc vial or the 7-cc vial without knowing what is going to happen at the end is inefficient. The way to make progress on this problem is start from the end.

Debriefing While there are numerous problems that cannot be easily solved by reasoning backwards, there are enough of them that the technique is one with which a good problem-solver is very familiar. The technique is useful for problems that have a specific end situation or position that has to be reached. The best way to develop the students' ability to independently recognize when the reason-backwards technique would be useful in solving a particular problem is to give the students lots of problems to solve without telling them how to solve them.

Pattern Recognition 7

Those who can remember their past are fortunate to reuse it.
— Raja Sooriamurthi

Our ability to recognize patterns is very useful in solving a variety of problems. Once we identify the pattern, it might be easier to suggest a solution – whether this might be to predict the next (or missing) symbol, number, action, or event (in the same way that fraud detection systems try to discover patterns in historical data and then use these patterns to predict which new transactions might be fraudulent[1]).

Before we move forwards with the material, it is worthwhile to emphasize that our ability to recognize patterns is of utmost importance. If we can identify a pattern, then we can build a model to find a solution. Marilyn Burns, in her book *I Hate Mathematics*, wrote: "*The password of mathematics is pattern.*" Indeed, in many branches of mathematics, we search for patterns that allow some generalizations.

Humans are very, very good at finding patterns – in fact, finding patterns in random data can be as unhelpful as finding a legitimate pattern is helpful! We search for patterns everywhere; we even recognize patterns in words (classes usually enjoy the slide with this text)[2]:

Aoccdrnig to a rscheearch at Cmabrigde Uinervtisy, it deosn't mttaer in waht oredr the ltteers in a wrod are, the olny iprmoetnt tihng is taht the frist and lsat ltteer be at the rghit pclae. The rset can be a toatl mses and you can sitll raed it wouthit porbelm. Tihs is bcuseae the huamn mnid deos not raed ervey lteter by istlef, but the wrod as a wlohe.

However, often the issue is to convince ourselves that the discovered pattern is the right one. If you look at your class as your students read the puzzle, you will often see a degree of satisfaction on the students' faces as they start to construct

---

[1] For more information on how patterns can help us to detect fraud, look for Benford's law, which uses an unexpectedly predictable distribution of numerical digits to detect possible fraud.
[2] This text circulated on the Internet in September 2003. See http://www.mrccbu.cam.ac.uk/~mattd/Cmabrigde/for translations in other languages

meaning from what is garbled nonsense. This joy of recognition can be one of the most satisfying *Aha!* moments to manufacture, when it works.

> **Teacher Tip**
> If there is any opportunity (e.g., there is some additional time available), it is highly recommended that the students watch (with the instructor) a video of famous Polya's lecture *"Let us teach guessing"* wherein Polya beautifully illustrates several problem-solving heuristics in the process of deriving a solution to the 5-plane problem.[3] Note that the maximum numbers of segments in the 3-dimensional space generated by zero, one, two, or three planes are 1, 2, 4, and 8, respectively, suggesting a wrong pattern.... This would emphasize one of the main points presented here – that the discovered patterns should be checked carefully.

Let's start with the following puzzle (it is actually a very good puzzle to move the students into appropriate "pattern recognition" mood):

Problem 7.1 Sequences of 4-digit numbers are assigned one 1-digit number; a few examples of such assignments are listed below:

8809 → 6
7111 → 0
2172 → 0
6666 → 4
1111 → 0
3213 → 0
7662 → 2
9313 → 1
0000 → 4
3333 → 0
8193 → 3
8096 → 5
7777 → 0
9999 → 4
7756 → 1
6855 → 3
9881 → 5
5531 → 0

Now, the challenge is to find the number assigned to 2581.

---

[3] The 5-plane problem requires finding the maximum numbers of segments that 5 planes can generate in the 3-dimensional space.

7 Pattern Recognition

> **Student Pitfall**
> Some people claim that young kids usually solve this puzzle within 10 minutes. However, the higher your education is, the longer time it takes to find the solution! The reason might be that educated people tend to analyze numbers – their parity (or divisibility by some number), existence of double or triple identical digits, growing or shrinking sequences of digits, etc. – whereas young kids would pay more attention to visual representations of symbols and their smaller components.

> **Teacher Tip**
> After a while consider giving a hint to your students: to pay more attention to visual representations of symbols and their smaller components. In other words, look at the number 7662 as a collection of circles, curves, and line segments. And from here to the solution is just one step.

Discussion 7.1 Let us start by looking at the sequences that are worth nothing. We can see that the numbers 1, 2, 3, 5, and 7 appear to be worth nothing as there is no combination of these that gives anything other than 0. With that in mind, look at some of the numbers that are worth 1. It appears that, somehow, 6 and 9 are both worth 1.

> **Student Pitfall**
> It is very tempting to leap ahead with a solution and make a clear statement as to what is happening before we have *checked* the solution. Students should always check their answers to see if it explains everything that they see before they conclude that they have the answer.

Discussion 7.1 (cont) Look at the number 7662 again: there are two circles in this sequence (lower parts of digit 6), and the number of circles (enclosed loops in this case) matches the assigned number for this sequence! Cleary, this is the case for all sequences – this is why 8809 is assigned to 6 (four circles in double 8, one circle for 0, one circle for 6), 6666 is assigned to 4, and 1111 is assigned to 0. And, of course, $2581 \to 2$.

Teacher Tip
Consider making a remark to the effect that the presented puzzle represented a pattern recognition case where all information necessary for solving it was included in the sequence – no external knowledge was required for discovering the solution. However, in most pattern recognition cases, some external knowledge is necessary to proceed.

The following problem illustrates how a puzzle can draw upon information that most people will recognize but that they won't necessarily be able to draw upon when asked to do so.

Problem 7.2 What is the missing letter (marked by the "?") in the sequence:

A ? D F G H J K L

Student Pitfall
Of course, we can analyze this sequence by characterizing each symbol by its features. For example, the letters **A**, **F**, **H**, and **K** each consists of 3 line segments. The letter **L** consists of only two line segments, whereas letters **D**, **G**, and **J** include some curves. Is there any pattern to this? Or is it important to be more specific and distinguish between longer and shorter segments? For example, the letters **A**, **F**, **H**, and **K** consist of 3 line segments: two long and one short. Is that useful?

Or it might be that we need a different approach based on numbers (a very natural thing to do). There is an obvious correspondence between letters and numbers (as **A** is the first letter, **B** is the second, etc.). So, we can translate the sequence in question into the following sequence of numbers:

1 ? 4 6 7 8 10 11 12

This is a growing sequence, and the growth is a pattern. If we believe that the pattern is genuine (i.e., it did not arise by chance), then we can conclude that the second number must be **2** or **3**; thus, the missing letter is **B** or **C**. But which of these two? Students will generally give both of these, and it's always worth asking *why* they have chosen a particular answer as it helps them to understand whether they have a reason or not.

7 Pattern Recognition

Discussion 7.2 Actually, none of them. The "obvious" answer is **S**, as the sequence

A S D F G H J K L

represents the middle row of letters on a computer keyboard!

> **Teacher Tip**
> Consider making a few remarks to the effect that the last puzzle illustrated nicely the fact that many pattern recognition activities are based on some a priori knowledge. In general, pattern recognition activities can be based either on information extracted from a sequence itself or on some a priori knowledge. The last puzzle was difficult as the information extracted from the presented sequence was not helpful, and it was not clear what type of external knowledge should be applied. It is also a very interesting puzzle to apply in the contemporary classroom as we have observed students who look at the puzzle and then go back to their (open) laptops. There is often a moment of mild surprise and then realization as they discover that the answer is right in front of them.
>
> You may also recall Problem 5.8, where the students familiar with atomic arrangements in crystal structures may get the solution to the "soccer ball" puzzle.

Problem 7.3 The following sequence of seven symbols (commonly known as the M-heart-8 sequence) is "meaningful" in the sense that it is not random:

What is the next symbol in the sequence?[4]

> **Student Pitfall**
> Many comments from the previous puzzle would apply here – usually students are confused, and the pattern is not that clear. Also, as it was the case with Problem 7.1, it has been empirically observed that children are better able to solve this puzzle than adults.

---

[4] This puzzle has appeared in pop culture playing a role in the movie *The Oxford Murders* and an episode of the TV show, *The Simpsons*. Interestingly, children are better able to solve this puzzle than adults.

> **Teacher Tip**
> There are many ways to analyze this sequence (i.e., many ways to search for a pattern). One possibility is to analyze the features of each symbol. If you have used the four-number puzzle from above, students may retain an idea of looking at the geometric elements of symbols in an effort to extract meaning. If we try to break these symbols down, what do we discover?
>   For example, the first symbol consists of four line segments, whereas the second symbol consists of one line segment and two curves. Following the occurrences of line segments and curves, we notice that the symbols consisting of only line segments appear on the first, fourth, and seventh positions. Does this mean that the next symbol consisting of only line segments will appear on the tenth position? If so, can we find a pattern in the number, length, and position of these line segments? This is much harder, because:
> - The first symbol consists of four line segments: two long and two short; the two long segments are vertical, and the two short segments run at 45 degree angles.
> - The fourth symbol consists of five line segments: three long and two medium; of the three long segments, two are vertical and one is horizontal, whereas the two medium segments run at 45-degree angles.
> - The seventh symbol consists of three long line segments: one horizontal and two at angles that are greater than 45 degrees. Even if we are convinced that the tenth symbol consists of only line segments, it would be impossible for us to determine the number, length, and orientation of these segments. Furthermore, it would be even harder for us to analyze the curves of the second, third, fifth, and sixth symbols!

Discussion 7.3 Note that mathematical notation of a sequence is

$$s[1], s[2], s[3], \ldots$$

where $s[i]$ indicates the ith symbol in the sequence. Thus, in any sequence, there is a clear correspondence between the symbols and natural numbers (i.e., the first symbol, second symbol, third symbol, etc.). In the above case, if we number all the symbols,

we may immediately notice that the sequence represents the initial sequence of natural numbers 1, 2, 3, etc., such that each number is displayed alongside its mirror image. With this observation, we should have no difficulty drawing the next symbol! This puzzle is now well known in popular culture due to its placement in an episode of the Fox Network's cartoon series *The Simpsons*. In that episode, much was made of a character's inability to solve the puzzle, when a number of other people around her could apparently solve it easily. This puzzle also appeared in the book/film *The Oxford Murders*.

So far, we have looked at fairly simple patterns, but it is very important to remember that hunting for patterns can result in finding patterns that were never intended. If you present students with patterns of head and tail flips, one of which is HHHTTHHHTT and one of which is HTHTHHTTHTH, they will see an obvious pattern in one, although if you introduce these as the result of coin flips, they are unlikely to think it repeatable. However, if you obscure the random nature of this, and its source, and present them with 1110011100 and 10101100101, they are far more likely to try and place a pattern over the top to predict the next element in the sequence, despite this being a 50/50 chance either way!

Returning to predictable, and nonrandom, patterns, we are going to look at a simple game of strategy where an understanding of patterns helps us to formulate a winning strategy. There are many references on the game *Nim*, and it is a game that senior students may enjoy playing in a puzzle club setting. We have already looked at one of *Nim* games – recall Problem 6.4 from the previous chapter. In this version, we "upgrade" the previous puzzle by adding more pebbles and changing slightly the rules.

Problem 7.4 Consider one of the easiest *Nim* games that consists of a single pile of 100 pebbles. Each player can take one, two, three, four, or five pebbles in a single move. The winner is the player who takes the last pebble. What is the winning strategy for the first player?

Teacher Tip
It might be a good moment to tell students more about the family of games called *Nim*. It is believed to be Chinese in origin, but the name *Nim* was given much later: at the beginning of the twentieth century. The objective for any *Nim* game is clear: to win any game, regardless of the opponent's strategy. There is an infinite number of possible *Nim* games – each game is defined by the number of piles, number of pebbles in each pile, and the rules of the game: how many pebbles (from how many piles) a player can take in a single move. Usually there are two players, **A** and **B**, who move alternatively. The winner (or loser) is the player who takes the last pebble.

Some *Nim* games are easy, some are harder, and some of them quite difficult to solve – and by "solving" a *Nim* game, we mean "finding a strategy for a player that would allow winning any game, regardless of the opponent's strategy."

The presented problem is ideal for class environment – pairs of students can practice their skills in playing this game before making any efforts in finding the best strategy how to play it.

> **Student Pitfall**
> Students often don't realize that any strategy must take into account their opponent playing at their *best*, rather than depending upon a foolish or naïve opponent. Some students also try to solve for a random opponent and make comments such as *"This will win such-and-such a percentage of the time."* A true winning strategy must win all of the time, for every opponent.

Discussion 7.4 Clearly, we need to discover "a pattern." To do so, as we did in solving Problem 6.4, we may use the strategy of reasoning backwards. If, after a number of moves, there is only one pebble left in the pile and the second player has his or her move, the first player would win the game. Also, if there are 6 pebbles left in the pile and the second player has his or her move, the first player would win the game – regardless of the number of pebbles taken by the second player, the first player can take a number legally that will reduce the number of pebbles to one. A straightforward reasoning would provide a general strategy here – the first player should reduce the number of pebbles to one of the following numbers: 6, 12, 18, 24, 30, 36, etc., in every move. Clearly, when 100 pebbles are in the pile, in the first move the first player should take 4 pebbles (of course, if the original number of pebbles in the pile was 96 or 102, the second player would have a winning strategy).

So the winning "strategy" for removing pebbles can be expressed as a "rule":

> *If there are n pebbles on the table, then remove p pebbles so the number $n - p$ is divisible by* 6, *if possible.*

This is a very important note, and we emphasize it: *a small change in the description of the game may result in a different winning strategy*. For example, assume now that the *winner* is the one who takes the *last* pebbles. What is the winning strategy for player **A** now, if one exists? We are still playing a very simple game of *Nim*, a single pile with the same number of pebbles in it, but we have altered the victory conditions in a significant manner.

> **Student Pitfall**
> A student may not pay close to attention to the rules and be surprised when they are told that different victory conditions are being used. It's good practice for students to write down the conditions to make sure that they understand them, especially when a familiar game has had a major rule change, such as a chess game where the objective is be the first player to be checkmated.[5] A student who is not paying attention to the change in rules will be very unpleasantly surprised when they say *"Checkmate!"* and realize that they've lost.

---

[5] Such a game is well beyond the scope of this course, but there are many online resources on Antichess, Loser's Chess, and Suicide Chess, all of which make rule changes to allow the game to be reasonably playable.

7 Pattern Recognition

Discussion 7.4 (cont) Some students, if playing this with physical stones, might try to adopt a strategy of not playing a move, to avoid entering one of these possible areas. However, this is when we can remind them that the puzzle is defined as removing one of these possible values of pebbles – which does not include zero!

> **Teacher Tip**
> This simple observation can be used to drive fairly deep discussion on why rules are important. If it was possible for someone to take nothing, then either player could guarantee a stalemate situation, and there would be no winning strategy. Returning to the rules of Antichess and its variants, there are often "enforced capture" rules that require players to remove an opponent's piece from the board if the opportunity arises, to force the players to move closer to a solution. (If we reach a situation where no one will move and we can move no further, we are *deadlocked*. If we reach a situation where people can move but we make no progress, as in Antichess, we are technically *livelocked* – the result is still useless but it's still moving!)

Discussion 7.4 (cont) We can only wish that the strategy for other games (e.g., chess) could be so simple and expressed by "if-then" rules! Pattern recognition is a very useful way to approach puzzles because it is applicable across many of the puzzle domains. As we've just seen, a pattern recognition approach to strategy helped us to formulate a winning strategy – or to work out if we couldn't! We can also use pattern recognition for probability problems as well.

> **Teacher Tip**
> If students are interested in *Nim* games, consider staying with this topic for a while. You may challenge your students with game-changing rules of making a move, e.g.:
> - At each move, the players can remove a number of pebbles, which must be a power of 2 (i.e., 1, 2, 4, 8, etc.).
> - At each move the players can remove one, three, or eight pebbles.
> - At each move the players can remove odd number of pebbles.
>
> Consider also departing from one pile of pebbles; the strategies of playing *Nim* with two or more piles of pebbles might be more complex. Consider the following:
> - There are three piles of pebbles on the table. The first pile contains two pebbles, the middle one contains three pebbles, and the last one contains four pebbles. There are two players, **A** and **B**, who move alternatively. Player **A** moves first. The rules of the game are the same for both players: at each move, they can remove one or two pebbles provided that they are from the same pile. The loser is the player who takes the last pebble. What is the winning strategy for player **A**, if one exists?

(continued)

- There are three piles of pebbles on the table containing 76, 65, and 48 pebbles, respectively. There are two players, **A** and **B**, who move alternatively. Player **A** moves first. The rules of the game are the same for both players: they can remove any number of pebbles at each move but only from one pile. The winner is the player who takes the last pebbles. What is the winning strategy for player **A**, if one exists?
- There are two piles of pebbles on a table. There are two players, **A** and **B**, who move alternatively. Player **A** moves first. The rules of the game are the same for both players: at each move, they can remove any number of pebbles provided they come from the same pile *or the same number of pebbles from both piles*. The winner is the player who takes the last pebble. What is the winning strategy for player **A**, if one exists?

Most of the above instances of *Nim* games are discussed in Chapter 11 of *Puzzle-based Learning* book.[6]

Problem 7.5 There are 512 tennis players, but two of them are twins. The typical tournament rules apply: there are 256 games in the first round, as two players play in each game. The winners advance to the second round, which consists of 128 games, and so forth. Assuming that each player has a 50/50 chance of winning any game against any opponent, what is the probability that the twins will play each other at some stage in the tournament?

Teacher Tip
Additional observation that should be shared with students at this point is that it might be a good idea to develop a strategy by reducing the problem into a simpler one (e.g., by reducing the number of available pebbles) and then gradually moving back to the original problem paying attention to the emerging pattern.

When we search for patterns in sequences and the search is based on information extracted from a sequence, it is often useful to start with a short sequence and gradually make it longer and longer to discover the pattern.

Clearly, there is a similarity here to the approach we took in Chap. 9 of starting with a small sequence and working our way up.

This puzzle illustrates this process in terms of a number of elements in a set (rather than a sequence). We have presented this puzzle with 512 players, but the general form is 2^n players, where there is always more than one player. (We must have an even number of tennis players, or someone will be left out.

(continued)

---

[6] Michalewicz Z, Michalewicz M (2008) Puzzle-based learning: an introduction to critical thinking, mathematics, and problem solving. Hybrid Publishers, Melbourne.

7 Pattern Recognition

> If we have only one player, that player wins automatically, as he or she has no opponent.) Going further, because we halve the number of players every time a sequence of matches is played, as one out of two people must lose, the number of players must be a power of 2. If this wasn't true, at some stage, we'll end up with an unmatched player. Once you've presented it as 512, step back, explain why it must be a power of 2, and then present this puzzle in a more general form as 2^n.

Discussion 7.5 To help frame the problem, we briefly introduce the tennis "draw," where players are chosen for each round. We start by assuming that we have enough players for at least one match – that is, 2, i.e., $n = 1$. In the first round, we randomly allocate pairs chosen from the available set. These pairs then play each other, and a single winner emerges from each match. Until we have a winner, this number must be even, and we then allocate pairs from this smaller set until everyone has an opponent. Once we have only 1 player remaining, we have a winner.

The formulation of this problem is quite clear – the task is to find the probability p of the meeting of two twins in the tournament, which is a function of n (the number of players is 2^n). But how can we calculate this probability? Let us start with smaller sets of players (i.e., small values of n) to see if a pattern would emerge:
- If $n = 1$ (i.e., there are $2^1 = 2$ players in the tournament), then $p = 1$ as they will meet for sure.
- If $n = 2$ (i.e., there are $2^2 = 4$ players in the tournament), then $p = 1/2$. This is because there is a 1/3 chance that the twins will be paired together in the first round and a 2/3 chance that they will play other opponents in the first round. In the latter case, there is a 1/4 chance that they will meet in the second round (as both of them must win their first round games), and there is a 1/2 chance for each of them to do that. So

$$p = 1/2, \text{ because } 1/3 \times 1 + 2/3 \times 1/4 = 1/2$$

- If $n = 3$ (i.e., there are $2^3 = 8$ players in the tournament), then we have three sub-cases to analyze:
  (i) The twins play each other in the first stage. Given that there are a total of 8 players, the probability that the twins will be paired in the first round is 1/7.
  (ii) The twins play each other in the second stage. Here we have two events we need to account for: (a) the twins should not have played each other in the first stage and (b) the twins meet in the second stage. For both of these to happen, the twins need to be in the first half in the first round and not have played each other, 2/7, and each twin needs to have won their game $(1/2 \times 1/2)$. So the total probability is $2/7 \times 1/2 \times 1/2 = 1/14$.

(iii) The twins play each other in the third stage. Here we have three events to account for: (a) the twins should not have played each other in the first stage, (b) they should not have played each other in the second stage, and (c) they play each other in the third stage. For (a) and (b) to occur, the twins need to be in different halves during the first stage (4/7). For (c) to happen, each twin has to win stage 1 and stage 2 which is $(1/2 \times 1/2) \times (1/2 \times 1/2) = 1/16$. So the probability for (iii) is $4/7 \times 1/16 = 1/28$.

Adding the probabilities for these three, we get $1/7 + 1/14 + 1/28 = 1/4$.

[During classroom discussions, we draw a decision tree which helps to figuratively convey the pairings and outcomes discussed above when $n = 1, 2,$ or 3.]

So
$p(1) = 1$
$p(2) = 1/2$
$p(3) = 1/4$

Thus, a reasonable assumption is that $p(n) = (1/2)^{n-1}$. Indeed, this is accurate and can easily be proved by the induction principle (it depends on the type of students in the class whether we present a simple proof or not).

> **Teacher Tip**
>
> This problem presents a good opportunity to challenge students further with some twists in the description of the original problem. For example, what would be the probability that twins will play each other in the tournament, if one twin is a better player (probability of winning any game is 60 %) and the other a weaker player (probability of winning any game is 40 %)? What would happen if we know (different) probabilities of winning a single game of every player in the tournament? Further, some players have a higher or lower probability of winning a game while playing against particular players – how can we approach the problem if for each pair of players we know the probability of winning/losing the game? Such discussion may lead to discovery that in such cases a derivation of analytical solution might be too difficult, so we can turn to simulation – and we cover this topic in Sect. 11.1.

The following puzzle is more advanced and is designed to illustrate how our reasoning on a pattern recognition-based solution is very dependent upon how many examples we look at.

Problem 7.6 n points are placed on a circle, and every point is connected by a line to every other point. Into how many pieces is the circle divided? The following figures illustrate the five initial cases (for $n = 1, 2, 3, 4, 5$):

7 Pattern Recognition

Discussion 7.6 The table below would lead us to believe that the number of pieces grows as 2^{n-1}, so adding the 6th point should result in 32 pieces:

n	Number of pieces
1	1
2	2
3	4
4	8
5	16

In actuality, we only get 31 pieces when adding the 6th point, and the right formula for the number of pieces is

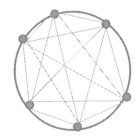

$$\frac{(n^4 - 6n^3 + 23n^2 - 18n + 24)}{24}$$

Student Pitfall
Usually students are convinced (after looking at the first five cases – and especially when they remember the previous puzzle on tennis players) that the number of pieces grows as 2^{n-1}. The number of pieces given for $n = 1$, 2, 3, 4, and 5 is very convincing and leaves very little room for imagination!

> **Teacher Tip**
> The take-home lesson (to be emphasized to the students) is to be careful not to jump to conclusions too quickly. This puzzle also introduced another key point – the discovered pattern must be checked carefully. Pattern recognition exercise just suggests a solution – whether this is the correct solution or not is up to us to determine.

Discussion 7.6 (cont) As human beings, we are great at seeing patterns – so good, in fact, that we can invent them when they do not exist! We can draw students' attention to a parallel between the last puzzle to some cases where a misleading pattern is created on purpose (e.g., to cover a crime, for counterintelligence, for competitive situations in the marketplace, or just for fun). One well-known example of this takes place in the short story *The ABC Murders* by Agatha Christie. In the story, a murder is committed in Andover, and the victim's name was Alice Ascher. Four days later, another murder is committed in Bexhill-on-Sea, and the victim was Betty Barnard. Four days after that, Carmichael Clarke was murdered in Churston. The point of the story was that the murderer's goal was just to kill Carmichael Clarke; the purpose of the previous two victims was to create a misleading pattern (useful for avoiding an investigation on the motive of the main crime).

Another enjoyable example of misleading circumstances (again, students are really interested in such stories, which also break a flow of the class material) is credited to José Raúl Capablanca, a Cuban-born world champion chess player (1921–1927). Apart from being referred to by many chess historians as the *Mozart of chess*, he also displayed an unusual sense of humor. One day he was on a train that was stopped in the middle of nowhere, and he was told it would take many hours to clear the tracks and continue the journey. While Capablanca was waiting, he was approached by a railman (who did not recognize Capablanca) and invited to a game of chess. Capablanca feigned that he did know how to play chess. The railman was not discouraged and offered a quick lesson. After a brief overview of the game, the railman said: "*Let's play. Since you are new to the game of chess and I am an experienced player, it would be only be fair if I play without my queen.*" They played ten games in which Capablanca enjoyed an advantage over his opponent, who played without the most powerful piece – his queen. Capablanca deliberately lost all these games. At the end of the tenth game, Capablanca said: "*After these ten games, I think I know what's most important in chess. Let's play another ten games, but this time I will remove my queen and you play with a full set.*" The railman was surprised (to say the least!) but obliged. They played another ten games, and this time Capablanca won all ten. He then said to the dazed railman: "*I knew it from the start: It is much easier to win chess if I play without a queen!*"

It is not surprising that many puzzles are constructed in such a way as to mislead us into the "wrong" pattern – this is, after all, the nature of the puzzle!

Reference

1. Michalewicz Z, Michalewicz M (2008) Puzzle-based learning: an introduction to critical thinking, mathematics, and problem solving. Hybrid Publishers, Melbourne

Enumerate and Eliminate

> *When you have eliminated the impossible, whatever remains, however improbable, must be the truth.*
> – Sherlock Holmes

One of the most powerful (and popular) problem-solving techniques applicable to many problems is a technique that is based upon the enumeration of all possible solutions and the systematic elimination of solutions which are "wrong." By repeating this process of elimination, we zoom in into the (usually relatively small) subset of possible solutions. This was one of the methods used frequently by Holmes, as seen in the opening quote!

In general, the process of enumerating all possible solutions for a given problem is very important in many aspects of problem-solving activities, from probability issues through logic and constraint satisfaction, to optimization problems. This chapter contains a small collection of problems/puzzles that were specifically selected to demonstrate to the student the basic steps behind the "enumeration and elimination" technique.

> **Student Pitfall**
> Many students do not understand how quickly the number of solutions for a puzzle can grow. While the often suggested answer of "*I'll look at all of them and pick the best*" may work for smaller problems, large problems rapidly become too big for students to solve in a reasonable time.

> **Teacher Tip**
> As we will discuss in Chap. 11, the whole area of picking one "good" solution from a very large number of "possible" solutions requires a combination of
>
> (continued)

124   8 Enumerate and Eliminate

> good thinking and a clever approach. Students often think that computers can do anything – a quick reminder that they cannot, because the problem is so big, is a good way to encourage them to think about the way that they are planning to solve a problem.

These problems do not have to be a dull inventory of possibilities. Consider, for example, the following puzzle (it is actually a very good puzzle to move the students into appropriate "enumerate and eliminate" mood).

Problem 8.1 Mr. Brown, Mr. White, and Mr. Green went to a hat shop where they all picked out hats of different colors. When they left the store, the man with a green hat said: *"Have you noticed that although our hat colors match our names, none of us has the same hat color as our name?"* Mr. Brown replied: *"Indeed, you are right! This is remarkable!"* What hat color did each man have?

Discussion 8.1 This is a good example of a problem that illustrates nicely the process of elimination – after all possibilities were enumerated. To do so, let's introduce the letters **B**, **W**, and **G** that would mark three colors: brown, white, and green, respectively. If we knew nothing apart from that fact that three men selected some colors for their hats from this list, the model would be:

$$
\begin{array}{ll}
\text{Mr. Brown:} & \textbf{B} \text{ or } \textbf{W} \text{ or } \textbf{G} \\
\text{Mr. White:} & \textbf{B} \text{ or } \textbf{W} \text{ or } \textbf{G} \\
\text{Mr. Green:} & \textbf{B} \text{ or } \textbf{W} \text{ or } \textbf{G}
\end{array}
$$

As none of the men's hat colors matched their name (this is the first fact, hence the first constraint), it is possible to eliminate a few potential assignments, and now the remaining possible arrangements are:

$$
\begin{array}{ll}
\text{Mr. Brown:} & \textbf{W} \text{ or } \textbf{G} \\
\text{Mr. White:} & \textbf{B} \text{ or } \textbf{G} \\
\text{Mr. Green:} & \textbf{B} \text{ or } \textbf{W}
\end{array}
$$

Now (as usual) we are ready for considering the other constraints. The man with the *green* hat made a remark that was answered by a *different* man, Mr. Brown. Therefore, the immediate inference is that Mr. Brown does not have a green hat! And this is all we need to solve this puzzle, as:

$$
\begin{array}{ll}
\text{Mr. Brown:} & \textbf{W} \\
\text{Mr. White:} & \textbf{B} \text{ or } \textbf{G} \\
\text{Mr. Green:} & \textbf{B} \text{ or } \textbf{W}
\end{array}
$$

implies:

8 Enumerate and Eliminate

Mr. Brown: **W**
Mr. White: **B** or **G**
Mr. Green: **B** (Because we know that Mr Brown is wearing a White hat)

which in turn implies:

Mr. Brown: **W**
Mr. White: **G** (Because Mr Green is wearing the Brown hat)
Mr. Green: **B**

So Mr. Brown has a white hat, Mr. White has a green hat, and Mr. Green has a delightfully brown hat.

> **Student Pitfall**
> Some students will struggle with this puzzle because they think that there is not enough information available. Indeed, this is an interesting puzzle with a "this is impossible!" flavor. However, this is also a very simple puzzle when we enumerate all possibilities.

> **Teacher Tip**
> In order to get the students to appreciate the value of the "enumerate and eliminate" technique, let them struggle with the problem on their own for some time. The purpose of presenting this problem to the student is not for the student to get the answer; the purpose is to develop the students' ability to independently solve problems. This is best accomplished by allowing the students to work on the problems independently.

Another important way that we can use enumeration is to work out all of the possibilities that are open to us. This is the foundation of probability: the ratio of things that interest us to the total number of things that are possible. For example, if we want to know the chances of heads on a single coin flip, we look at the ratio of number of ways a head can come up (1) to the number of possible outcomes (heads or tails, hence 2). Therefore, the chances of flipping a head is 1/2 – or 50 %.

Enumeration can also help in understanding the problem as it may lead towards the correct solution, as the following puzzle illustrates.

Problem 8.2 Let us consider the case of two bears – one white and one black. We assume that among all bears, the two sexes and two colors (black or white) are equally likely and we wish to answer three simple and similar questions:
1. What is the probability that both bears are males?

2. What is the probability that both bears are males if you were told that at least one of them is male?
3. What is the probability that both bears are males if you were told that the white one is male?

Discussion 8.2 The first question is: What is the probability that both bears are males? Let us represent the bears using symbols, as we can easily put these into the text or draw them on a board. We pick a very simple representation, using alphabetic characters. There are four equally likely possibilities for the two bears:

$$(m\ m), (m\ f), (f\ m), \text{ and } (f\ f),$$

where the first is the white bear and the second is the black bear. Thus, *(m m)* is two male bears and *(m f)* is a male bear and a female bear. Why do we have *(m f)* and *(f m)*? We interpret the possible outcomes listed above in a way that the *first* letter represents the gender of the *white* bear and the *second* letter represents the gender of the *black* one. In other words, the outcome $(f\ m)$ describes the situation where the white bear is female and the black one is male, whereas $(m f)$ describes the situation where the white bear is male and the black one is female. (This distinction would be significant in answering the remaining two questions.) Clearly, the probability that both bears are male is 1/4, as the case *(m m)* is one of the four possible cases listed above.

Now, let us consider the second question: What is the probability that both bears are males if you were told that at least one of them is male? Again, we can easily find the answer if we enumerate carefully all possible outcomes, which are:

$$(f\ m), (m\ f), \text{ and } (m\ m)$$

as the case (ff) is excluded. So the probability that both bears are males is 1/3, as this case is one of three possible cases listed above. Why is *(ff)* excluded? Because we are told that at least one of them is male, which means that we don't even have to consider the possibility that both are female.

> **Teacher Tip**
> If students struggle with this, you can restate the question as "*If you are told that one of the two bears is a male, what is the possibility that both bears are female?*" The answer is, obviously, zero and this is why we can ignore this case as options that have no likelihood of occurring have no bearing on calculating the success or otherwise of an outcome. For example, when flipping a coin, we don't have to take into account the possibility of the coin being carried off by an eagle to be dropped into an active volcano.

8 Enumerate and Eliminate

Discussion 8.2 (cont) Now, let us consider the final question, which students often find the most challenging: What is the probability that both bears are males if you were told that the white one is male? As before, we can easily find the answer by carefully enumerating all possible outcomes, which are:

$$(m\ f) \text{ and } (m\ m)$$

as the other cases are excluded. So the probability that both bears are male is 1/2, as this case is one of only two possible cases listed above. This draws upon our representation of the bears as (white black).

> **Student Pitfall**
> For most students, intuition usually fails in distinguishing between cases 2 and 3. If we know that one of the bears is male, why should it matter whether he is white or not? In such cases, intuitive reasoning might be quite dangerous – and we can avoid many traps and arrive at the correct answer, if we just enumerate all the possible outcomes.

> **Teacher Tip**
> Encourage students to enumerate all possibilities – they should see that such simple enumeration helped a lot! From the list of all possible solutions, it was sufficient to eliminate some solutions (e.g., the solution $(f f)$ when we were told that at least one of them is male) to arrive at the correct answer. As before we can restate the question, to highlight the problem in reasoning, as *"If the white bear is male, what is the chance that the white bear isn't male?"* (which would admit the $(f\ m)$ option.) The answer is, fairly obviously, *"No chance at all,"* which clearly illustrates why (again) we don't have to include this option in our final calculations.

The next puzzle builds upon the ideas that we have been describing in the bear puzzle but increases the difficulty.

Problem 8.3 You have three bags made of a dark and opaque material. One of these bags contains two black marbles. One contains two white marbles. One contains a marble of each color. You draw from one of the bags at random and examine the marble you extract. You see that is black. What is the probability that the remaining marble in the bag is black?

> **Teacher Tip**
> This puzzle is great for a classroom environment, as it is very easy to arrange for some experiments between pairs of students. The puzzle also illustrates the power of enumeration. If students struggle with this, consider asking them what the chances of drawing a black marble from a bag are, if they haven't drawn anything out.

> **Student Pitfall**
> As before, for many students it is hard to resist the temptation of intuitively answering that the probability is 50 %. The reasoning behind this answer is that as we draw a black marble, this excludes the bag with two white marbles. Then we know that the bag we selected has either one black marble or one white marble left in it, as we have already removed one black. So, in the former case (the selected bag had two black marbles), we will have a black marble in the bag, and in the latter case (this was the mixed bag), we will have a white marble to pick next. It seems that because either of these two cases is equally likely, the probability of drawing a black marble out is 50 %.

Discussion 8.3 The mistake in this reasoning is that we have not enumerated all the possibilities. We know that we are only working with two bags, as we can legitimately exclude the bag with two white marbles the moment we see the black. However, there are three possible ways that we could draw a black marble out of a bag. We could be picking the black marble out of the mixed bag (1 way), we could be picking a black marble out of the all-black bag (1 way), or, as this is the stumbling block, we could be picking the *other* black marble out of the all-black bag (1 way) – for a total of 3 ways to draw a black. For two of these ways, the all-black bag, the other draw will also be a black marble, hence two success out of three possibilities gives us a 2/3 chance of drawing another black marble – not 50 %.

> **Teacher Tip**
> Students may need to try this themselves, with bags and marbles, and you can label the black marbles in the all-black bag as B1 and B2. This will make it clear, when students draw one or the other, that they really do have two different ways to draw that next black marble.

Discussion 8.3 (cont) We can go further in using enumeration to answer apparently quite complicated probability puzzles. Students are often interested in games of chance and where better to find a game of chance than in a casino? However,

8 Enumerate and Eliminate   129

rather than requiring a grounding in Bayes' theorem, we can answer this question just by writing down the lists of things that could happen and then work out which ones are the interesting ones for this question.

> **Teacher Tip**
> A very similar puzzle (that can be used to check whether the students really understood the last puzzle) goes like this:
>   There are three cards in a bag. The first card has the symbol **X** written on both sides, the second card has the symbol **O** written on both sides, and the third card has an **X** on one side and an **O** on the other. You draw one card at random and examine one side of this card. You see an **X**. What is the probability that there is also an **X** on the other side?

Problem 8.4 The dice game *Chuck-a-Luck* is played in some casinos. The rules are extremely simple: you bet $1 on a number from one to six. Three six-sided dice are then rolled. If your number does not appear on any of the dice, you lose your money. If the number appears once, you get your money back. If your number appears two or three times, you get $2 or $3, respectively. What percentage of the money bet at this game is the casino expected to win?

Discussion 8.4 The method of finding the answer is straightforward. Again, it requires the listing of all possible "elementary events." Indeed, we can easily list all $6 \times 6 \times 6 = 216$ possible outcomes of a three dice roll and, for each of them, get the profit or loss (i.e., $-\$1$, $\$1$, $\$2$, or $\$3$).

> **Teacher Tip**
> Enumerating a space this large can be time-consuming, but if you do decide to do the whole thing, consider assigning student groups to the problem, which will require the students to work collaboratively. Having two teams for each set will also set up a small competition – and allow you to check that both obtained the same answer!

Discussion 8.4 (cont) For example, if we bet on the number 3, there would be:
- 1 outcome with all 3s, namely, (3, 3, 3), when we win $3
- 15 outcomes with two 3s, for example, (3, 3, 4) or (3, 6, 3), when we win $2
- 75 outcomes with one 3, for example, (3, 4, 6) or (1, 3, 5), when we win $1
- 125 outcomes with no 3s, for example, (1, 2, 5) or (6, 6, 5), when we lose $1

Each of these outcomes happens with the probability of 1/216. Thus, our expected gain, when we bet $1, is:

$$\frac{(1 \times \$3 + 15 \times \$2 + 75 \times \$1 - 125 \times \$1)}{216} = \frac{-17}{216} \approx -0.08$$

That means that we can expect to lose 8 cents every time we bet $1. We can easily convince ourselves that the game is unfair – we can even argue that the mere fact that the game is offered in a casino serves as sufficient proof! For younger students, we don't even need to do the division as we can see that the result is going to be negative. If our expected return is less than zero, then we can clearly see what is going to happen to our money!

> **Teacher Tip**
> It would be worthwhile to tell students that here is also a different enumeration method applicable in analyzing various games, which is based on a situation where many players bet on *all* possible outcomes. In the case of the last game, one can reason as follows: assume that there are six players and each bet $1 on a different number. Now, there are three possible outcomes:
> - All three numbers on the dice are different. In this case, there are three winners (who each get $1) and three losers (who lose their bets), and the house moves the $1 bets of the losers to the winners.
> - Two numbers are the same and one is different. In this case, there are two winners (one gets $2 and the other one gets $1) and four losers who collectively lose $4, and the house gains $1.
> - Three numbers are the same. In this case, there is one winner (who gets $3) and five losers who collectively lose $5, and the house gains $2.
>
> On average, it is impossible for the house to lose. A similar argument can be made for other games: say, we bet $1 on the roulette wheel. (For roulette, it may be important to remind students of the 0 and 00 positions on the wheel. Having one or two outcomes where no one except the house wins is essential if casinos are going to offer bets such as red or black, which would otherwise be 50/50 outcomes.)

The following puzzle can be quite challenging and often requires a reasonable amount of preparation and scaffolding for students. As we note, many students will assume that this problem has no solution. Our most successful versions of this puzzle in classroom delivery have always employed a reasonable amount of theatricality, to maintain student interest while they struggle with the apparent absurdity of the questions!

Problem 8.5 Mr. Smith and Mr. Jones met on the street after not seeing each other for many years. Mr. Jones is a school teacher and Mr. Smith is a mathematician. After a few minutes, their conversation turned into family matters:

Mr. Jones: All three of my sons celebrate their birthday today. Can you tell me how old each one is? After all, you are a mathematician...

Mr. Smith: Yes, of course, but you have to tell me something about them.

8 Enumerate and Eliminate

Mr. Jones: The product of their ages is 36.
Mr. Smith: I need more information...
Mr. Jones: The sum of their ages is equal to the number of windows in the building next to us...
Mr. Smith: Still I need more information...
Mr. Jones: My oldest son has blue eyes.
Mr. Smith: This is sufficient!

And Mr. Smith gave the correct answer: the ages of Mr. Jones' three sons. What are the ages of these three sons?

> **Student Pitfall**
> Usually students are confused here – they pay attention to "blue eyes" rather than "oldest son" – and they may feel that this is an "impossible" puzzle. Usually students do not know where to start in solving such puzzles and, as a result, many will give up rather than start.

> **Teacher Tip**
> After some time when students struggle on their own, consider giving them a hint: *"Enumerate the possibilities."* Note that in this puzzle, there is no probability component (as it was the case with earlier puzzles) – here, the constraints would drive the elimination process. Of course, the task is to find a solution that satisfies all these constraints. We can enumerate all possible potential solutions and gradually eliminate those which do not satisfy the constraints. Note also that the misleading information in this puzzle was the last statement, *"My oldest son has blue eyes,"* as most of us pay attention to the word *blue* rather than to the *oldest*, and we do not see the relevance of this constraint to the problem. This is often the case in solving real-world problems, where during the model building phase we try to identify the important information and reject irrelevant ones. In this puzzle, as in real life, the process of identifying the relevant information is not that straightforward!

Discussion 8.5 Clearly, we can enumerate all possible answers, and from this point we can run the elimination process. It is very clear that a solution can be represented by three positive integer numbers: x, y, and z. Why integers? That's a good question. Here, we are depending upon the human convention to answer questions about age with a whole number. Above a certain age, no one says that they are "something and a half" and we are depending upon that implicitly. (Some students may ask about this, so be ready to answer whether the answers are in "years," integers.) Why are they positive? Because we never talk about people being "-3" or "-100." There are also some additional assumptions that we can make. For example, the ranges of these three variables must be between 1 and 36 (because multiplied together, they produce 36). Further, we can assume that $x \geq y \geq z$, as there must be some age order among three boys.

At this stage we can enumerate all possible answers: at this stage any triplet (x, y, z), where $1 \leq z \leq y \leq x \leq 36$, would do. For example, one may consider $x = 25$, $y = 16$, and $z = 3$. Of course, this "solution" is no good, as the problem-specific constraints are not satisfied – so we can eliminate it from our considerations. (This is an infeasible solution, one which does not meet our requirements.) Fortunately, we can eliminate many of the potential solutions without having to carry out a lot of calculation, as the above dialog between Mr. Smith and Mr. Jones contains additional statements, and we can use these to narrow our search and find a solution (or solutions) that would make the above dialog meaningful. So, let us start.

The first piece of information Mr. Smith got from Mr. Jones was that the product of his sons' ages is 36. This is very helpful, as there are only eight sets of 3 integer numbers, x, y, z, where none of these numbers are greater than 36, whose product is 36. These are:

$36 \times 1 \times 1 = 36$
$18 \times 2 \times 1 = 36$
$12 \times 3 \times 1 = 36$
$9 \times 4 \times 1 = 36$
$9 \times 2 \times 2 = 36$
$6 \times 6 \times 1 = 36$
$6 \times 3 \times 2 = 36$
$4 \times 3 \times 3 = 36$

At this stage, Mr. Smith narrowed his search to these eight possibilities. The second piece of information was a bit mysterious: *"The sum of their ages is equal to the number of windows in the building next to us...."* Indeed, we do not know what the building next to Mr. Jones and Smith looks like, and so we do not know the number of windows it has. But Mr. Smith, who was standing next to the building, knew the number of windows, and the answer he got was *"The sum of their ages is equal to such and such number."* So, what was the number of windows? What was this "such and such number" which represents the total of ages of Mr. Jones' three sons? To answer this question, let us calculate all the totals for the eight possible cases we have identified above:

$36 + 1 + 1 = 38$
$18 + 2 + 1 = 21$
$12 + 3 + 1 = 16$
$9 + 4 + 1 = 14$
$9 + 2 + 2 = 13$
$6 + 6 + 1 = 13$
$6 + 3 + 2 = 11$
$4 + 3 + 3 = 10$

8 Enumerate and Eliminate

Now everything should be clear. If the number of windows was, say, 16, Mr. Smith would not have any difficulties in identifying the ages of all three sons: they would be 12, 3, and 1. And if the number of windows was 11, again, Mr. Smith would have easily identified the ages of all three sons (6, 3, and 2). This is also the case if the number of windows was 38, 21, 14, or 10; in all these cases, Mr. Smith would have immediately given the answer. However, Mr. Smith was still not sure and asked for additional information. The only reason for his request for more information was that the number of windows in the house next to them was 13, and Mr. Smith needed to distinguish between two possible cases: ages of 9, 2, and 2 and 6, 6, and 1; this is why he asked for additional information!

So the number of windows was 13 and the additional clarification "*My oldest son has blue eyes*" eliminated one of the two possible solutions, as there is no *oldest* in the 6, 6, and 1 case.[1] So the answer is 9, 2, and 2.

Another excellent use of enumeration is in answering logic puzzles, where we try and put together a set of statements and, based upon whether they are true or false, answer a question or solve a puzzle. The next puzzle takes a logical approach to enumeration and makes an apparently paradoxical statement[2] much easier to understand!

Problem 8.6 There are two tribes on the island (tribes **A** and **B**), and it is widely known that these tribes consist only of liars or truth-tellers. That is, for each tribe, if one member is a liar, they are all liars, and if one member is a truth-teller, then they are all truth-tellers. No one knows what each tribe consists of. Both tribes could be liars, both could be truth-tellers, or they could be a mix. No one is sure! One day, one member from each tribe was invited to make a statement (a statement made by a member from tribe **A** is marked as S(**A**)):
S(**A**): Exactly one of us is lying.
S(**B**): At least one of us is telling the truth.
Who is telling the truth and who is lying?

> **Student Pitfall**
> Some students will have no idea where to start with this because in the absence of any concrete evidence, they will try to work this out in their heads and not be able to resolve it. Writing down an attempt at an enumeration is vital here as it forces them to remember how they have set the tribes up. As we will see, a simple table-based representation makes it very easy to work out who is who.

---

[1] Take these explanations with a grain of salt – this puzzle is half a joke. Some students may not interpret "my oldest has blue eyes," meaning that the oldest is *not* a twin. One of the authors of this book (Ed Meyer) has twins in his family, and Johnny, the older of the two, does have blue eyes!

[2] A paradox is an apparently contradictory statement.

Discussion 8.6 This is a good example of a problem that illustrates nicely the process of enumeration and elimination in logic puzzles. The most straightforward approach for this and many other logic puzzles of this type would be to enumerate all possibilities and then do some reasoning. Let's start by introducing some labels of each tribe: L for liars and T for truth-tellers. With two tribes **A** and **B**, and two possibilities for each tribe (L or T), there are four possible arrangements of "liars" and "truth-tellers." Each line starts with a row number and then we have a column for both tribes **A** and **B**. We then list all of the possibilities – both tribes are truth-tellers (T) (line 1) down to both tribes are liars (L) (line 4). This is known, formally, as a *truth table*, because it lists all of the possible truth values for these tribes. Truth tables are used extensively in computer science, mathematics, logic, and engineering.

	A	B
1	T	T
2	T	L
3	L	T
4	L	L

Now we need to evaluate the truthfulness of each statement made by invited members from each tribe. This is easy if we look carefully at the statements and assess them in light of what each line tells us about the tribe. Let's look at line 1 for S(**A**), "*Exactly one of us is lying.*" If both **A** and **B** are truth-tellers, then the statement "*One of us is lying*" is actually a lie (so we label it as L in the table). Similarly, S(**B**) is labeled as T because, if both are truth-tellers, then at least one will be telling the truth.

> **Teacher Tip**
> Some students will get stuck at this point because they will, quite reasonably, argue that a truth-teller telling a lie about telling a lie is the truth, and then their heads will explode. You will need to make sure that students understand that you need to enumerate all of the possibilities here to show everything that is going on. For more advanced students, you can discuss the possibility of halting an enumeration when it's apparent that it is nonsense, as in line 1.

Then we can fill in the rest of the table.

	A	B	S(**A**)	S(**B**)
1	T	T	L	T
2	T	L	T	T
3	L	T	T	T
4	L	L	L	L

8 Enumerate and Eliminate

The third and the final step in our reasoning process is to carefully check each row of the table and analyze its meaning, so we can determine whether each row describes a possible or impossible scenario. We can do this by seeing if lies are told by liars and truths are told by truth-tellers – a simple matter of comparing the values in columns across a given line. Let's look at these rows one by one:

- The first row of the table represents a situation where both tribes are truth-tellers. This is clearly impossible, as the statement S(A) is a lie, which a truth-telling member of tribe **A** could not say. We can eliminate this case from our considerations.

	A	B	S(A)	S(B)	
1	**T**	**T**	**L**	**T**	**A and S(A) don't match**
2	T	L	T	T	
3	L	T	T	T	
4	L	L	L	L	

- The second row of the table is also impossible, as a lying member of **B** is apparently telling the truth (S(**B**) is true).

	A	B	S(A)	S(B)	
1	T	T	L	T	
2	**T**	**L**	**T**	**T**	**B and S(B) don't match**
3	L	T	T	T	
4	L	L	L	L	

- The third row of the table is impossible as well, as a lying member of **A** tells the truth (S(**A**) is true).

	A	B	S(A)	S(B)	
1	T	T	L	T	
2	T	L	T	T	
3	**L**	**T**	**T**	**T**	**A and S(A) don't match**
4	L	L	L	L	

- In the final row, both members of these two tribes are liars and both of their statements are false. This is a possible scenario and is the only one left.

So we can conclude that the fourth row describes the only possible scenario, so the solution is that both tribes consist of liars!

For harder instances of puzzles that involve liars and truth-tellers, see Problem 13.4 and Problem 13.5.

Many of the puzzles that we have discussed in this chapter can, at first glance, appear to be quite challenging or, in some cases, totally unsolvable. Looking at all of the possible ways that things *could* happen and then working out, from the information given, what you can exclude is a great way to take a problem and start working towards the correct solution.

Simplify! 9

I couldn't repair your brakes, so I made your horn louder.
– Steven Wright

Much of being a good problem-solver is utilizing clever strategies to make the solution to the problem more accessible. One of the most useful of these strategies is to simplify the problem. There are a number of different simplifications that will lead to progress towards the solution. One of these is to simply restate or rephrase the problem in terms that are more understandable. In mathematics and computer science, you will use techniques described as *divide-and-conquer, decrease-and-conquer*, and *transform-and-conquer*, but we're going to talk about them here under the general banner of "simplify."

Another way to simplify a problem is to solve a simplified version of a more difficult problem. This often allows connections and insights into the solution of the problem as stated. It's really important, however, to make sure that you are still solving the same problem. There is an old joke about a man who leaps out of a plane with a parachute and an instructor. They pass through 10,000 feet and the instructor says *"Open the chute!"* The man does nothing. They pass through 5,000 feet and the instructor screams *"Open the chute!"* The man still does nothing. They get to around 100 feet and the instructor says *"Are you mad!"* The man replies *"No, I'm going to wait until I get to 6 feet because I can jump down from there."*

> **Student Pitfall**
> Students will often try to solve a simpler problem, rather than a simpler version of the same problem. Watch carefully for excessive assumptions that make the problem trivial. Students can become very defensive when this is addressed so be ready to explain why a given assumption is a step too far. (Like waiting until 6 feet so you can jump!)

Yet another way to simplify a problem is to assume a value for missing information and then do the problem with that information. This often leads to progress towards the solution.

> **Student Pitfall**
> Many challenging problems seem completely intractable at first glance. It is not uncommon for students to dislike problems that are unfamiliar to them. In fact, some completely shut down when presented with a new problem, often saying something like, "*How do you expect us to do the problem if you don't show us how?*" We have also heard, "*It's the teacher's job to show us how to do the problems.*" As the teacher, your goal should be to convince the students that solving new problems is a key skill for success in today's society. The formal term for this, in terms of Perry's classification of how students gain knowledge, is *dualism*. Students require an authority and, if you won't dole out easy answers the moment they get stuck, may make the mistake of assuming that your lack of cooperation indicates a lack of knowledge! One of the key skills a teacher has to develop, when using puzzle-based techniques, is a way to get the class working on the problems on their own and resisting the urge to give out early answers.

The unfamiliarity and challenge of a large and intimidating problem can cause students to "freeze up" for two reasons: firstly, because they require too much thought, and secondly because they require too much action. Cognitive and kinetic load combined can be very discouraging for students. Not only does a simpler version of a problem allow students to reach solutions more easily, it also limits the amount of time that has to be spent in the search for the solution and reduces any associated physical activity as well.

> **Teacher Tip**
> Pitching the load at the right level for your class will greatly increase the number of active and engaged students that you will work with. Set the bar too low and problems become trivial, with little satisfaction gained by solving them. Set the bar too high, and load issues mean that frustration and disappointment will dominate your class. The younger the class, the simpler the problem, but, very importantly, the converse is not true! An older class may not have the background, patience, or training to sit and experiment with a large complex problem for 30 minutes. Watch carefully for signs of fatigue, boredom, or general disengagement and use this to try and match the load of the puzzles to the capacity of your class.

Simplification can take several forms. In some cases, we can reduce the problem to less confronting number of possibilities. In other cases, a simple change of

9 Simplify!

representation can yield results. Simplification is a powerful and useful mathematical technique that we will not address here mathematically, but we will provide some basic guidelines for applying it generally:
1. Can we reduce the problem to a much smaller problem and still solve the same problem? (This is called *instance simplification*.)
2. Can we turn complicated or hard to understand concepts into simpler ones? (Can we make a *representation change*?)
3. Can we move the problem from one that we don't know how to solve into something that we do know how to solve? (Can we carry out *problem reduction*?)

Our fearless skydiver from the joke was (badly) applying our third principle of simplification. Rather than deal with the bigger problem, he opted to solve a problem he already understood and could solve. Physics, of course, had other ideas.

This chapter contains a collection of problems that were specifically designed to demonstrate to the student that simplification is a powerful problem-solving tool.

We start with a classic that can certainly cause consternation in students that lack the ability to simplify.

Problem 9.1 A man stands in front of a picture of a man and says, "*Brothers and sisters I have none, but this man's father is my father's son.*" Who is in the picture?

Discussion 9.1 As the problem is stated, it usually requires multiple readings before the student can even begin to solve the problem.

Teacher Tip
Many students will guess at the answer to this problem simply because there are not many possibilities. You will often get students quickly shouting out answers like stepbrother, uncle, himself, his son, his father, and perhaps others. Try to get the students to understand that guessing the answer does not develop problem-solving ability. No answer should be accepted without clear explanation – even if it is a correct guess.

This one becomes a lot easier to understand when the last phrase, "my father's son," is replaced by the word "me."

So, the simplified version of the same problem is, "brothers and sisters I have none, but this man's father is *me*." This can even be simplified further by considering only the phrase, "This man's father is me." With this simplification, it is clear that the man in the picture is his son. Without this simplification, the problem is difficult for the students to wrap their heads around.

Problem 9.2 A young woman is in the second day of a job interview for the position of scientist at a multinational chemical company. She is in a conference room with three senior scientists at the company who want to test her problem-solving ability. The senior scientist shows her a standard deck of 52 cards in which 19 of the cards are face up. He shuffles the deck several times and then blindfolds her (with a pair of safety goggles that have been painted black). He then shuffles the deck a couple more times and then places it in her hands, informing her that the challenge is to separate the deck into two piles, each with the same number of cards that are face up. She thought in silence for a couple of minutes, barely moving a muscle. Then, her fingers started slowly moving the cards across the top of the deck. She then placed two piles on the large oak table – each with the same number of face-up cards. How did she do it?

Discussion 9.2 This is a great example of a problem that seems impossible but yields readily to a person who knows how to simplify.

When considering a deck with 19 face-up cards and 33 face-down cards, it is very difficult to make any headway on this problem. Let's tackle a much simpler version of the same problem. How about a deck with two cards, one of which is face up? The only way to make two decks with two cards is with one card in each deck. When there are two cards on the table, one of which is face up and the other of which is face down, it is not hard to see that flipping over either card will produce two piles each with the same number of face-up cards. When one card is flipped, either both are face up or both are face down. In both cases, the number of face-up cards in each pile is the same.

This relatively simple observation actually represents significant progress towards the solution that would not have been made while trying to tackle the problem as stated.

Now let's consider a total of three cards, one of which is face up. We can make two piles by placing one card in the first pile and two cards in the second pile. One of the three is face up and, blindfolded, we don't know which one. However, flipping over the lone card in the first pile will ensure that the two piles have the same number of face-up cards.

It might also be prudent to consider also a case with three cards, two of which are face up. To figure out what to do here, split the three cards into two piles as we did before. There are two possibilities, either the lone card is face up or the lone card is face down. Here we need to flip the pile with the two cards in it. This will ensure that both piles have either no face-up cards or both have one face-up card.

Now let's consider a standard 52-card deck with only one face-up card. The same principle applies. We put one card in the first pile and the rest in the second pile. By flipping the lone card in the first pile, we ensure that there will be the same number of face-up cards in each pile. If the lone card in the first pile happened to be the lone face-up card, then all the cards will be face down and both piles will have zero cards face up. If the card in the pile was face down, then both piles will have only one card that is face up.

As a last stepping-stone towards the solution, let's consider a 52-card deck with *two* face-up cards. Let's start by separating the deck into a pile of two cards and a pile of 50 cards. It is possible that the two-card pile contains the two face-up cards, one face-up card and one face-down card, or two face-down cards. In each of these

9 Simplify! 141

three cases, flipping over the two-card pile will result in two piles with the same number of face-up cards.

So, the young woman in the job interview counted off the top 19 cards, flipped those 19 cards and placed them on the table. Then she placed the remaining 33 cards on the table next to it.

This is a true story, and she got the job.

> **Teacher Tip**
> It would be a good idea for the students to confirm that this would work for a range of possible face-up cards among the 19 that the candidate flipped. For example, if 9 of the 19 cards in the first deck were face up, there would be 10 face-up cards in the second deck. Flipping the entire 19-card deck will result in ten face-up cards in the first deck as well.

Problem 9.3 A child leaves his home and rides his bike to the store, averaging 15 km/h. When he comes out of the store, he sees that he has a flat tire and walks the bike back home at a constant speed of 5 km/h. Assuming that he takes the same path, what is his average speed for the back-and-forth trip?

Discussion 9.3 This puzzler will stymie many students because there is a variable missing – the distance between the child's home and the store. An experienced problem-solver will simply make up a value for the missing parameter to move the problem forwards. Further, the experienced problem-solver will select a value that is very convenient. In this case, the distance of 15 km from the home to the store is a good choice.

> **Student Pitfall**
> Students have a tendency to plus numbers into equations to get answers. The equation relating average speed, distance traveled, and time taken is $x = vt$. The students have only the average speed and, without the distance or the time, many will claim that the problem can't be done.

With an assumed distance of 15 km, the average speed can be determined by dividing the total distance of 30 km by the time it takes the child to bike there and walk the bike back. At 15 km/h, it will take the child 1 hour to get to the store, and at 5 km/h, it will take the child three hours to walk back. So the average speed is:

$$v = \frac{30 \text{ km}}{4 \text{ hours}} = 7.5 \text{ kph}$$

Does this work for all distances? Let's use the variable x as the distance from the home to the store. Therefore, the total distance traveled is $2x$. The time it takes for the child to bike to the store is $x/15$ km/h, and the time it takes for the child to make the return trip is $x/5$ km/h. So, the total time traveled is:

$$t = \frac{x}{15 \text{ kph}} + \frac{x}{5 \text{ kph}} = \frac{4x}{15 \text{ kph}}$$

The average speed is the distance traveled divided by the time taken:

$$\bar{v} = \frac{2x}{4x/15 \text{ kph}} = \frac{30 \text{ kph}}{4} = 7.5 \text{ kph}$$

So, the average is 7.5 km/h irrespective of the distance between the home and the store.

Making up values for missing information won't always lead to the correct answer, but it is a valuable problem-solving technique because it often can provide insights into the solution to the problem.

Problem 9.4 There is a 2-meter long, one-lane vine on which there are 100 tiny ants. Sixty are traveling to the left and 40 are traveling to the right. Each ant moves at a constant speed of 2 cm/second. When ants collide on the vine, they immediately reverse direction. If they reach either end of the vine, they keep going away from the vine. What is the longest time you will have to wait before all of the ants have left the vine?

Discussion 9.4 After a quick read of this problem, the students may start a classroom coup, as it appears exceedingly difficult. However, a clever simplification renders it almost trivial. The key is to realize that there is no difference between ants going through one another when they collide and two ants reversing directions when they collide. Once this simplification is made, the problem is reduced to the question, "What is the maximum time it would take for an ant to walk off a 2-meter long vine if it was traveling at a speed of 2 cm/second?" The answer is 100 seconds and that is the answer to the problem as posed.

Problem 9.5 Ten pirates have plundered a ship and discovered 100 gold pieces. They need to divide the loot among themselves. They want to be fair and abide by the law of the sea: to the strongest go the spoils. They have an arm-wrestling match to determine how strong each pirate is and then sort themselves from weakest to strongest. No two pirates are equally strong so there is no doubt about the order. We can label the pirates from weakest to strongest as P1, P2, and so forth, up to P10. The pirates also believe in democracy, and so they allow the strongest pirate to make a proposal about the division, and everyone votes on it, including the proposer. If 50 % or more of the pirates vote in favor, then the proposal is accepted and implemented. Otherwise, the proposer is thrown overboard, into the shark-infested waters, and the procedure is repeated with the next strongest pirate.

9 Simplify! 143

All pirates like gold (a lot!), but they hate sharks even more than they like gold. So any one of them would rather stay on board the ship and get no gold than be thrown overboard to the sharks. All the pirates are rational, and they know that if they damage any of the gold pieces (e.g., by trying to divide them into smaller pieces), then the bullion will lose almost all of its value. Finally, the pirates cannot agree to share pieces, as they do not trust each other.

What proposal should the strongest pirate propose to get the most gold?

Discussion 9.5 The strongest pirate may start with something along the lines of "we get 10 coins each." While a group of students voting for this might support the arrangement, it's important to remind them that pirates are *greedy*, and, while they are sensibly scared of sharks, any pirate who reasonably thinks he/she can get more gold will vote against the proposal. When reminded of this, the strongest pirate may then keep more gold and divide the rest (without fractions) to the rest but it's rare for students to get the solution immediately.

> **Teacher Tip**
> This problem can be very rewarding if you split the students up into groups of the right size and get them to work through the solution. (If you have smaller groups, then, once you are familiar with the concepts, you could vary the number of pirates or gold to reflect the number of students you have.)

Discussion 9.5 (cont) Working through the solution is often easiest if we start with the simplest case: a single pirate. In this case, the single pirate is the strongest pirate, and assuming he votes for his own strategy of keeping everything, he walks away with 100 gold pieces.

So, let us start at the second simplest case, when there are just two pirates, P1 and P2. In that case, the strategy of the strongest pirate, P2, is obvious: propose 100 gold pieces for himself, and none for P1. His vote would carry 50 % of the vote necessary for the acceptance of the proposal and he would be one rich pirate!

> **Student Pitfall**
> The biggest mistake most students make is that when they extend to two pirates, they forget that a successful vote only requires 50 % of the pirates to agree.

Discussion 9.5 (cont) Now we can consider the case with three pirates. Note that pirate P1 knows (and P3 knows that P1 knows!) that if P3's proposal is turned down, the procedure would proceed to the two-pirate stage where P1 gets nothing. So P1 would vote for absolutely any proposal from P3 that gets him *something*. Knowing then that the optimal strategy for P3 is to use a minimal amount of gold to bribe P1 to secure his vote, P3 should propose 99 gold pieces for himself, 0 for P2, and 1 gold piece for P1.

The strategy of P4 in the scenario with four pirates is similar. As he needs 50 % of the vote, he needs a vote of one additional pirate. Again, he should use a minimum amount of gold to secure this vote, so his proposal is 99 gold pieces for himself, 0 for P3, 1 gold piece for P2, and 0 for P1. Of course, P2 would be happy to vote for this proposal; otherwise, P4 is thrown overboard, the procedure reduces to three pirates, and P2 gets nothing.

Now, the strategy of P5 in the scenario of five pirates is just slightly different. He needs two additional votes from his fellows. Thus, he proposes 98 gold pieces for himself, 0 for P4, 1 gold piece for P3, 0 for P2, and 1 gold piece for P1. Clearly, the votes of P3 and P1 are secure, because in the four-pirate scenario, they would get nothing.

It is straightforward now to design a proposal for P6 in a six-pirate scenario, for P7 in a seven-pirate scenario, etc. In particular, the proposal for P10 is: 96 gold pieces for himself, 1 gold piece for each of the pirates P8, P6, P4, and P2, and none for the rest. This solves the small version of the puzzle. It is good to be the strongest pirate, at least when there is a small number of pirates and a lot of gold.

Teacher Tip
At this stage, consider taking the students to "the next level" and move to the larger version of this puzzle, leaving all the assumptions as they were but increasing the number of pirates to 500. The same pattern emerges, but there is a catch, because it only works only up to the 200th pirate. P200 will offer 1 gold piece for himself, 1 gold piece for each even-numbered pirate, and none for the rest. And that is when the fun starts in this larger version of the problem.

Discussion 9.5 (cont) P201 still can follow the previous strategy except that he runs out of gold and he proposes nothing for himself. So he proposes 1 gold piece for each odd-numbered pirate from P199 to P1. In that way, he gets nothing but at least he stays on board and avoids being eaten by sharks.

P202 also gets nothing. He has to give all 100 gold pieces to 100 pirates and stay dry. The selection of these pirates is not unique, as there are 101 pirates who are willing to accept the gold (pirates who do not get anything in the 201-pirate scenario), so there are 101 ways to distribute these bribes.

What about the 203-pirate scenario? This pirate must get 102 votes for his proposal including his own vote and he does not have enough gold pieces to give to 101 of his fellow pirates. So P203 will go overboard regardless of what he proposes! Too bad for him.

This is important for P204 though, as he knows that P203 would vote for anything to save his life! So P204 can count on P203 no matter what he proposes. That makes his task easy, as he can count on P203, himself, and 100 fellows that get

9 Simplify!   145

a gold piece each, so he can secure 102 votes. Again, the recipients of the gold should be among the 101 pirates who would receive nothing under P202's proposal.

The pirate P205 in the 205-pirate scenario faces an impossible task. He cannot count on P203 or P204 for support: each will vote against him to save themselves. So P205 will be thrown overboard no matter what he proposes. The moral is do not be the strongest in a group of 205 democratic pirates. The same fate awaits P206: he can be sure of P205's vote, but that is all he can count on, so overboard he goes. Similarly, P207 faces a soggy end to his existence, as he needs 104 votes: his own, 100 from the gold, and 3 additional followers. He can get votes from P205 and P206, but these are not enough, so overboard he goes.

The fate of pirate P208 is different, as he also needs 104 votes, but P205, P206, and P207 will vote for him to save their lives! With his own vote and 100 votes, his proposal will be accepted and he will survive. Of course, the recipients of his gold must be among those who would get nothing under P204's proposal: the even-numbered pirates P2 through P200, and then P201, P203, and P204.

Now, we can see the pattern, which continues indefinitely. Pirates who are capable of making successful proposals (even though they get no gold from their proposals, but at least they get to stay on the ship) are separated from one another by ever longer sequences of pirates who would be thrown overboard no matter what they propose! So the pirates who can make a successful proposal are P201, P202, P204, P208, P216, P232, P264, P328, P456, and so on (i.e., pirates whose number equals 200 plus a power of 2).

It is also easy to see which pirates receive the gold. As we saw before, the solution is not unique, but one way to do this is for P201 to offer gold to the odd-numbered pirates P1 through P199, for P202 to offer gold to the even-numbered pirates P2 through P200, for P204 to the odd numbers, for P208 to the even numbers, and so on, alternating between even and odd.

So, as the puzzle clearly illustrates, being the strongest and having a chance to put forwards the first proposal is not always the best (unless, of course, the number of pirates is quite small; sometimes, it is good to be a big fish in a small pond!).

The last problem in this section is one in which the simplification strategy is almost essential as it reduces the difficulty of the problem by at least an order of magnitude.

Problem 9.6 Imagine a village of one-eyed aliens who are very logical, but have an unusual cultural tradition regarding the color of their eye – which is known to be either brown or blue. If any member of the community can logically deduce their eye color, they have to leave the village forever after making the announcement that they have figured out the color of their eye (at the daily meeting, which is mandatory for everyone in the village). The village is isolated with no contact with the outside world and they have no mirrors or reflecting surfaces. Let's say that there are 5 blue-eyed aliens and 25 brown-eyed aliens in this village. One day, a visiting anthropologist addresses them at their daily meeting and says, "*My word, your blue and brown eyes are beautiful!*" There was a collective gasp among the aliens because she mentioned eye color. It would be natural to think that no harm was done as a result of this announcement because every member of the community already knew this – any blue-eyed alien could see 4 blue eyes and 25 brown eyes, and any brown-eyed alien could see 5 blue eyes and 24 brown eyes. However, the village will be empty within a week. Why?

Discussion 9.6 This puzzler stymies a person without a lot of experience solving problems. After all, everyone in the community already knew that there were both blue-eyed and brown-eyed members. Without experience in tackling new, unstructured puzzles, it is a challenge to make any progress whatsoever on this classic.

> **Teacher Tip**
> In order to convince the student the value of simplification, it is best to let them struggle with the problem as it stands. You can attempt to solve the problem as an entire class in a brainstorming session or you can break the class up into smaller groups. They are very unlikely to make any progress on this one without simplifying it (or googling it on their PDA).

This problem will not easily yield to a direct attack. To slay this one, it helps to start with a much simpler version. What if there was only one blue-eyed alien? Here the answer is straightforward. The lone alien with a blue eye must know he is blue as soon as the anthropologist reveals that they have both blue and brown eye color among them. The reason is, of course, that the alien with a blue eye sees no one else with a blue eye. Since he knows his eye color, he must announce this fact at the daily meeting the next day and say his goodbyes. As soon as he leaves, all the other members know they are brown because the only way the lone blue-eyed alien would know he is blue immediately after the anthropologist's remark is if he saw no other alien with a blue eye. Therefore, the rest of the community leaves on the second day.

Of course, we don't have the solution to the original problem yet because with only one alien with a blue eye. Nevertheless, we have made progress. Now, what if there were only two aliens with a blue eye?

Addressing the situation in which there are 2 blue-eyed aliens and, say, 28 with brown eyes, no one would leave at the first daily meeting after the anthropologist makes her announcement. However, both blue-eyed aliens are expecting that the other blue-eyed alien will leave because they see only one blue eye. When the two blue-eyed aliens see that the other blue-eyed alien did not leave on the first day, they conclude that they must have a blue eye as well and they leave together on the second day. When the two blue-eyed aliens leave on the second day, the rest of the community realizes that they must have a brown eye and they all leave on the third day.

With three blue-eyed aliens and the rest brown-eyed, the whole process is delayed by a single day. Let's consider the point of view of a blue-eyed alien when there are a total of three aliens with blue eyes. He could guess that he has a brown eye because he is looking at 27 brown eyes and 2 blue eyes. He would also reason that the two blue-eyed aliens he can see would not be able to figure out that they have a blue eye until the second day. However, when they do not leave on the second day, he must conclude that the only reason they did not leave is that he has a blue eye as well. The other two blue-eyed aliens reason similarly and the three leave together on the third day. On the fourth day, all the brown-eyed aliens abandon the island because they realize their eye must be brown.

So, in the original problem, the answer is that the 5 blue-eyed people leave on the fifth day and the remaining 25 brown-eyed people leave the day after.

Teacher Tip
Consider to challenge the students with the following question: "What *new* information from the anthropologist doomed the community?" This is discussed in Problem 15.7.

Debriefing The simplification technique should be a tool in every problem-solver's toolbox. The simplification technique is a very useful technique to apply when the problem is too difficult to attack as it is stated.

Perform a Gedanken: "What If?" and "So What?"

10

He who thinks little errs much.
– Leonardo da Vinci

A *gedanken* (from the German) is a thought experiment, a hypothesis that is evaluated in the mind. There can be many reasons to perform a gedanken. One is that the actual experiment is too difficult or even impossible to perform. Many great discoveries are made with or start from a gedanken. Albert Einstein wondered how a light beam would look if he could travel right beside it, and this led to the development of special relativity. In 300 B.C., Euclid proved that there was infinitude of primes just by thinking and wondering, which is perhaps the most impressive gedanken of all. Euclid simply wondered, *"What if there was a highest prime number?"*

Another reason to perform a gedanken is simply because there is no obvious path to the solution to a problem. There are many problems that students will find intractable at first glance, and one of the key contributions of Puzzle-based Learning is to encourage students to move past this, often illusory, obstacle. In many cases, the problem can be tackled by performing a quick thought experiment.

There is a significant overlap between performing a gedanken and other problem-solving techniques as many can be performed solely in one's head. There are also different complexity levels of a gedanken. There is the simple, *"What if it started at four?"* and there is, *"What if we eliminated minimum wage?"* Some gedankens are a lot more complicated than others, and it is as important to find the right approach to the gedanken as it is to undertake the exercise in the first place.

An underappreciated aspect of problem-solving is to question every aspect of the puzzle as it is presented, to understand the importance and relevance of each component. Puzzles are notoriously full of false leads, dead ends, and red herrings, so the questioning process is vital to reducing the puzzle down to only the relevant details. This is where the notion of "So what?" becomes an important component of a gedanken. By looking at every statement of fact, or apparent statement of fact, in a puzzle description and asking "So what?" we are forcing ourselves to explicitly include or exclude this detail as part of thought experiment, and these two small words become a vital component of the gedanken.

The ability to perform a gedanken will help the students to make decisions in everyday life. Correctly carried out, the gedanken will give students insight and develop their ability to anticipate and understand the consequences of their actions. As we have noted, creativity is one of the hardest aspects to develop, and the gedanken framework gives us a model that we can share through cognitive apprenticeship.

Let's look at several examples.

Problem 10.1 Samantha is 20 years old and thus twice as old as Allison was when Samantha was as old as Allison is today. How old is Allison?

> **Student Pitfall**
> It is very likely that many students will respond quickly with the answer "ten years old," because they are using only their System 1 thinking. This problem requires a careful reading and rereading of the problem as well as a depth of focus.

Discussion 10.1 Students can try to use equations here, and there is nothing wrong with that. However, a simple gedanken will lead to the answer relatively quickly. But before we begin the gedanken, let's use the "Simplify!" technique and assume that Samantha was born in the year 1990 and it is now 2010.

A typical starting point for this problem is a gedanken that asks, *"What if Allison was ten?"* Well, the question states, "...when Samantha was as old as Allison is today." The relevant question now is, how old was Allison when Samantha was ten? Samantha was 10 years old in the year 2000. So, Allison was zero when Samantha was ten. If Allison is ten years old in 2010, she must have been born in 2000. Zero is not one-half of twenty, so the guess that Allison was ten years old is too low.

Let's guess that Allison is fifteen years old, so she was born in the year 1995. Since Samantha is twenty, she was as old as Allison is today five years ago, in 2005. So, Allison was ten years old when Samantha was as old as Allison is today, and ten is indeed one-half of twenty. Therefore, Allison is 15 years old. (As in earlier chapters, we could solve this with a system of equations using algebraic variables.)

> **Teacher Tip**
> If you give this as a one-problem, 10-minute quiz, it is likely that a few students will write down "ten years old" quickly and then stop thinking about the problem. This intellectual laziness is not desired. The strategy to combat it depends upon the individual student. Sometimes it is best to collect their paper when they want to hand it in early, and other times it is best to

(continued)

10 Perform a Gedanken: "What If?" and "So What?" 151

encourage them to keep thinking about the problem. The ultimate goal is to get the students to engage their System 2 thought process, and this requires a depth of focus and a shutting down of external stimuli. It is natural for students to resist this because it makes them vulnerable. As a teacher and leader in the classroom, it is necessary to provide an atmosphere where the students can easily slip into state of deep thought.

Problem 10.2 A roller coaster starts at rest at the top of a long straight hill. It speeds up uniformly and reaches the bottom at a speed of 60 miles/hour in six seconds. Where was it going 30 miles/hour: one-fourth of the way down, halfway down, or three-fourths of the way down?

Student Pitfall
Many students will want an equation to use to solve this one. Some might even protest that they have not taken physics and hence are not yet trained to solve a problem like this.

So What?
The hill is long and straight. Is this important? No, the type of vehicle and the fact that it's on a hill are irrelevant. The key phrase is "speeds up uniformly" as we'll see below.

Discussion 10.2 No physics training is necessary to solve this one. A simple gedanken will reveal the solution. To start, compare the distance traveled by the coaster from 5 to 6 seconds and the distance traveled by the coaster during the first second. It should be fairly clear that the distance traveled in the first second is much smaller simply because the coaster is not going very fast at the top of the hill. At the bottom of the hill, the coaster takes a long distance to speed up another 10 miles per hour simply because it is going so fast. So, right away it seems like the answer must be one-fourth of the way down the hill. However, we can calculate it knowing only that the distance traveled by an object is its average speed multiplied the time it was traveling that averaged speed.

During the first three seconds, the coaster goes from 0 to 30 mph at a constant rate, averaging 15 mph. During the final three seconds, the coaster goes from 30 to 60 mph at a constant rate, averaging 45 mph. Since the average speed during the final three seconds is three times as fast, the distance traveled is three times as far. Therefore, the solution is, again, one-quarter of the way down the hill.

There are a number of familiar phenomena that are analogous, such as a drop of rain falling from a gutter or a car accelerating on an entrance ramp to a highway. The distance traveled at the beginning of the acceleration is smaller than at the end simply because the object is moving faster. Students are usually misled by any simple intuitive puzzle that works with inverse relationships or nonuniform distances. Gedankens offer a rich opportunity to allow students to practice their thinking on these more challenging areas.

Problem 10.3 Consider a metal washer, a metal disk that has a hole in the middle. It is a well-known fact that metal expands when it is heated. That is, the atoms get a little bit further apart on the average. So, when a metal washer is heated, what happens to the size of the hole?

Discussion 10.3 This is a challenging puzzle, and the fact that the actual expansion is very small contributes to this fact.

> **Student Pitfall**
> Many times, when students do not know the answer and can't look it up or use an equation, they are stymied. They lack the experience in tackling new problems and hence have not had the opportunity to develop their wondering skills, that is, their ability to perform a few gedankens to gain insights towards the solution.

> **So What?**
> The metal washer has a hole in it. So what? Do we need to know the size of the hole? Students will often get stuck on requiring an exact specification of the size of the hole and, if this is a concern, provide one (as a fraction of diameter) as it makes no difference to the problem at hand. In fact, the size of the hole should be irrelevant, which is a desirable "So what?" outcome for more advanced students.

Discussion 10.3 (cont) A good problem-solver will mentally go through a variety of scenarios in order to shed light on the question, as in the following example.

10 Perform a Gedanken: "What If?" and "So What?"

What if instead of a washer, you considered a metal disk? In place of the hole, a circle was drawn on the disk with a felt-tip marker. When the disk is heated, what happens to the diameter of the drawn circle?

Or, what if a smaller disk that had the same diameter of the hole and was made of the same metal was placed inside the hole in the washer and they were heated together? Doesn't the solid inner disk have to expand?

What if a marching band made a big "O" on the field and the director then asked each member of the band to get a foot farther from their nearest neighbor? Which way would a band member on the inner ring have to move to get further from his neighbors?

Consider a long metal rod that is bent into a ring instead of a washer. When this large-diameter ring is heated, does the hole get bigger or smaller?

What if the ring was made of a single chain of atoms? If the atoms get farther apart, does the hole get bigger or smaller?

It should be clear by now that the hole gets larger when the washer is heated. The problem was solved – not by attacking it directly but by attacking related problems whose answer might be easier to see.

As a historical note, before the invention of rubber, wood was used to make the wheels of wagons and carriages. To protect the wood surface on which the wheels rolled and to keep the wooden spokes from working loose, a flat iron ring was placed around the outside of the wheel. The iron ring had to be heated so that it would expand enough (about one centimeter) to be placed around the wheel. As it was cooled, its diameter decreased making it a very tight fit. Similar technology is used in the coopering of barrels for wine and spirits. Your students may already know of one or both of these examples but may not have realized the application of this knowledge to the problem at hand.

The next problem builds on Problem 8.3 – three bags, two marbles (black or white) in each of them.

Problem 10.4 There are three bags that each contains two marbles. One contains a black and a white marble. One contains two white marbles, and the third contains two black marbles. Your goal is to avoid a black marble when you choose the marble that counts. You are allowed to sample one marble from one bag. The marble happens to be white. Now you have to select the marble that counts, and your goal is to avoid a black marble. Do you select a marble from the same bag or choose a different bag?

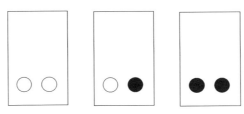

Discussion 10.4 This problem demonstrates the utility of performing a gedanken because so many students get the wrong answer by incorrectly applying the fundamental concept of probability.

> **Student Pitfall**
> It is very common for a student – especially one who knows some probability formulas – to come up with the wrong answer to this problem. A common mistake is to assume that the probability of selecting a white marble went down because a white marble was removed. Another common mistake is that the probability of selecting another white marble from the bag that the first one came from is 50 %.

> **So What?**
> You drew a white marble. So what? So you cannot have drawn from the bag containing two black marbles. This is a fundamental application of "So what?" because this is the "trick," for want of a better word, of the entire puzzle.

Discussion 10.4 (cont) To perform a gedanken, we simply wonder, "*What if I performed this experiment six million times?*" Well, each of the six marbles is equally likely so each one would get selected about one million times. In about three of the six million, the marble will be black so we don't count those. Now let's look at the three white marbles. In two of the three million instances in which a white marble was selected, the other marble in the bag is white, and in one of the three million instances in which a marble was selected, the other marble in the bag is a black one. Therefore, the probability of selecting a white marble from the bag from which a white marble was already removed is 2/3. Therefore, the correct strategy is to select the marble that counts from the same bag from which the first white marble was chosen.

> **Teacher Tip**
> The application of a thought experiment to probability problems is tremendously underutilized. It not only provides a direct route to the answer, it provides the students with a fundamental underpinning of the concept of probability. If you do this problem with your class, we recommend letting them try to solve it – either individually or in small groups – before teaching them about performing a gedanken. We also recommend asking the students questions that were posed in the discussion section, specifically the following:

(continued)

1. "If we sampled a marble from one of the bags six million times, how many times would a black marble be selected?"
2. "If we sampled a marble from one of the bags six million times, how many times would a white marble be selected?"
3. "In how many of the three million times a white marble was selected was the other marble in the bag white?"
4. "If a white marble is selected, what is the probability that the other marble in the bag is white?"

Problem 10.5 The three students with the highest test average in Mr. Johnson's fifth grade class get to play a logic game for the honor of being named "Most Clever." To start, Mr. Johnson seats them facing the class and shows them a total of five hats – three red and two white. He then blindfolds all three students, puts a hat on each student's head, and hides the remaining two. It turns out that he put red hats on all three students and hid the two white hats. He then announces that he will, in turn, remove the blindfolds of each student so each can see the color of the other two hats and then ask if the student can figure out the color of their own hat. The top student gets to go first. Her blindfold is removed and she announces that she can't tell the color of her hat based on the color of the other two hats. The second student's blindfold is removed, and – after some thought – she announces that she can't tell the color of her hat. As the teacher is about to remove the third student's blindfold, she announces, "*I got it!*" How could she figure out what color her hat was before her blindfold was taken off?

Teacher Tip
This is a *great* experiment to perform in front of the class. Five hats should be easy to find at a thrift store, and, of course, any two colors will do. It is usually best to give the three volunteers seats, because the discussion can last a long time. Blindfolds are not necessary; you can ask the student to close their eyes when putting on their hat and then poll the students one at a time. If the third student doesn't know, go back to the first student and keep circling through. It is hard to get the rest of the class involved with the process because they inevitably give it away when proposing a theory, saying, for example, "*Well, Joe can see two red hats, so...*"

So What?
All of the details of the students are, of course, irrelevant here. Some students will still get bogged down by the fact that the students have an "order."

Discussion 10.5 The third student could see nothing so the only information she has is that the first two students could not logically deduce the color of their hats based on all the information that was available.

Rather than wait for something to happen, the third student was performing a gedanken. She wondered, *"What if my hat was white?"* Once this thought process is started, it leads directly and inevitably to the answer. The key is to start it.

The only way that the first person could know the color of his hat is if he is looking at two white hats. So, once the first person says, *"I can't tell,"* the other two know that he is not looking at two white hats. When the second person's blindfold is removed, she can't deduce the color of her hat based on the color of the other two hats *and* the knowledge that the first person couldn't tell.

Now let's return to the gedanken. The third student, still blindfolded, asks herself, *"What if my hat were white?"* She answers in her mind thusly; if my hat were white, then the second person would know she is red because she knows that the first person was not looking at two white hats.

> **Teacher Tip**
> This conclusion might not be obvious to the students. To make it clearer, replace the red hat on the third student with a white hat and start over. Once the first person can't tell, it should be clear that the second person now knows he can't have a white hat and therefore must be red.

You can conclude the activity with the advice, "Don't just sit there, *THINK*!"

Here is a nice problem that can be used in a variety of settings. It can be used to start a class or it can be used as filler if there are ten minutes left in class or it can be used if there is a coveted prize available. Once it was used as a game at a birthday party. It is similar in nature to Icebreaker 3 (Chap. 3) as there is no "correct" answer and the success of a student depends on choices made by other students.

Problem 10.6 Write down an integer between 0 and 100 that will be 4/5 of the average of all the numbers written in a group.

> **Teacher Tip**
> Have small pieces of paper available for the students to write down their name and their number; they don't need an entire sheet of paper for this exercise. You can explain what you are going to do with the numbers to help the students achieve a better understanding of their goal. *"I'm going to add up all the numbers and divide by the number of students to get the average. Then I'm going to multiply that number by 4/5. Whoever is closest to that result is the winner."*

10 Perform a Gedanken: "What If?" and "So What?"   157

> **So What?**
> There is a subtle aspect to this puzzle, which is that everyone is being given the same information. So what? It means that your peers are now trying to guess what you would do, based on what they think that they would do. As will be noted below, everyone knowing that people are aiming for 4/5 is very different from everyone being asked to provide a number from 0 to 100 and one person, secretly, being asked to guess the 4/5 average.

Discussion 10.6 If you have a thoughtful class, it may take them some time to decide what number to write because they will be performing a series of gedankens. At first, they will think, *"Well the average of all the integers from 0 to 100 is 50, so perhaps I should write down 40."* But then they should realize that the other students are not choosing their numbers randomly. Everyone else has the same instructions and is trying to get 4/5 of the class average. So, if everyone else writes down 40, then perhaps 32 is the best number. In our experience, the longer it takes for a student to hand in their number, the lower it will be – in most cases anyway.

Typically the calculated result for this puzzle is between 20 to 40. The more advanced the class, the lower this calculated number tends to be.

While this problem has no "correct" answer, it provides an opportunity for the student to think hard, and that is the overriding goal of the course.

> **Teacher Tip**
> This is a nice problem to try on consecutive class days. If you are going to try this, give the problem at the end of class one day and then start the next class by announcing the winner. Then, before a discussion, have the students do it again. Will they write down a number that is 4/5 of the winner of the first game or will they write down 4/5 of the previous winner? As long as the students are thinking, you are doing your job.

Debriefing Many problems require an incubation period in which the solver mentally asks questions that will provide stepping stones to the answer. That is, the solver is "wrapping his/her head around the problem." The more experienced the problem-solver, the more relevant and revealing the questions. If the student thinks hard for 15 minutes and comes up with nothing, it is not a wasted effort. Thinking hard for 15 minutes and not getting anywhere is just as good for getting into good mental shape as running on a treadmill for 15 minutes and not getting anywhere is for getting into physical shape.

As a teacher, the more you can do to allow and encourage this process, the better problem-solvers your students will become.

Staying on Track When confronted with "What if?" and "So what?" some students will go off on quite extreme tangents, which can be counterproductive to keeping the class solving, especially if the students become defensive about their directions. One of the problems with a more open-ended course such as Puzzle-based Learning is that less able students sometimes cannot differentiate between a fruitful exploration and a wild stab in the dark. As students are exposed to more and more valid stepping stones and develop their mastery by observing you and each other and carrying out the process over and over again, these excursions should start to become more useful.

As a teaching technique for managing this, it often helps to turn the question back onto the person who is going off track. If the "What if?" can't happen, then get them to work this out. If the "So what?" has an implication that immediately negates itself, ask them "So what?" to help them get this sorted out in their own mind.

Simulation and Optimization 11

> *It is better to be rich, healthy, and happy, than poor, sick, and unhappy.*
> – Zbigniew Michalewicz

Many real-world problems are so complex that it is impossible to conduct a full theoretical analysis. In such cases, we can turn to *simulation* – we make experiments and carefully record the results. We have already suggested simulation when we discussed Problem 7.5, where different tennis players might have different probabilities of winning their games against different opponents, and we have to determine the probability of twins playing against each other in the tournament.

A simulation is an imitation of something real (whether a process, state of affairs, etc.). In other words, the word simulation is defined as the imitation of the functioning of one system or process by means of the functioning of another (e.g., a simulation of an industrial process). The act of simulating something real generally requires representing its certain key characteristics or behaviors. Simulation can be used in many areas, from human systems to safety engineering, in order to gain insight into their functioning. Simulation can provide insights into the effects of alternative conditions and courses of action. Simulations are useful when other types of analysis are too difficult (e.g., they require solving thousands of differential equations).

Some problems fall in the category of *optimization* problems. These problems require finding the *best* solution among *many* possible solutions. There is hardly a real-world problem without an optimization component. For example, how should we get to a particular destination in the *shortest* possible time? How should we schedule orders on a production line to *minimize* the production cost? How should we cut components from a piece of metal to *minimize* the waste? and so on.

In this chapter we discuss some aspects of simulation and provide a few diverse examples of optimization problems.

11.1 Simulation

A simulation uses a model of a real system to reproduce the behavior that we would expect to see from a real system. In most cases, when we say simulation, we actually refer to a *computer simulation*, where we write a computer program that will behave in the same way as some real-world system. For example, if we wanted to see what happened if we flipped a coin a million times, we could either put aside 12 days to flip the coin nonstop (we assume here that you are able to complete the flip – and record the result – in one second) or write a computer program that simulated the real system. A simple model would be to choose a number from 0 and 1 and record a 0 as a head and a 1 as a tail. Then we perform this operation a million times at machine speeds, which would take about a millisecond. Again, it is important to emphasize the need for *replicating* the simulation process to reduce the effect of stochastic variation.[1]

Simulation models can be very handy for performing experiments that involve probability, because of an approach known as *Monte Carlo* methods[2] (Monte pronounced "Mont-ee"). If we run the same experiment over and over again, we are effectively taking a sample over the random values and this repeated random sampling over the simulation would be a *Monte Carlo experiment*. One of the characteristics of Monte Carlo approach is that we can obtain an answer without calculating it directly. If students employ this kind of simulation to answer a question that has an exact, or formulaic, answer, they will not be able to give you the precise answer, just the approximation over as many tests as they ran. For example, if we wanted to know the chances of two coins being flipped and both coming up heads, we could run a Monte Carlo experiment with a million flips of two coins and record when they were both heads. This would give us an approximate answer, close to 0.25. However, we can easily calculate this with simple probability by saying that the chances of one coin coming up heads is 0.5, as is the chances of the other coin coming up heads, and given that one coin being heads has no impact on the other coin, we can multiply the two probabilities together to get the answer. 0.5 times 0.5 is 0.25. From a teaching perspective such as this, a simulation can be an excellent way to validate a solution – rather than to arrive at one. The Monty Hall problem (Problem 5.5), which is often challenging for students, is a common target for simulation as it quickly becomes apparent that the counterintuitive mathematical solution is correct.

---

[1] The term "stochastic" implies the presence of a random variable. In particular, *stochastic variation* is variation in which at least one of the elements is a random variable.

[2] The term *Monte Carlo method* (defined as a technique that involves using random numbers and probability to solve problems) was coined by Stanislaw Ulam and Nicholas Metropolis in reference to games of chance, which are a popular attraction in Monte Carlo, Monaco (Metropolis N, Ulam S (1949) The Monte Carlo method. J Am Stat Assoc 44:335–341). The concept of Monte Carlo simulation is quite general and the technique has universal applicability to a variety of problems in economics, environmental sciences, nuclear physics, chemistry, logistics, etc.

11.1 Simulation

Let's start with a simple example.

Problem 11.1 A boy is often late for school. When approached by his teacher, he explained that it is not his fault. Then he provided some details. His father takes him from home to the bus stop every morning. The bus is supposed to leave at 8:00 am, but this departure time is only approximate. The bus arrives at the stop anytime between 7:58 and 8:02 and immediately departs. The boy and his father aim at arriving at the bus stop at 8:00; however, due to variable traffic conditions, they arrive anytime between 7:55 and 8:01. This is why the boy misses the bus so often.

Can you determine how often the boy is late for school?

> **Teacher Tip**
> This is a good opportunity to develop the students' intuition by asking them to estimate the answer. Note that this problem can be solved in multiple steps by starting with simple models that allow enumeration of possible cases, through probabilistic considerations, to simulation. Finally, this puzzle connects the material presented earlier in this part of the book: Chap. 5 (understanding the problem and, in particular, building a model), Chap. 8 (enumeration), and Chap. 9 (simplification).

Discussion 11.1 Let's start with a simple model. Clearly, there are two important variables to consider:

x: arrival/departure time of the bus
y: arrival time of the boy

Note that the problem definition does not distinguish between the arrival time and the departure time of the bus. If the boy is not there when the bus arrives, the boy misses the bus.

Also, the problem description makes it very clear that the arrival time x of the bus may happen anytime between 7:58 and 8:02 (i.e., $7:58 \leq x \leq 8:02$), where any particular arrival time in this segment is equally likely, and the arrival time y of the boy may happen anytime between 7:55 and 8:01 (i.e., $7:55 \leq y \leq 8:01$), where any particular arrival time in this segment is equally likely. In other words, we assume a uniform distribution of arrival times in intervals 7:58, 8:02 and 7:54, 8:01 for the bus and the boy, respectively.

With this understanding it is easy to formulate the objective of this puzzle. If we know the arrival time x of the bus and the arrival time y of the boy, then the boy is late only if $x < y$ (if $x \geq y$, either the bus and the boy arrive exactly at the same time ($x = y$) or the boy arrives earlier than the bus ($x > y$); in both cases the boy boards the bus). So to determine how often the boy is late for school is equivalent to determining how often $x < y$. Indeed, what is the probability that $x < y$ for randomly generated x and y from their appropriate intervals?

Let's simplify the problem. Say, we consider only arrival times (whether for the bus or the boy) at one minute intervals. In other words, the bus can arrive at any of the following times:

7:58, 7:59, 8:00, 8:01, and 8:02

There are 5 possible arrival times for the bus. The boy, on the other hand, can arrive at any of the following times:

7:55, 7:56, 7:57, 7:58, 7:59, 8:00, and 8:01

There are 7 possible arrival times for the boy. From this enumeration, we can conclude that there are $5 \times 7 = 35$ possible arrival times for the bus and the boy. When matching up these various arrival times, how many pairs would lead to a situation that the boy misses the bus? Well, let us count. If the boy arrives at 7:55, 7:56, 7:57, or 7:58, he will catch the bus. If he arrives at 7:59, he will miss the bus only if the bus arrives at 7:58 (one out of 35 possible cases). If he arrives at 8:00, he will miss the bus only if the bus arrives at 7:58 or 7:59 (2 out of 35 possible cases). Finally, if he arrives at 8:01, he will miss the bus only if the bus arrives at 7:58, 7:59, or 8:00 (3 out of 35 possible cases). So altogether, the boy will miss the bus in 6 out of 35 possible cases (i.e., in slightly more than 17 % of cases). So the probability that the boy is late is around 17 %.

However, are we sure that this result is accurate? After all, we have made an additional assumption (to simplify our model, hence to simplify the process of getting to the solution) that the arrivals of the bus and the boy happen at full minutes (no seconds). Of course, there is no justification for such an assumption – we did this to make the calculations easier.

So, let us be a bit more precise and allow arrival of the bus and the boy with some finer granularity. Say, we consider only arrival times (whether for the bus of the boy) at 10 seconds intervals. In other words, the bus can arrive at any of the following times:

7:58:00, 7:58:10, 7:58:20, 7:58:30, 7:58:40, 7:58:50,
7:59:00, 7:59:10, 7:59:20, 7:59:30, 7:59:40, 7:59:50,
8:00:00, 8:00:10, 8:00:20, 8:00:30, 8:00:40, 8:00:50,
8:01:00, 8:01:10, 8:01:20, 8:01:30, 8:01:40, 8:01:50, and 8:02:00.

There are now 25 possible arrival times for the bus. The boy, on the other hand, can arrive at any of the following times:

7:55:00, 7:55:10, 7:55:20, 7:55:30, 7:55:40, 7:55:50,
7:56:00, 7:56:10, 7:56:20, 7:56:30, 7:56:40, 7:56:50,
7:57:00, 7:57:10, 7:57:20, 7:57:30, 7:57:40, 7:57:50,
7:58:00, 7:58:10, 7:58:20, 7:58:30, 7:58:40, 7:58:50,
7:59:00, 7:59:10, 7:59:20, 7:59:30, 7:59:40, 7:59:50,
8:00:00, 8:00:10, 8:00:20, 8:00:30, 8:00:40, 8:00:50, and 8:01:00.

There are now 37 possible arrival times for the boy.

From this enumeration we can conclude, as before, that there are $25 \times 37 = 925$ possible arrival times for the bus and the boy. When matching up these various arrival times, how many pairs would lead to a situation that the boy misses the bus? Well, let us count.

If the boy arrives anytime between 7:55:00 and 7:58:00, he will catch the bus. If he arrives at 7:58:10, he will miss the bus only if the bus arrives at 7:58:00 (one possible

11.1 Simulation

case out of 925 cases). If he arrives at 7:58:20, he will miss the bus only if the bus arrives at 7:58:00 or 7:58:10 (two out of 925 possible cases), and so on. Finally, if he arrives at 8:01:00, he will miss the bus only if the bus arrives at 7:58:00, 7:58:10, 7:58:20, ..., 8:00:50 (eighteen out of 925 possible cases). So altogether, the boy will miss the bus in $1+2+3+\cdots+18=81$ out of 925 possible cases, i.e., approximately 18.5 % of cases. So the probability that the boy is late is around 18.5 %.

Which of these two answers is "better" (i.e., more precise)? It seems that the latter one, as our model was more precise: we considered arrival times of every 10 seconds rather than every minute. What would happen if we consider an even finer granularity of arrivals occurring every second? Most likely, in such a scenario, the answer would be even more precise.

Now we can go after the general solution.

As we know, the two variables of the problem are:

x: arrival/departure time of the bus, $7:58 \leq x \leq 8:02$
y: arrival time of the boy, $7:55 \leq y \leq 8:01$

The boy is late for the bus if $x < y$. We can represent the overall problem graphically (this model of the problem is very elegant – recall Sect. 5.3 of this book): the coordinates, x and y, mark the arrival times of the bus and the boy; the rectangle within the boundaries of these variables ($7:58 \leq x \leq 8:02$ and $7:55 \leq y \leq 8:01$) defines the area of all possible events. Any point within the rectangle (e.g., point A, below) defines a particular event: the arrival of the bus at $x = 7:59:11$ and the boy at $y = 7:57:04$ (of course, these times can be even more precise). The thick line $x = y$ divides the rectangle into two areas:

(a) Area, where $x < y$ (dark part), when the boy is late for the bus
(b) Area, where $x > y$ (light part), when the boy is on time

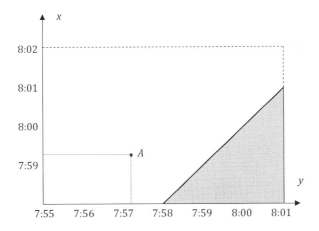

As any point within the rectangle can occur with equal likelihood, it is sufficient to find the ratio between the dark area of the rectangle and the area of the whole rectangle to calculate the probability of the boy being late. This is easy:

- The dark area: $3 \times 3 / 2 = 4.5$
- The whole rectangle: $4 \times 6 = 24$

Thus, the probability of the boy being late is exactly

$$\frac{4.5}{24} = 18.75\%$$

So our earlier estimations, where we used simpler models of the problem, 17 and 18.5 %, were not that bad.

There is also a way for arriving at a solution by simulation. We can generate a large number N of random points A from the intervals $7{:}58 \leq x \leq 8{:}02$ and $7{:}55 \leq y \leq 8{:}0$, with a point A having two coordinates, x and y,

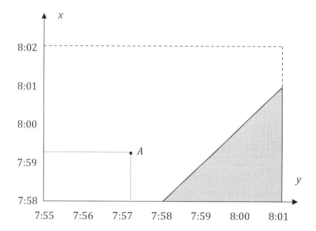

and keep track how many of these points would fall into dark area (i.e., $x < y$). Say, we generated 100,000 random points (x, y) out of which 18,738 fell into the dark area, a value we represent with the label k. Thus the probability of the boy being late is

$$\frac{k}{N} = \frac{18{,}736}{100{,}000} = 18.738 \%$$

Of course, for a smaller number of generated points, the precision of the estimate would most likely deteriorate.

Do we need a simulation model in such simple case? Indeed, there is no need as we can arrive at the precise answer by applying the basics of probability, as we already did a bit earlier. However, if the problem is complicated (e.g., the distribution of arrival times of the bus and the boy are not uniform and quite complex), the derivation of the solution might not be so straightforward. Then the usefulness of simulation is apparent.

Consider the following:

Problem 11.2 There is a crowd of people in a room. Each person is asked to select some other person in the room (without telling anyone). At a signal, every person

11.1 Simulation

tries to approach the selected person and stand behind his/her back. What pattern would emerge when "the dust settles"?

> **Teacher Tip**
> This is a good case to check students' intuition by asking them to describe possible patterns. Note also that if classroom environment is appropriate, we can run a real simulation by involving all students who follow the rules of the exercise. In this simple example, it would be easy to assemble a group of students and impose the behavioral rule of randomly selecting another person and standing behind them. This setup would allow us to observe the emerging pattern. Note that we can impose rules that are much more complex, such as requiring each person to take a position *in between* two selected persons. What pattern would emerge when "the dust settles"? Clearly, if the interaction rules are very complex, it will be difficult to find the solution (i.e., the emerging pattern) without simulation.

Discussion 11.2 This problem is a bit harder to analyze: it takes a while to visualize the pattern that might emerge in this scenario. Eventually we would get one or more clusters of people (by a "cluster" we understand a connected chain of people where there is a connection from person A to B if person A selected B). Each cluster is a circle (two or more individuals arranged in a circular fashion, where each individual selected the next one in a clockwise – or counterclockwise manner) with possible tails; a sample cluster (where the circle consists of six individuals marked by small dark circles and arrows represent the selection decision: one person selected "another") is displayed below:

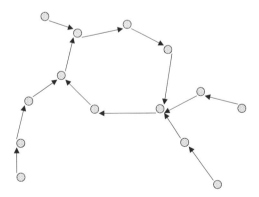

As the above example illustrated, not all simulations need computers. The CS Unplugged program[3] makes the very good point that you can provide a really

---
[3] http://csunplugged.org/

interesting puzzle and problem environment by simulating computers with everyday objects. In the same way that you don't need a pirate ship to solve the pirate puzzle, you don't need to write a computer program to produce the results of running the pirate puzzle – you can just run it in class and use students to stand in for the pirates.

> **Teacher Tip**
> While some students are happy to click repeatedly on the screen for little reward, many others will get frustrated quickly. A computer simulation that requires a lot of pointing, clicking, or typing may prove to be too much *kinetic load* for students on top of the existing cognitive load of the puzzle. Similarly, an in-class exercise that says "flip a coin thirty times" is less likely to succeed than one that requires only ten flips. Testing this with a small group first is a great way to see how students will react. (Remember, however, that anyone who volunteers for such a test is more likely to work for longer on the problem!)

There are several issues to discuss with students when considering simulation as a technique to demonstrate or verify a puzzle outcome. Much in the same way that you can trust the results from a calculator too much, when it's only as good as the accuracy of key presses and correctness of process, simulations can provide a believable solution to a problem and still not be correct or appropriate. It is important to emphasize that any simulation is only as good as the model that we use to build it. One of the biggest pitfalls in simulation is producing a bad model, which in turn leads to a bad simulation. A model that either ignores key factors or gets them wrong is going to produce a result that reinforces your belief in the model – but both the simulation and the model are actually incorrect! Listing all of the aspects of a puzzle to be able to build a correct model is very challenging – students must *really* understand the problem. They cannot have any ambiguities in their models before they produce accurate simulations. While this is a potential problem, it is also a strong encouragement to get students involved in simulating. If a student can produce a solid simulation that produces the right answer, then they probably understand the puzzle properly!

The above discussion on the importance of the simulation models leads to two important activities: *verification* and *validation*. It is worthwhile to emphasize that all models (including simulation models) are just *representations* of real problems and they always leave something out. If all aspects of the real world were represent in a model, then the model would be as complex and unwieldy as the real world itself. As a result, we work with simplifications of how things really are. On the other hand, we have to preserve the important characteristics of the problem we are trying to solve – otherwise the derived solution might be meaningless.

So verification is concerned with building the *model right*. During the verification phase, we compare the conceptual model to the computer representation and ask: Is the model implemented correctly in the computer? Are the input parameters and logical structure of the model correctly represented? Validation, on the other hand, is concerned with building the *right model*. During the validation phase, we want to determine that the model is an accurate representation of the real problem, situation, scenario, etc. Validation is usually achieved by calibrating the model, which is an iterative process of comparing the model to the real problem, situation, scenario, etc. and using the discrepancies and insights to improve the model. This process is repeated until the model accuracy is judged to be acceptable.

Let us return to general discussion on computer simulations. As indicated earlier, a computer simulation is a software program that attempts to imitate a real-world problem or scenario and provide predictions on possible outcomes. Every software program (including simulation software) requires a certain number of inputs; simulation software usually includes a few equations that use those inputs to give us a set of outputs (called also *response variables*). Many computer simulations are *deterministic*, meaning that we get the same results no matter how many times we recalculate. A classic example of this is simulation of compound interest, which always gives us the same result (for the same investment amount and interest rate). In such cases, we talk about deterministic model of a problem.

Sometimes models include variables that have a known range of values, but an uncertain exact value at any particular time. This is true for economics (e.g., the variables include interest rates, currency exchange values, and stock market prices, which have a known range of values that are constantly changing), logistics (e.g., the variables include inventory levels, production schedules, and transportation schedules, which also have a known range of values that are constantly changing), etc.

In such cases, we can use a *Monte Carlo simulation*. As discussed at the beginning of this chapter, the idea behind the Monte Carlo simulation is quite simple: by sampling the values of a model's variables from their (predefined) probability distributions, many scenarios are generated and the outcome is calculated. In other words, Monte Carlo simulation is just one of many methods for analyzing how random variation, lack of knowledge, or error affects the sensitivity, performance, or reliability of the system that is being modeled.

The best way to explain the Monte Carlo simulation is through a simple puzzle that involves just one variable:

Problem 11.3 Calculate (with some precision) the number π.

Discussion 11.3 Because it is not straightforward to calculate the exact length of the circumference of a circle, we can approach this problem by building a simulation model. We know that area A of a circle is expressed by

$$A = \pi r^2$$

where r represents half of the diameter of the circle:

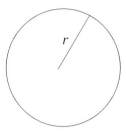

Then we can use a Monte Carlo simulation to approximate area A. Let us circumscribe this circle with a square:

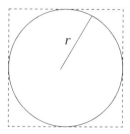

Area S of this square is

$$(2 \times r) \times (2 \times r) = 4 \times r^2$$

and the ratio between the area of the circle and the area of the square is

$$\frac{A}{S} = \frac{(\pi \times r^2)}{(4 \times r^2)} = \frac{\pi}{4}$$

Now we are ready for the Monte Carlo simulation. Imagine throwing darts at a square target with a circle inside (same as the above figure). Each dart lands somewhere inside the square: the coordinates of a throw are x and y (which are the horizontal and vertical coordinates, respectively). If the center of the circle is positioned at point $(0, 0)$, then the x and y coordinates can take any values from $-r$ to r. We can simulate a "throw" by generating two random numbers from this range (one for x and the other for y) and then calculate how many throws landed inside the circle. A throw is inside the circle if the distance between the center of the circle $(0, 0)$ and the position of the dart (x, y) is within the radius r:

$$x^2 + y^2 \leq r^2$$

Say, we simulated 10,000 throws (by generating 10,000 pairs of random numbers from the range from $-r$ to r) and the result of the simulation was that 7,854 darts landed inside the circle, while the remaining 2,146 darts landed inside the square (but outside the circle). This completes the Monte Carlo simulation, and we are ready to estimate the value of π. As the number 0.7854 (7,854/10,000) approximates the ratio A/S, and

$$\frac{A}{S} = \frac{\pi}{4}$$

then π is simply $4 \times 0.7854 = 3.1416$ (a pretty good approximation of 3.14159 after 10,000 throws). Of course, the larger the number of throws, the better the approximation.

> **Teacher Tip**
> There are a few issues to discuss with students when considering Monte Carlo simulation. Where we use computers, computers use *pseudorandom* number generators, and while these are reasonably reliable for most applications, they are not truly random numbers. While a detailed discussion of the issues are beyond the scope of this book, if the quality and unpredictability of the random numbers generated are going to be important to the simulation, one should look into *random number generation algorithms* and *seed values*.
>
> Further, it's a common misconception (the gambler's fallacy) that random numbers must show up during the span that someone is watching – because they are somehow "due." While it is true that a fair coin, flipped 100 times, is unlikely to come up as "heads" each time, there is no reason why it can't. The coin flips do not depend upon each other, and the coin doesn't *know* nor *remember* what has happened before. A good simulation that uses randomness has to take place over a large enough set of events that "runs", such as the long run of heads, take place over a sufficiently great set, that we still get the averages we expect. As a rule of thumb, computer simulations should run for hundreds of thousands to millions of times, depending on what you are modeling.
>
> If you throw a fair six-sided die six times, there is no guarantee that you will see all six numbers. (The chances of getting a simulation with six throws that gives you the correct expected probability of 1/6 for each side are actually less than 2 %!) When you are working with simulation, it can be very rewarding for students to see the influence of allowing the simulation to go on for longer and watch as values *converge* to their expected values.

For additional exercises on simulation, Chap. 12 on probabilistic reasoning provides a wealth of appropriate puzzles. Whether we consider different arrangements of lining boys and girls (Problem 12.1), draw/putt marbles from/to bags (Problems 12.2 and 12.14), search for "radar notes" on paper currencies (Problem 12.3), resolve some lost baggage issues (Problem 12.4), look for some distributions of cards (Problems 12.5 and 12.10), throw coins on parallel lines (Problem 12.6), play "cash wheel" (Problem 12.7), compete in shooting contest (Problem 12.8), select your date (Problem 12.9), select seat in the airplane (Problem 12.11), deal with emergence of butterflies (Problem 12.12), or roll dice (Problem 12.13), a simulation would be an ideal tool to approximate the answer (probability of some event). The thing to remember is that simulation would return experimental probability that approaches the theoretical probability as the number of trials increases.

11.2 Optimization

In this section we introduce optimization problems. These problems require finding the *best* solution among *many* possible solutions. The main characteristic of optimization problems is that in optimization problems each potential solution is assigned a *quality measure* (sometimes called the *evaluation function*) that allows us to compare the quality of different solutions. For example, each schedule of orders on a production line has a cost, which can serve as the quality measure of a schedule, which, in turn, is viewed as a possible solution: the lower the cost, the better the schedule. Optimization problems are also defined by a number of variables, and each variable has a domain of possible values. There might be also a number of constraints. However, the objective is to find a solution (a unique solution or the set of all solutions) that not only satisfies all the constraints (i.e., *feasible* solution) but also has the best quality measure. In other words, we have to find the *best feasible* solution.

We will begin with a delightful puzzle that introduces the topic of optimization and provides material for further discussion.

Problem 11.4 Four travelers (**A**, **B**, **C**, and **D**) have to cross a bridge over a deep ravine. It is a very dark night and the travelers only have one oil lamp. The lamp is essential for successfully crossing the ravine because the bridge is very old and has plenty of holes and loose boards. What is worse, its construction is quite weak and it can only support two men at any time.

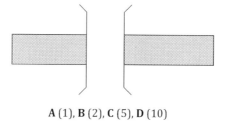

A (1), **B** (2), **C** (5), **D** (10)

It turns out that each traveler needs a different amount of time to cross the bridge. **A** is young and fast and only needs a minute to cross the bridge. **D**, on the other hand, is an old man who recently had a hip replacement and will need 10 minutes to get across the bridge. **B** and **C** need two and five minutes, respectively. And since each traveler needs the lamp to cross, it is the slower man in a pair who determines the total time required to make the crossing.

The question is how should the men schedule (i.e., organize) themselves to cross the bridge in the shortest possible time?

> **Teacher Tip**
> It might be good to discuss some types of solutions (possibly suggested by the process of lateral thinking), which include the following: traveler **A** gets across the bridge and throws the light to his companions on the other side, or traveler

(continued)

11.2 Optimization

> A (being the fastest) carries traveler **D** on his back, or the travelers manufacture the second lamp from some components of the original lamp. All these considerations/questions clarify the nature of the problem and help the students to focus on the real issues. We should be able to convince students that this is a genuine scheduling problem (no tricks are allowed) and only two travelers can be on the bridge at any one time. They can cross either individually or in pairs, but since they also need the lamp to ensure that they do not fall through a hole in the bridge, we know that it would be impossible for any one individual to carry the lamp from the starting side without having a partner go with them (to take the lamp back). So no matter what the solution, it has to involve a sequence of pairs traveling across the bridge. The question is: Which pairs?

Discussion 11.4 We can summarize our discussion so far. From the problem description it is clear that the only way four travelers can cross the bridge is when they follow this sequence of steps:
1. Two of them cross the bridge (with the light).
2. One returns (with the light).
3. Another pair crosses the bridge (with the light).
4. One returns (with the light).
5. The final pair crosses the bridge (with the light).

Of course, we can ignore some unproductive sequences, for example, where traveler **A** walks aimlessly back and forth or two travelers cross the bridge and then cross back together again.

The above sequence of five steps defines the structure of the solution that takes into account the constraints – thus it represents a model of the problem. Our only task is to specify *who* is crossing the bridge at what stage.

> **Student Pitfall**
> Many students[4] follow their intuition that the fastest traveler A should walk back and forth with the light and offer the following solution:
>
> | | | |
> |---|---|---|
> | 1. | A and B cross the bridge | Time: 2 minutes |
> | 2. | A returns | Time: 1 minute |
> | 3. | A and C cross the bridge | Time: 5 minutes |
> | 4. | A returns | Time: 1 minute |
> | 5. | A and D cross the bridge | Time: 10 minutes |
>
> Total time: 19 minutes
>
> In other words, the quickest traveler, **A**, is sent across the bridge with each man in turn. A could carry the lamp. So A and B could go across together – this
>
> (continued)

---
[4] The authors' experience indicates that the term *many* corresponds to over 90 % of individuals.

> would take two minutes – then **A** would return with the lamp, which would take an additional minute. Then **A** could go across with **C** and come back to do the same thing again with **D**. In all, this would require 19 minutes. This is a solution based on our intuition that the quickest traveler should always carry the lamp back to minimize the total time.

Discussion 11.4 (cont) But is there is a better way for them to accomplish their task? Is the above solution "optimal"? Well, to answer this question, we should consider other possibilities. Another appealing option is to send the two slowest travelers (**C** and **D**) together. When this possibility is offered to people working on this problem, they usually answer that they had thought about that possibility; however, it is no good, as one of these two slowest guys would have to return, and all the time gained by pairing them together at the beginning would disappear. This is a good point, but is it really necessary to send one of these guys back? Indeed, on second thought we can avoid this by scheduling the travelers as follows:

1.	**A** and **B** cross the bridge	Time: 2 minutes
2.	**A** returns	Time: 1 minute
3.	**C** and **D** cross the bridge	Time: 10 minutes
4.	**B** returns	Time: 2 minutes
5.	**A** and **B** cross the bridge	Time: 2 minutes

Total time: 17 minutes

So, after all it was possible to cut 2 minutes from our first (intuitive) solution. Note that these 2 minutes represent more than a 10 % improvement over the first solution – and this is an impressive number. If we could cut costs in a large manufacturing company by 10 % by more effectively scheduling the production orders, then this would be something! The above solution is optimal: they cannot cross the bridge in a shorter time than 17 minutes.

> **Teacher Tip**
> There are many similar puzzles to this one. For example, six travelers approach the same bridge and their respective times for crossing the bridge are 1, 3, 4, 6, 8, and 9 minutes. Again, what is the best way to schedule them
>
> (continued)

11.2 Optimization

> to minimize the crossing time? Or suppose there are seven travelers with crossing times of 1, 2, 6, 7, 8, 9, and 10 minutes, but in this case the bridge is stronger and can handle three travelers at a time. Both of these additional puzzles illustrate interesting points that are worthwhile to discuss further with students. The first of these two puzzles shows that a particular pattern (pairing the slowest travelers together) may or may not lead to the optimal solution. The second problem shows that the fact that the bridge can support three travelers at the time does not mean that we *have to* send three travelers at every crossing to get the best solution.

Discussion 11.4 (cont) These bridge-crossing puzzles are good examples of optimization problems, where we search for a solution that maximizes or minimizes some measure.[5] As indicated earlier, there usually are *many* possible solutions for an optimization problem. The set of all solutions is called the *search space*, which is further divided into feasible and infeasible solutions (i.e., solutions that satisfy and do not satisfy the constraints, respectively). Again, each solution has a *quality measure* (*evaluation function*) that allows us to compare the quality of different solutions. The main question for any optimization technique[6] is: How to search through the very large set of possible solutions to find the best solution (in terms of the quality measure) in the shortest number of steps?

The point is that very often the number of possible solutions is enormous. Even if the number of possible solutions is "only" 10^{30}, evaluating all of these solutions is simply impossible. If we had a fast computer capable of evaluating 1,000 solutions per second and if we started our calculations around 14 billion years ago (at the Big Bang), today we would have searched less than 1 % of all 10^{30} possible solutions! And this is the main challenge for all optimization techniques – how to find the best solution while testing only a very limited subset.

Let us return to Problem 11.4 and discuss it from the perspective of the search space and evaluation function. Remember that after the problem was analyzed, we found out that the structure (the representation) of the solution to this puzzle is this:
1. Two of them cross the bridge (with the light).
2. One returns (with the light).
3. Another pair crosses the bridge (with the light).
4. One returns (with the light).
5. The final pair crosses the bridge (with the light).

---

[5] In this text we restrict our attention to single-objective optimization problems, where we try to maximize or minimize a single objective (like the crossing time for the four travelers over the bridge).

[6] *Optimization technique* and *search technique* are considered synonymous. The search for the best feasible solution is both an optimization problem and a search problem.

Note that the constraints are already incorporated into this representation, as we require that only a "legitimate" traveler can take part in crossing the bridge. For example, the solution
1. **A** and **C** cross the bridge
2. **B** returns
3. **B** and **D** cross the bridge
4. **B** returns
5. **A** and **D** cross the bridge

does not make much sense, as it violates the constraints and is therefore infeasible. The above representation allows us to exclude such solutions, so we restrict the search space to only feasible solutions (the process of separating feasible and infeasible solutions is not always easy).

The above representation implies the size of the search space as we can enumerate all the feasible solutions. There are 6 possible pairs to be sent across the bridge in the first step (**A** and **B**, **A** and **C**, **A** and **D**, **B** and **C**, **B** and **D**, **C** and **D**). In each of these cases, there would be two possible choices for a traveler to return with the light in step 2 (e.g., if **A** and **B** cross the bridge, either **A** or **B** must return with the light). At this stage, another pair selected out of three available travelers would cross the bridge (step 3), and there would be three possible pairs. Then one of the three travelers "on the other side" would return with the light – again, this would give us three possibilities (step 4). Then the final pair (no choice here) crosses the bridge for the last time (step 5).

Thus, the total number of possible (feasible) solutions is

$$6 \times 2 \times 3 \times 3 \times 1 = 108$$

where each number corresponds to the number of possible choices at each step (i.e., 6 choices at step 1, 2 choices at step 2, etc.). In other words, there are 108 possible combinations here (and we will talk about combinations in more detail in the following chapter).

Note two important things:
- Each possible arrangement, for example,

1. **A** and **B** cross the bridge
2. **A** returns
3. **A** and **C** cross the bridge
4. **A** returns
5. **A** and **D** cross the bridge

or

1. **B** and **C** cross the bridge
2. **B** returns
3. **A** and **D** cross the bridge
4. **A** returns
5. **A** and **B** cross the bridge

11.2 Optimization

represents just one solution out of 108 possible solutions. The quality measure of each solution is the total crossing time for all four travelers. The two solutions given above each have their own quality measures of 19 and 20, respectively. Hence, the first solution (out of these two) is a better one.

So Problem 11.4 was indeed very simple: we could have listed all 108 possible solutions, calculated their quality measures, and selected the best!

> **Teacher Tip**
> The above discussion leads us to the following observation: when you are modeling an optimization problem, it is worthwhile to also think in terms of what the solution looks like and how it can be represented. For example, is the solution a sequence of actions (as was the case for the four travelers)? Or does the optimization problem call for the best number, the best sequence, the best arrangement, or the best strategy? It is important to keep in mind how the solution is represented in our model. Note that the structure of the solution will imply the search space (i.e., the set of all possible solutions) together with the feasible and infeasible parts (i.e., solutions that satisfy and do not satisfy all the constraints, respectively). It is important to discuss these issues with students at this stage.

Problem 11.5 Suppose we have to build a road from city **A** to city **B**, but these cities are separated by a river. We would like to minimize the length of the road between these cities and the bridge must be constructed perpendicular to the banks of the river:

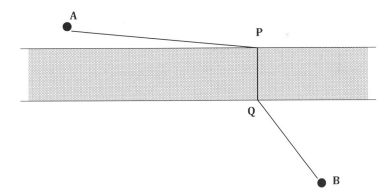

Now, the question is, Where to build the bridge to minimize the total length of the road?

Discussion 11.5 It is very easy to illustrate the problem on a diagram. The river is represented by two parallel lines and the bridge must be constructed perpendicular to the banks of the river. Our intuition is not of much use, as there is no obvious placement for the bridge. Should we leave it where it is? Or rather, move it a little bit to the left or right?

The solution that we are after is to find the location of the bridge such that the combined length of three segments

$$\mathbf{AP} + \mathbf{PQ} + \mathbf{QB}$$

is minimal. Note also that the length of the bridge, **PQ**, is always the same.

> **Teacher Tip**
> There are many methods for solving this problem but some of them require a significant amount of calculations. We should allow students to explore some of them. For example, we can assume some coordinates (x_A, y_A) and (x_B, y_B) for cities **A** and **B**, respectively. We can also assume that the river flows horizontally and is bound by y_t and y_b (where $y_A > y_t > y_b > y_B$). Then, we can build a formula for the length of the connection between cities **A** and **B** that is a function of an angle α (see figure below) and find the minimum length.

> **Student Pitfall**
> Students have difficulties in concentrating on the important parts of the model and reject the "noise" – those elements that obstruct our path to the goal.

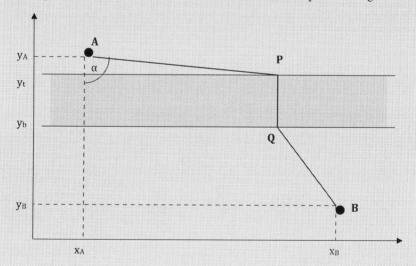

> For example, $\mathbf{AP} = (y_A - y_t)\cos(\alpha)$, where the only variable is the angle α.

11.2 Optimization

Discussion 11.5 (cont) Let us assume that there is no river. The river is reduced to a line (of width zero) and city **B** is moved upwards to **B′** by the distance equal to the original width of the river:

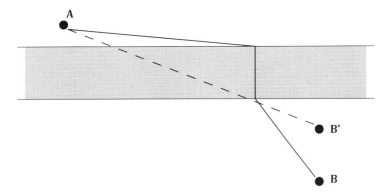

This problem is extremely easy to solve: a straight line between **A** and **B′** gives a solution!

This solution also solves our original problem. The line between city **A** and **B′** crosses the bank of the river at some point, say, **P**. This is the point where the bridge should originate; the terminal point of the bridge (i.e., the point of the other side of the river) is **Q**.

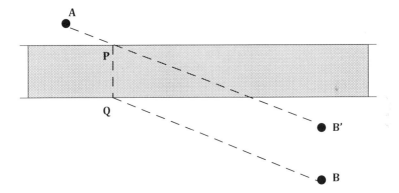

The segments **QB** and **AB′** are parallel so the total distance

$$\mathbf{AP} + \mathbf{PQ} + \mathbf{QB}$$

is the shortest possible as the segment **PQ** is constant (the length of the bridge) and **QB** is equal to **PB′**. So the best way to solve the problem of getting across the river is to abstract the river out of existence! In other words – the inclusion of unnecessary objects (variables) in the model may unnecessarily complicate the process of solving the model.

Problem 11.6 A rectangular chocolate bar consists of $m \times n$ small rectangles and you wish to break it into its constituent parts. At each step, you can only pick up one piece and break it along any of its vertical or horizontal lines. How should you break the chocolate bar using the minimum number of steps (breaks)?

Discussion 11.6 We can easily visualize the problem. The rectangular chocolate bar (9×13) is given below:

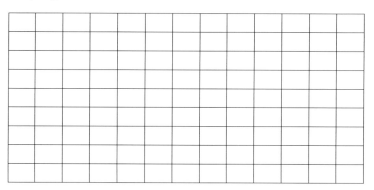

Clearly, we can break it in many ways. For example, in the first step we can break the chocolate into two pieces along the third (inside) vertical line:

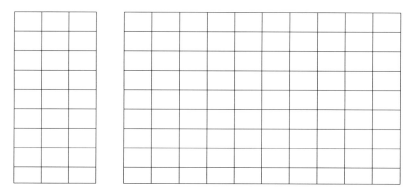

We can then break the left piece along the 6th (inside) horizontal line; this step would result in three pieces of chocolate:

11.2 Optimization

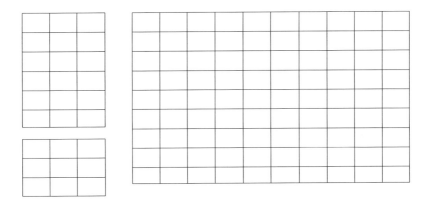

We can continue this process until we have a collection of all $9 \times 13 = 117$ separate small rectangles. How many breaks did we make?

Student Pitfall

Students have difficulties in solving this puzzle, as they try to find the best strategy based just on their intuition. Some of them focus on the "long" (horizontal) lines to make the process as efficient (at least at the beginning) as possible. Other students concentrate on "short" (vertical) lines, reasoning that it would pay off later. Yet others investigate some alternating strategy (one horizontal break followed by one vertical break).

Teacher Tip

It is important to present students the key question: What is a solution here? The solution is a strategy: a sequence of breaks that would lead to the final arrangement (only the smallest pieces are left). However, it is not trivial to *represent* a strategy. One possibility would be to index all the horizontal lines available for breaks by $h_1, h_2, \ldots, h_{m-1}$, and all the vertical lines – by $v_1, v_2, \ldots, v_{n-1}$. Furthermore, we have to index the pieces of chocolate that emerge after each break. Then a strategy for breaking the chocolate bar would be a sequence of recommendations; for example,

"break kth piece along the line h_j."

Of course, some care should be taken to ensure that the horizontal line h_j is included in the kth piece, which need not be the case (e.g., a small 3×4 piece from the top-left corner of the chocolate bar does not include the 7th horizontal line for breaks). Students should appreciate the complexity of this task!

Discussion 11.6 (cont) Actually, this puzzle is quite trivial. Note that each step of breaking a piece results in replacing one (larger) piece by two additional (and smaller) pieces, which are added to the collection. So we start with a single piece – the whole chocolate bar. After the first break we get 2 pieces, after the second break we get 3 pieces, after the kth break we get $k+1$ pieces of chocolate. The number of chocolate pieces *at any stage* is always one larger than the number of executed breaks.

As we have to continue the breaking process till we get 117 pieces (or, in general, $m \times n$ pieces, each of them being the smallest rectangle), the number of required breaks is 116 (or $m \times n - 1$). Interestingly, there is nothing to optimize! Any breaking strategy would result in the same number of steps!

This puzzle illustrates the unusual situation where there are many possible solutions (many different strategies of breaking a chocolate bar into individual pieces); however, the quality measure of each solution is the same!

Problem 11.7 Suppose you wish to know which floors in a 36-story building are safe to drop eggs from and which will cause the eggs to break on landing (using a special container for the eggs). We eliminate chance and possible differences between different eggs (e.g., one egg breaks when dropped from the 7th floor and another egg survives a drop from the 20th floor) by making a few (reasonable!) assumptions:
- An egg that survives a drop can be used again (no harm is done and the egg is not weaker).
- A broken egg cannot be used again for any experiment.
- The effect of a fall is the same for all eggs.
- If an egg breaks when dropped from some floor, it would break also if dropped from a higher floor.
- If an egg survives a fall when dropped from some floor, it would survive also if dropped from a lower floor.
- There are no preexisting assumptions concerning when the egg will break. It is possible that a drop from the first floor in the special container would break an egg. It is also possible that a drop from the 36th floor in the special container would not break an egg.

Now, if only one egg is available for experimentation, we have no choice. To obtain the required result, we have to start by dropping the egg from the first floor. If it breaks, we know the answer. If it survives, we drop it from the second floor and continue upwards until the egg breaks. The worst-case scenario would require 36 drops to determine the egg-breaking floor.

Now, suppose we have two eggs. What is the least number of egg drops required to determine the egg-breaking floor? Note that the method should work in all cases.

> **Student Pitfall**
> Most students when given this puzzle try to start somewhere in the middle of the building, e.g., dropping the first egg from the half-height. Most likely they

(continued)

11.2 Optimization

use, more or less consciously, a technique called *binary search*, where we divide a sample in half (or as close to half as possible) and – based on the outcome – proceed further. Clearly, this technique will not provide us with a good solution here. If we drop the first egg from, say, the 18th floor, there are two possible outcomes:

- The egg breaks. In this case, we have to move slowly and test every floor starting from floor 1. In the worst case, we drop the second egg 17 times to determine the egg-breaking floor, making the total of 18 drops.
- The egg does not break. We have to examine remaining 18 floors; however, we have still two eggs for experimentation.

These two cases suggest that we should start lower than the 18th floor: if the first egg breaks, we will have a shorter segment of floors to experiment with; if the first egg does not break, the segment would be longer but we would have still two eggs for experiments.

Teacher Tip

An important issue to discuss with students is the question how to represent a solution? One of the ways to represent a solution for this optimization problem is to represent our decisions (of where to drop the eggs) in the form of a tree. Each node is represented by a number that corresponds to a particular floor level, and the two lines leaving the node represent the two cases where the egg breaks or survives the drop:

This part of the tree represents a drop from floor level a. The thick line indicates that the egg breaks and the thin line that it does not. In the former case, we can drop the second egg from level b, whereas in the latter case we drop the first egg from level c.

Discussion 11.7 So, where should we start? In the case of this puzzle, let us present the solution first and then discuss how to arrive at this solution. Note that the diagram below shows a method that would determine the egg-breaking floor in no more than 8 droppings:

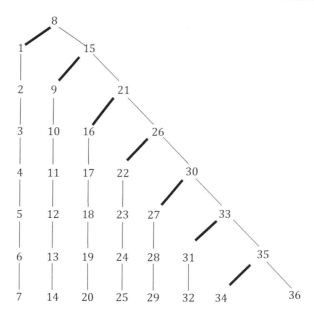

Again, the thick lines indicate cases where the first egg breaks and the thin lines when it does not. The whole diagram can be interpreted as follows. We start by dropping the first egg from the 8th floor. If it breaks (left branch marked by the thick line), we have very little choice, as only one egg is left for experimentation and we have to experiment with floors 1, 2, ..., 7 (in that order). So we have to start with the lowest floor, gradually moving up one by one. In the worst-case scenario, we need all seven attempts to determine that the egg-breaking level is 7 or 8. If the first egg does not break, we repeat the experiment from the 15th floor. Again, if it breaks, we have to examine floors 9, 10, ..., 14 (in that order); if not, we move up to the 21st floor. It is clear that in the worst-case scenario, we need 8 attempts to determine the egg-breaking floor.

The trick in finding this solution lies in our ability to build a *balanced* tree – balanced in the sense that the nodes that have two branches are of similar (ideally, identical) lengths (i.e., "left" length is the same as the "right" length). Note that to accomplish this task, we have to proceed in a systematic way that is determined by the rightmost branch of the tree. The last drop of the first egg (if it survives that long) must be made from the 36th floor, the second-to-last drop from the 35th floor, the one before from the 33rd floor, and so on. Hence, moving from the top of the building downwards, the gaps between floors become larger by one. Note that
$$1+2+3+4+5+6+7+8 = 36$$
which gives us the following information:
- Eight attempts should be sufficient to determine the egg-breaking floor.
- We should start from the 8th floor.

The rest (i.e., the construction of the tree) is straightforward.

11.2 Optimization

So let us solve the same problem for a 100-story building! Again, note that
$1+2+3+4+5+6+7+8+9+10+11+12+13 = 91$
and
$1+2+3+4+5+6+7+8+9+10+11+12+13+14 = 105$
which means that 13 attempts are not sufficient (in the worst-case scenario) and 14 attempts will do the trick. To better understand the above equalities and their significance for this egg-dropping puzzle, let us construct the decision tree for this 100-story building (starting from the top of the building).

If the first egg survives the first 12 drops (we will define what these drops are later), the 13th, second-to-last drop, would be from floor 99. If the egg breaks, we would test floor 98 in the final, 14th attempt. If the egg survives, we would test floor 100 in the 14th attempt:

If the first egg survives the first 11 drops (again, we will define them later), the 12th drop would be from floor 97. The reason is that if the egg breaks, we would test floors 95 and 96 in the 13th and 14th attempts, respectively. If the egg survives, we would test floor 99 in the 13th attempt:

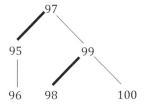

If the first egg survives the first 10 drops (again, we will define them later), the 11th drop would be from floor 94. The reason is that if the egg breaks, we would test floors 91, 92, and 93 in the 12th, 13th, and 14th attempts, respectively. If the egg survives, we would test floor 97 in the 12th attempt:

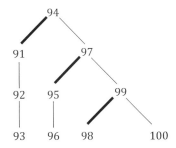

It is a very trivial exercise to complete this decision tree. The numbers on the higher levels at the rightmost branch of the decision tree, from bottom to top, would be
100, 99, 97, 94, 90, 85, 79, 72, 64, 55, 45, 34, 22, and 9
exactly 14 levels. Note that the number of levels would not change if the building is a bit higher (up to 105 floors, as $1+2+3+4+5+6+7+8+9+10+11+12+13+14 = 105$). In the case of 105 floors, the numbers on the higher levels at the rightmost branch of the decision tree, from bottom to top, would be
105, 104, 102, 99, 95, 90, 84, 77, 69, 60, 50, 39, 27, and 14

> **Teacher Tip**
> And now, having full understanding of the solution process for this puzzle, we can challenge our students to design a decision tree for the case where 3 eggs are available for experimentation.

Debriefing A class on optimization methods may require some summary (or overview) to make some general points that students would remember. One of the points would be to indicate that there are some methods (like *dynamic programming* or *branch-and-bound method*) that would guarantee optimal solutions, but they are not much more efficient that the enumerative search. Another option would be to use approximate methods that would provide good (but not optimal) solutions. Here we can choose from classic and intuitive *greedy algorithms* (where we start from a random city and, at each iteration of the process of constructing the solution, we select the closest unvisited city), through *linear/integer programming*, to modern heuristic methods like *tabu search*, *simulated annealing*, or *genetic algorithms*. For more information on this topic, the reader is referred to *How to Solve It: Modern Heuristics* by Z. Michalewicz and D.B. Fogel (Springer, 2004). Also, in Part III of this book, there are a few puzzles (e.g., Problems 13.7, 14.15, 15.2) with an optimization flavor.

Reference

1. Metropolis N, Ulam S (1949) The Monte Carlo method. J Am Stat Assoc 44:335–341

Part III
Challenges

It is not enough just to be aware of problem-solving techniques; students must develop the skill to use the techniques to solve problems. The only way to develop this skill is to solve problems.

In this part of the book, we present a range of challenging problems that will give the students practice applying the techniques presented in Part II.

As mentioned many times previously in this book, it is better to go over one problem carefully and thoroughly rather than try to get through a large number of problems. For this reason, we recommend being very selective when deciding which problems to use in the classroom.

The first chapter in this part of the book (Chap. 12) contains problems involving probabilities. These problems are designed to develop the probabilistic intuition of the students as well as their ability to enumerate all the possibilities. As a side benefit, the students will get practice manipulating equations, recognizing patterns, and using the factorial function. We also explore continuous probability distributions and summing infinite series.

Following this, there is a chapter on logical reasoning (Chap. 13). Here, we challenge the student to think about what others know and when they know it. This will develop their empathy. They will also have to be able to retake inventory multiple times when solving some of the problems because new knowledge is being generated. Students also will get practice performing gedankens by asking themselves "what if..." and by utilizing "if–then" logical statements.

The chapter on logical reasoning is followed by a series of problems involving geometric reasoning (Chap. 14), that is, problems involving two- and three-dimensional shapes. We start with a few problems involving the pentominos and continue with problems involving the Pythagorean Theorem. In this section, students will get to practice drawing a diagram, building a model, and working with manipulatives. There are also problems involving dissection of shapes and optimization.

The final chapter in the book (Chap. 15) contains a collection of problems that may challenge the best students. In some of our classes, we present a page of these on the first day of class for those students that need an extra challenge. They are not officially "assigned," but every couple of weeks or so, we ask if anyone has thought about or has made progress on any of the problems. These problems might be good candidates for special projects or grand challenges to present at a poster session or any "evidence of scholarship" event that is used at your school or university.

The problems you present to your students should be chosen carefully and perhaps even customized for them. It is best to look through the entire chapter before selecting which one to use. If you are comfortable with letting the student choose, this can work as well. In the beginning of the class period, you can present two or three problems and have the students pick one to work on that day.

Although you know your students best, we recommend that they be allowed to decide how to attack the problem independently.

> **Teacher Tip**
> In a typical classroom, there is often a wide range of experience, ability, and formal training among the students. If a student or two need a greater challenge, there are plenty of problems in this part from which to choose. Not everyone in the class has to be working on the same problem.

If the goal of the course is to develop independent problem-solvers, a good metric for your success as a cognitive apprentice is how often the students look to you for guidance or hints. If they completely ignore you and work independently and successfully, you have done well.

The problems in this part of the book are presented in a format that starts with a statement of the problem. This is followed by a summary of the strategies that may be utilized to solve the problem. While every problem should include the pre-solving strategy *understand the problem*, we only included it here if the problem needs special attention to understand what is being asked.

This is followed by a discussion of the problem in which the answer is revealed and then a debriefing. Teacher Tips and Student Pitfalls are given throughout the problem sets, as needed.

Probabilistic Reasoning 12

All knowledge degenerates into probability.

– David Hume

Probability theory is the branch of mathematics that deals with estimating or calculating the degrees of likelihood. If it is *impossible* that a particular event would happen, it is given a probability of zero. If it is *certain* that a particular event would happen, it is given a probability of one. The probabilities of other events (expressed as fractions or decimals) lie between zero and one.

In his book *Entertaining Mathematical Puzzles,* Martin Gardner talks about the strong connection between real-life decisions and probability:

> "Everything we do, everything that happens around us, obeys the laws of probability. We can no more escape them than we can escape gravity. The phone rings. We answer it because we think someone is calling our number, but there is always a chance that the caller dialed the wrong number by mistake. We turn on a faucet because we believe it is probable that water will come out of it, but maybe it won't. 'Probability,' a philosopher once said, 'is the very guide of life.' We are all gamblers who go through life making countless bets on the outcome of countless actions."

While probability has many applications in everyday life, the problems in everyday life are not well defined and often involve some educated guesses and estimations. For example, what is the probability of a football team winning its next game? What is the probability that a 60-year-old male will live to his eightieth birthday? What is the probability that the price of gold will go up 5 % over the next year? What is the probability that it will rain tomorrow? What is the probability that the polar bear will be extinct by the year 2050? These are all good probability problems, and a lot of people make a living trying to answer them.

To answer complicated questions like the ones posed above requires a sound foundation in probabilistic thinking. This is what we will try to develop in this chapter. With a strong foundation in probabilistic thinking, the student will be better equipped to tackle real-world problems in the future.

In general, humans tend to have poor intuition about the probability of an event occurring. One reason for this is that we have evolved to recognize patterns as a survival mechanism. As a result, we see patterns everywhere – even in randomness. We see patterns in nature[1] and in numbers – recall also comments we made in the introductory part of Chap. 7.

This has been well documented in the literature,[2] and you can do some simple exercises in class to demonstrate this to your students. One experiment is to say, "*I have here coin that was a gift from a friend. I'm going to flip it over and over again and announce the results to the class. Raise your hand when you think something strange is going on.*" Even though the coin is fair, it won't be long before most of the students have their hands up.

We have also given quizzes with 200 blank spaces with the only instructions to fill in the blanks with either a one or a zero in order from left to right by modeling a 50/50 coin toss. In other words, the students have to simulate the results of a 50/50 event in their heads. Almost invariably, the students have a natural aversion to putting down long strings of ones or zeros – thinking that a long string of ones or zeros is not random. That is, once they write down three or four zeros or ones in a row, they have a natural tendency to end the sequence, thinking, "This is not randomness." The result of any truly random event – by definition – is independent of the results of previous events.

One way to "grade" this quiz is to count (or have the students count) the number of "switches" from 0 to 1 or from 1 to 0. There should be about 100 of these switches, but the class average is usually between 110 and 120, with many of the human-generated sequences having no strings of either ones or zeros of length five or more. It is actually quite remarkable that a person familiar with probability would be able to differentiate, in most cases, a human attempt at generating a random sequence from a true random sequence.

Another way to demonstrate that patterns occur naturally in randomness is to present sequences found in the expansion of π. For example, starting at the 50,366,472nd digit after the decimal point in π, there is a sequence "31415926," which is the first eight digits of π. Amazing coincidence? No. If enough randomly generated sequences are examined, some patterns will be recognizable and hence not appear to be random. The students should understand that randomness means "without bias," not "without pattern" (see also discussion for Problem 7.6).

The main purpose of this chapter is not to make the student more familiar with probability formulas and terminology; it is to develop the student's probabilistic intuition and their ability to effectively and efficiently tackle new problems. For this reason, there are no formal protocols to commit to memory and no formulas to

---

[1] Patterns in clouds and rock formations are good examples. For a striking example of a natural pattern found in satellite images of the earth, use the Internet to check out the "Indian Head" feature found in Canada, located at coordinates (50°00′ 38.20″ N, 110° 06′ 48.32″ W).

[2] For an entertaining and eye-opening book on this subject involving the stock market, see Nassim Taleb's book, *Fooled by Randomness*, Random House, 1994.

remember. There will be a minimum of mathematical terminology and symbols, and the emphasis will be on pure problem-solving. The only math skills needed in this section is to be able to solve simple equations and to be able to manipulate fractions.

Probability will simply be the "playing field" upon which we develop the students' thinking and reasoning skills. Probability was selected for inclusion in this book because it offers a wide range of simply stated problems that will reward the student with hours of challenging and enjoyable mental exercise. In other words, the problems are relatively easy to understand but can be very challenging to solve. Further, the amount of formal training needed to be able to solve the problems is minimal.

To understand the difference between training a student to calculate answers and training a student to be a skilled problem-solver, consider the two solutions to the following problem in probability:

> *A committee of ten people is going to select one person among them to be chairperson and another to be the secretary. How many different possible combinations of chairperson and secretary are there?*

Solution 1 To help think about the problem, let's frame it better by naming everyone on the committee. For convenience, we'll pick names that begin with the letters A through J. How about Amanda, Bert, Cheryl, David, Edward, Frank, Glenda, Hillary, Ian, and Jocelyn? If Amanda is chairperson, how many different secretaries can there be? Since there are nine people remaining from whom to choose a secretary, there are nine different combinations of chairperson and secretary in which Amanda is the chairperson. There are also nine such combinations when Bert is chairperson, nine when Cheryl is chairperson, etc. Since there are nine combinations when each person on the committee is the chairperson, there are 90 possible ways to choose a chairperson and a secretary from ten people. The answer is 90.

Solution 2 The formula for the number of ways to choose r items from n objects in which the order matters is

$$P(n,r) = \frac{n!}{(n-r)!}$$

Plugging in $n = 10$ and $r = 2$ into the formula gives the answer 90.

On reflection, however, how much thinking went into the second solution? Is this the formula for permutations or combinations? (For those who don't know the answer, welcome to the world of the general student who confuses these all the time. And, for the record, it's a permutation, because the order matters.)

We believe that teaching students to reason through the problem as demonstrated in first solution is much better for developing the student's thinking skills because the student is actually thinking. Students who are trained to calculate answers using the second technique without developing the underpinning of the main concept of probability are vulnerable to two potential pitfalls. The first is that they can't remember the formula, and second is that they misapply the formula.

If the student is a little bit unsure about the answer to the above example, it is probably a very good exercise for the student to write out all 90 permutations (there's that word again, but now we can put in a *thinking* context, rather than a mathematical one) of chairperson and secretary. Then the students can *SEE* where the 90 possibilities come from. A table such as this tells the complete story, with the members of the committee represented by the first letter of their name.

		Chairperson									
		A	B	C	D	E	F	G	H	I	J
	A	–	BA	CA	DA	EA	FA	GA	HA	IA	JA
	B	AB	–	CB	DB	EB	FB	GB	HB	IB	JB
	C	AC	BC	–	DC	EC	FC	GC	HC	IC	JC
	D	AD	BD	CD	–	ED	FD	GD	HD	ID	JD
Secretary	E	AE	BE	CE	DE	–	FE	GE	HE	IE	JE
	F	AF	BF	CF	DF	EF	–	GF	HF	IF	JF
	G	AG	BG	CG	DG	EG	FG	–	HG	IG	JG
	H	AH	BH	CH	DH	EH	FH	GH	–	IH	JH
	I	AI	BI	CI	DI	EI	FI	GI	HI	–	JI
	J	AJ	BJ	CJ	DJ	EJ	FJ	GJ	HJ	IJ	–

The table clearly illustrates that there are 90 possibilities. Basically, there are ten people and two different "slots" to fill. There are ten possibilities for the first slot and nine possibilities for the second slot. This makes a total of 90 possible different combinations of president and secretary.

If all possibilities were equally likely – as they would be if all ten names were drawn randomly from a hat with the first being assigned the role of chairman and the second being assigned the role of secretary – the probability of each combination occurring would be 1/90.

This demonstrates the one basic, overarching concept in probabilistic thinking. It is simply that

> *The probability of particular outcome is the number of ways it can occur divided by the total number of possible outcomes.*

Let's look at another example – the rolling of a single six-sided die. There are six possible outcomes, and there is only one way to roll, say, a four. So, the probability of getting a four is one-sixth on any roll. (From this point on, we'll assume that any dice we refer to are six sided, unless we explicitly state otherwise.)

Now let's consider two rolls of the same die. The first roll can be any of six outcomes, and the second roll can be any of six outcomes. This is a total of 36 possible outcomes, and each of the 36 is equally likely. These are listed in the table below, with a dash separating the two rolls.

12 Probabilistic Reasoning

The 36 possible outcomes when rolling two dice					
1-1	2-1	3-1	4-1	5-1	6-1
1-2	2-2	3-2	4-2	5-2	6-2
1-3	2-3	3-3	4-3	5-3	6-3
1-4	2-4	3-4	4-4	5-4	6-4
1-5	2-5	3-5	4-5	5-5	6-5
1-6	2-6	3-6	4-6	5-6	6-6

With this table, we can answer a number of questions involving outcomes of a roll of two dice. For example, what is the likelihood of the sum on the two dice being eleven?

Well, there are two different ways to roll an eleven, 5-6 and 6-5. So the probability of rolling an eleven is 2 out of 36.[3] That is, two of the thirty-six possibilities sum to eleven, which is the same as 1 out of 18.

Similarly, the odds of rolling a seven with two dice is 6 out of 36 because six of the thirty-six possible rolls sum to seven.

The probability of a particular sum of two rolled dice can also be calculated by considering the rolls *individually*. To start, let's calculate the probability of an eleven. To get an eleven, the first roll must be either a 5 or a 6. The probability of this occurring is 2 out of 6, which is the same as one-third. The second die must be a 6 if the first roll was a 5, and it must be a 5 if the first roll is a 6. The probability of getting the number that you need on the second die to make a total of eleven is 1 out of 6. The probability that both of these events occur is one-third times one-sixth, which is one-eighteenth.

This *multiplication principle,* which we will use quite often in this chapter, is a basic concept in probability theory. The multiplication principle states that if some choice can be made in *r* different ways and some subsequent choice in *s* different ways, then there are $r \times s$ different ways these choices can be made in total.

This principle was clearly evident in the example that involved choosing a president and secretary and in the example involving the rolling of two dice.

The important assumption here is that these choices are independent of each other – otherwise the multiplication principle cannot be applied. A simple example to illustrate the multiplication principle is when determining the total number of combinations of shirts and pants we can wear if we have 5 shirts and 3 pairs of trousers. Since each of the five shirts can be matched with each of the three trousers, there are $3 \times 5 = 15$ fifteen different combinations. This is intuitive and simple and used frequently when enumerating the possibilities.

---

[3] You can get 15-1 odds on an eleven appearing on a single roll of two dice in a casino (at a craps table). This means that a $1 bet on an eleven being rolled on any particular roll will win $15 and you get your $1 back. So, if you "invest" $1 on each of 36 rolls at these odds (betting that a total eleven will appear on the two dice), your average return will be $32 (winning $15 and getting your $1 bet back on two of the 36 rolls).

The fact that the probability of two unrelated, independent events occurring is the product of the probabilities of each event occurring individually is not a rule that should be remembered, it is a concept that should be understood.

> **Student Pitfall**
> The definition of independent events in probability can be confusing to students because of their familiarity of the word "independent" in common usage. Independent can mean separate. So a student might think that the results of rolling two *separate* dice are independent but the results of rolling the same die twice are not – it's the same die. What independent means in probability is that one outcome does not depend upon the other. The fact that result of the first roll is independent of the result of the second roll means than no information about the result of the second roll can be inferred from the result of the first roll.
>
> The same thing is true of the outcomes from a roulette wheel. Many casinos have an electronic chart available that gives the results of the last twenty spins of the wheel. Bettors are free to perform any statistical analysis they want on these recent results, but the result of the next spin is independent of the past results.[4]

We can return to the dice for an illustrative example of the fact that the probability of two independent events occurring is the product of their individual probabilities. The chance of rolling a 12 with a pair of dice is one-sixth times one-sixth, which is 1/36. The fact that a six was rolled first does not change the probability of a six being rolled on the next roll: the rolls are independent.

Similarly, the probability of selecting a black ace with a random choice from a 52-card deck can be calculated by multiplying the probability that a randomly chosen card is black (one-half) by the probability that a randomly chosen card is an ace (one-thirteenth). The result is 1/26. The same result can be achieved by noting that there are two black aces among 52 cards, so the probability of getting a black ace from a random selection of a full deck is 2/52 or 1/26.

> **Teacher Tip**
> If a student ever says, "*I forget whether I'm supposed to add the probabilities or multiply the probabilities,*" you could take the role of information provider and respond by telling the student the rule, or you could be a help the student to become an independent thinker by responding, "*It's OK that you have forgotten, now you have a wonderful opportunity to learn something. Let's do some problems and figure it out.*" The authors enthusiastically endorse the latter.

---

[4] Not all casino games share this property – for example, in blackjack, the probability distribution of getting particular type of card (high or low) changes on every hand.

12 Probabilistic Reasoning

Problem 12.1 Three boys and two girls are in line at lunch. There is no preference for any child to be next to any other. That is, the position of one girl in line does not depend upon the position of the other girl. What is the probability that the two girls are next to each other?

Strategies Utilized Understand the problem. Take inventory. Enumerate the possibilities. Draw a diagram.

> **Teacher Tip**
> It is a good idea to develop the students' probabilistic "instinct" whenever possible. A great method to do this involves asking the students to give a rough estimate of the likelihood of an occurrence using fractions or percentages. This is also a good way to keep the students' attention and get them involved. You can make a quick table of estimates on the board and later see which student is closest, or you can take an electronic poll if such equipment is available. It might be best to give the students a list of choices from which to select their response. For example, offer them a choice of all the percentages from 10 to 90 % in increments of 10 %. For this problem, estimates usually range from 20 to 50 %.

Discussion 12.1 When trying to understand the problem, the students may have questions. Try to avoid answering authoritatively; use the opportunity to reason through the question with the student that asked or perhaps with the entire class.

To solve this one, let's simply sketch all the possibilities in an organized fashion. To simplify, we'll use the "key" B for boy and G for girl. Whenever enumerating the possibilities, it is very easy to skip one if they are not "swept through" in an organized fashion. In fact, students who write down the sequences in a haphazard fashion will invariably miss one or two of the possibilities. In this example, we start with all the boys on the left and systematically work them through to the right, being careful not to skip any of the possible combinations:

BBBGG
BBGBG
BBGGB
BGBBG
BGBGB
BGGBB
GBBBG
GBBGB
GBGBB
GGBBB

It looks like there are ten possibilities, and in four of the ten, the girls are next to each other. If each of these ten possibilities is equally likely, there is a 40 % chance that the girls will be next to each other. (It may be important to remind students that we are not distinguishing any of the boys from each other or the girls from each other. If we had named the boys and girls and then asked, *"What is the probability that Sophie stands next to Jim?"* then there are far more possibilities. Why? Because BGGBB, for example, then becomes $B_1G_1G_2B_2B_3$ or $B_2G_2G_1B_1B_3$ and so on.)

Of course, there are many problems in probability where making an exhaustive list of all the possibilities is not reasonable and examples of these are presented in this chapter. However, making a list whenever possible is a great way for the students to develop a fundamental understanding of probabilistic thinking, and, even with a big problem, a list of *some* of the possibilities will guide their thinking and help them frame the problem.

> **Teacher Trap**
> Resist the temptation to tell the students the formula for getting the total number of combinations of two different things. Remember, the goal is not to finish as many problems as possible; it is to develop the students' ability to solve problems independently. If you think the students will be interested, you can encourage them to try to determine the number of ways, for example, 6 heads and 4 tails can be ordered in ten tosses of a coin.

Debriefing 12.1 This problem clearly demonstrates the main concept of probability: the probability of a particular outcome is the number of ways it can occur divided by the total number of possibilities assuming that each is equally likely. There are ten possible arrangements – each of which is equally likely to occur – and in four of the ten, the girls are next to each other. This makes the probability that the girls are next to each other 4 out of 10, which is 40 %, which is 2/5.

We have already looked at some puzzles (e.g., Problem 8.3 or Problem 10.4) where there were some bags with some marbles. Have a look at another one.

Problem 12.2 There are four identical bags. One has two white marbles. One has two black marbles, and two have one of each (see below).

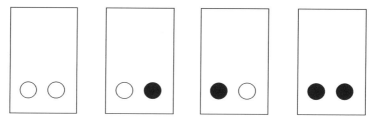

One of the bags is selected randomly and one marble is drawn from it, also randomly. The marble is black. What is the probability that the other marble in the bag is white?

12   Probabilistic Reasoning   195

Strategies Utilized   Simplify. Enumerate the possibilities. Draw a diagram. Build a model. Perform a gedanken.

> **Teacher Tip**
> This problem offers a great opportunity to develop the communication skills of the students. Start the problem-solving process by asking the students to silently think about the answer on their own. The time you allow the students to think on their own should be dependent on their ability to do so. If they are all still thinking or figuring with their heads down, try not to interrupt. You should be able to get a good read on how hard the students are thinking by looking at them. Are they thinking hard or are they looking at you for some help? When you feel that it is time to move on, ask some of the students to reveal their answers. The majority should guess either one-half or two-thirds. Now get the class to vote on the answer by first asking the students who think that the answer is two-thirds to raise their hands and then asking the students who think that the answer is one-half to raise their hands. Here, again, electronic polling is useful.
>
> The goal now is to reach a consensus. There are a couple of ways to do this. If the class is relatively small, you might ask everyone who thinks that the answer is one-half to stand up and go to one side of the room and everyone who thinks the answer is two-thirds to go to the other side of the room. Then the groups argue their position back and forth. This activity may need periodic intervention from the teacher, as it can get loud. If any students change their mind, they walk to the other side of the classroom. The exercise is over when all the students are on the same side of the room. When this happens, they all should be on the "one-half" side. If they are not, this is not a bad thing; it actually provides a great opportunity for you to guide the development of their probabilistic reasoning skills.
>
> If the class is larger or tends to get too rowdy, divide the class into groups of three in which there is a not a consensus on an answer. If there are too many of one answer, you may have to use groups of four. The goal of each individual group is to reach a consensus via logical argument. Once each group reaches a consensus, they can present their result to the class – perhaps even noting whether the initial majority in the group of three was the "winner."

Discussion 12.2   This is a problem given at job interviews to determine whether the candidate is a thinker or a person who simply plugs things into formulas without thinking. The employer is very likely looking for the former of these two. The correct answer will become clear with careful, logical, probabilistic thought.

A clear way to see the answer is to perform a gedanken. That is, mentally perform the experiment a large number of times to see what fraction of the time the other marble in the bag is white when the first marble selected from it is black. To aid in our gedanken, let's number the bags 1–4 from left to right as shown in the figure.

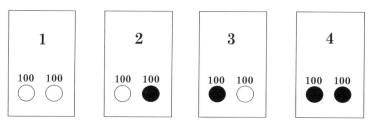

Now let's think about what would happen if we performed the test, say, 800 times. We choose 800 because each marble will be chosen 100 times on the average, and this makes the math easier. We could also have chosen 8 million, 8, 40, or any other multiple of eight. If we wanted to practice dealing with decimals and using a calculator, we would mentally perform the test, say, 47 times.

When performing a gedanken with 800 trials, each bag will be selected about 200 times and each marble will be selected about 100 times simply because each of the four bags and each of the eight marbles are equally likely to be chosen.

The 200 times that bag #1 is selected are not counted in the results because a black marble can't be selected from this bag. Of the 200 times that bag #2 is chosen, a black marble will be drawn about half the time. Of the 100 times that a black marble is chosen from bag #2, the other marble will be white in all 100 instances. Similarly, of the 100 times that a black marble is chosen from bag #3, the other marble will be white in all 100 instances. Bag #4 contains two black marbles, and it, like each of the other bags, will be chosen about 200 of the 800 times. In all 200 instances, the first marble chosen from the bag will be black and the other marble in the bag will be black.

With this information, we can now determine the probability that the second marble in the bag is white given that the first marble selected from the bag is black. Of the 400 times that a black marble was chosen first, the other marble in the bag will be white 200 times on the average, for a probability of one-half. So, the answer to the problem is one-half.

Teacher Tip
The instructor should emphasize that this problem can be solved merely with clear, logical thought. No formal training in probability theory or use of a formula is necessary. By drawing a picture and modeling the situation, the answer becomes clear. This is the essence of good problem-solving technique – the ability to take a direct path to the answer by a series of logical steps.

Discussion 12.2 (cont) This is a relatively simple experiment to do. If there are no marbles and bags handy, consider modeling it with eight playing cards – four of a black suit and four of a red suit. Make four piles of two cards, with all the cards face down. One pile contains two red cards, one contains two black cards, and two piles contain one black and one red card. Have someone select one of the four piles randomly and then draw a single card randomly from that pile. If the card is red, start over. If the card is black, look at the other card in the deck and record whether it is red or black. Many of these experiments can be performed simultaneously in small groups. With a lot of data, it should become clear that the probability that the other card in the pair is red given that the first card from the pile was black is 50 %. This is a great lesson in modeling and it should make an impression on the students – especially those who were "sure" that the answer was two-thirds.

Debriefing 12.2 It is worth discussing why two-thirds is a popular wrong answer. Students tend to reason, *"There are three bags with at least one black marble and each is equally likely to get chosen. In two of the three bags is the 'other' marble in the bag is white."*

While it is true that each bag is equally likely to get chosen for the draw of the *first* marble, it is not true that each bag gets included in the results an equal number of times. Bag #1 never gets included in the results, bag #2 and bag #3 get included half the time they are chosen (when the black marble is selected first), and bag #4 is included 100 % of the time it is chosen. So, the results of a draw involving bag #4 are twice as likely to be included in the results as the result of a draw from bag #2 or bag #3.

Problem 12.3 A "radar note" is a piece of paper currency in which the serial number reads the same forwards as it does backwards. US currency has eight-digit serial numbers. An online search for images of "radar note" will reveal many examples from numerous countries – some of which are for sale for a premium price. Examples of 8-digit radar serial numbers include 67444476, 12344321, and 90844809. Assuming that all serial numbers are equally likely, what is the probability that a randomly selected US dollar bill is a radar note?[5]

Strategies Utilized Enumerate the possibilities. Simplify. Perform a gedanken.

Discussion 12.3 It seems as if making a list of all the possible radar notes would be a daunting task. However, a good problem-solver will discover more than one way to get to the solution relatively quickly.

---

[5] It would be nice if you had an actual radar note to pass around.

> **Teacher Tip**
> Many times it is difficult to watch the students struggle – especially if they piteously ask for a little clue. If you give the students hints to solve the problem, you are robbing them of the opportunity to figure it out for themselves. Remember, the goal is not to get the students to know the answer; the goal is to develop the students' ability to think independently to solve new problems. If a student starts at 00000000 and continues thusly 00011000, 00022000, 00033000, and so on, they might just have the flash of insight needed to solve the problem.

Discussion 12.3 (cont) One way to tackle this problem is to first determine the total number of possible 8-digit serial numbers and then determine the total number of 8-digit serial numbers that read the same forwards as backwards. The ratio of the latter to the former is the probability of a randomly selected bill being a radar note.

Assuming that the serial numbers range from 00000000 to 99999999 (although apparently the 00000000 serial number is not used), there are 1,000,000,000 possible serial numbers. The second bit of information we need is the total number of possible radar notes. The condition that a bill is a radar note is that the first number matches the last number, the second number matches the seventh number, the third number matches the sixth number, and the fourth number matches the fifth number.

So, if the first four digits are 7463, the last four digits must be 3647. Similarly, if the first four digits are 2209, the last four digits must be 9022. We can see that for each combination of the first four digits from 0000 through 9999, there is a unique set of the remaining four digits that will make a radar note. In other words, there is exactly one radar note for each set of the first four digits. Since there are 10,000 possible different first four digits (0000 through 9999), there are 10,000 possible radar notes. So the probability of a radar note is 10,000 possible radar notes divided by the 1,000,000,000 possible notes, which is one in ten-thousand. The answer is one in ten-thousand.

Another, perhaps more straightforward, way to see this is to ask, "*What is the probability that the first digit matches the eighth digit?*" The answer is one out of ten. In fact, the chance that the any digit matches any other digit is one-tenth. Therefore, the chance that the first digit matches the eighth digit *and* the second digit matches the seventh digit *and* the third digit matches the sixth digit *and* the fourth digit matches the fifth digit is one-tenth times one-tenth times one-tenth times one-tenth, which is one in ten thousand, as expected:

12 Probabilistic Reasoning

$$P = \frac{1}{10} \times \frac{1}{10} \times \frac{1}{10} \times \frac{1}{10} = \frac{1}{10,000}$$

Debriefing 12.3 This problem demonstrates the fact that the probability of two independent events occurring is the product of their probabilities, which we called the multiplication principle in the introduction to this chapter. This is not a rule to follow, but a concept to understand. It comes directly from the main concept of probability.

Also, it might be worthwhile to connect this with earlier material (Sect. 11.1) on simulation. It is possible to generate many random sequences of 8-digit numbers and check which of these read the same forwards as backwards. Such simulation would allow us to estimate the true probability (the higher number of generated sequences, the better estimation).

Problem 12.4 On a particular flight, a passenger's baggage, consisting of two black suitcases, gets misdirected. The passenger goes to the lost baggage office at the airport to see if he can locate his bags. The clerk takes the passenger's name and goes to search for the missing bags. He finds three that meet the description of the bags given by the passenger. He also finds the passenger's name on one of the three bags. As he can only carry two at a time, he brings the identified bag and one of the other two. To open the office door and reenter, the clerk puts down the bags and then brings the bags into the room one at a time. The passenger recognizes that the first bag he sees is his. What is the probability that the second bag is his as well?

Strategies Utilized Perform a gedanken. Build a model. Enumerate the possibilities.

Discussion 12.4 It should be clear that there are only two possible situations – either the second suitcase is his or it isn't. However, there might be some disagreement regarding the probability that the second suitcase is his.

If there is a significant disagreement among the students, it gives them the opportunity to develop their problem-solving skills. Students can work individually or in small groups, or you can choose to have a class discussion. The size and maturity level of the class will determine the best choice, but feel free to change strategies if the one being used is not working. As soon as the discussion reaches either a consensus or a stalemate, it is time to discover the answer.

Teacher Tip

Often, it is better to do the experiment *before* revealing the correct answer to the problem. This way the students will be performing the experiment with some anticipation, and they will get to practice interpreting the results. To quote an experienced mathematics teacher, "*Intuition is not born, it is built.*"
– Samantha Meyer.

Student Pitfall
Whenever there are only two choices, students tend to reason, "Well it is or it isn't, so therefore it must be 50-50." You can disabuse them of this notion, by saying something like, "Well, if you buy a lottery ticket you will either win or you won't – but that doesn't make it a 50-50 chance."

Discussion 12.4 (cont) The problem-solving process for this one might begin with an enumeration of the possibilities. To better frame the problem, let's identify the three suitcases. S1 is the passenger's suitcase that the clerk identified as his. S2 is the passenger's suitcase that has no identification and S3 is not the passenger's suitcase. To start, we can enumerate the six possibilities:

First	Second	Third
S1	S2	S3
S1	S3	S2
S2	S1	S3
S2	S3	S1
S3	S1	S2
S3	S2	S1

The column headers are defined as follows: First is the first suitcase seen by the passenger, Second is the other suitcase that was brought to the passenger, and Third is the suitcase that was not brought to the passenger.

The last two possibilities listed above can be eliminated because the first suitcase was one of the two owned by the passenger and S3 was not owned by the passenger. The fourth possibility in the table can be eliminated because the clerk would not leave S1 behind because he knows it is the passenger's suitcase. In two of the three remaining possibilities, the other suitcase is the passenger's. The answer is 2/3.

The answer of 2/3 will not be accepted by many of the students. This is a good thing. Nobel Prize winning physicist Niels Bohr once said, "*It is good that we have come to a paradox, because now we have a chance to learn something.*"

A great way for the students to learn is to perform an experiment to get some actual data. This problem can be modeled with three coins. The coins that represent the passenger's suitcases can be heads up, and the coin representing the other suitcase can be represented by a tails-up coin. The student playing the clerk takes one of the heads-up coins; this coin represents the suitcase that is known to be the passenger's. Then the clerk selects one of the other two coins randomly, perhaps by closing his/her eyes. Then the clerk selects one of these two selected coins to show to the student playing the role of the passenger. If this coin is heads up, record whether the other coin is heads up as well.

Note that this is the only data point that needs to be taken. The recorder can simply have a sheet of paper with two columns, one labeled "heads" and the other labeled "tails." Emphasize to the students that nothing is recorded if the first coin shown is tails. After about 25 experiments in which the first coin shown to the passenger is heads, it should start to look like the answer is not 50/50. To emphasize this fact, all the results of the groups in the class can be combined. One recent result was 384 heads and 189 tails. It is very unlikely that this is the result of a 50/50 probability.

> **Teacher Tip**
> This might be a good point at which to discuss the difference between experimental probability (simulation; see Sect. 11.1) and theoretical probability. If the student performs ten trials of the above model and gets six heads for the second coin selected, the experimental probability is 60 %. It is not clear from this experiment whether the theoretical probability is 1/2 or 2/3. What the students should understand is that the experimental probability approaches the theoretical probability as the number of trials increases. This can be demonstrated clearly if each group performs ten trials. The results of each group will vary considerably, but the combined results should produce an experimental probability that is closer to 2/3.

Discussion 12.4 (cont) Once the experimental results start to reveal that the answer is not 50/50, the students should start to wonder why this is so. Many times, actually doing the experiment will trigger the engagement of the students' System 2, and they will start to question the 50/50 answer.

Another way to see the answer is to perform a gedanken where we repeat the experiment 1,000 times. In roughly 500 of the trials there will be two heads chosen by the clerk, and in roughly 500 of the trials there will be a head and a tail. Every time he chooses two heads, the first coin shown to the passenger will be a head and the other coin will be a head as well. When the clerk has a head and a tail, the head will be selected in roughly 250 of these. When this occurs, the other coin will be a tail in all 250 occurrences. So, of the 750 trials in which a head is shown first, the other coin will be a head in about 500 of them. 500/750 is the same as two-thirds. The answer is two-thirds.

Debriefing 12.4 Many times, experiments are performed (we call this process simulation; see Sect. 11.1) to verify a theory or hypothesis. However, sometimes experiments are performed just to see what happens, with the experimenters forming their theories based upon the results of the experiment. It is a good idea to utilize both techniques. In our experience, we have found that the students would rather perform experiments than listen to a lecture. As a side benefit, we have found that when the students get to know each other, they are much more likely to contribute in class discussions. More advanced students, and especially those with computing skills, can write much larger-scale simulations of this problem to see what the pattern is.

Problem 12.5 What is the probability of being dealt four deuces (cards with a numerical value of 2) in five-card poker using a standard, well-shuffled deck of cards? In other words, if five cards are selected randomly from a standard deck of 52 playing cards, what is the probability that the five cards will contain all four deuces?

Strategies Utilized Understand the problem. Enumerate the possibilities. Build a model.

> **Student Pitfall**
> It is easy to get overwhelmed by such a problem and believe that it is too hard. The numbers involved are indeed large, but this is a great opportunity to tackle a challenging problem and build confidence.

Discussion 12.5 There are formulas to calculate this, but a good thinker with an understanding of the main concept of probability might start by asking, "How many possible five-card hands are there?" One way to think about it is to calculate the probability of getting any particular hand. After all, every single five-card hand is equally likely. That is, the following two hands are equally likely:
- Ace of spades, king of spades, queen of spaces, jack of spades, and ten of spades
- Two of clubs, five of diamonds, seven of spades, nine of spades, and jack of clubs

So, to get the possibility of any particular hand, we can calculate the probability of getting dealt a royal flush in spades. A flush, in poker, is when every card in the hand comes from the same suit. Hence, a flush in spades means that every card in your hand is a spade. A royal flush occurs when every card in the hand is from the same suit *and* the cards happen to be the ace, king, queen, jack, and ten. The first card can be any of the five that make up a spade royal flush. The chance of this happening is 5/52. There are four cards remaining that will complete a spade royal flush, and there are 51 cards in the deck. This makes the chance that the second card drawn is also part of the spade royal flush equal to 4/51. Skipping ahead to the fifth card, we see that there is only one card remaining that will complete the spade royal and the chance of getting it is 1/48. So, the probability of getting dealt a spade royal flush from a well-shuffled deck is as follows:

$$P = \frac{5}{52} \times \frac{4}{51} \times \frac{3}{50} \times \frac{2}{49} \times \frac{1}{48} = \frac{1}{2,598,960}$$

Since the probability of getting *ANY* particular hand is 1 out of 2,598,960 and all hands are equally likely, it must be true that there are 2,598,960 different poker

12   Probabilistic Reasoning

hands. Now that we have the total number of poker hands, we can think about the total number of hands that contain four deuces.

> **Teacher Tip**
> If the students are having trouble thinking about this problem, you can ask them to start by determining the number of different hands that contain four deuces. You can also guide them in the determination of the total number of possible hands by asking them to consider using the simplify technique. For example, consider a deck of only six cards. How many different five-card hands can be made with six cards? How about with ten cards? Can the students calculate the probability of randomly drawing all five black cards from a well-shuffled group of five black and five red cards? If possible, give the students decks of cards to work with.

Discussion 12.5 (cont) The four-deuce hand contains the four deuces and any one other of the 48 cards in the deck. Therefore, there must be 48 different four-deuce hands. Continuing down this logical path, the probability of getting dealt four deuces must be

$$P = \frac{48}{2,598,960}$$

which is 1 in 54,145. The answer is 1 in 54,145.

Debriefing 12.5 This problem is yet another that continues to demonstrate the main concept of probabilistic thinking. Here we did not have the time or the space to list the 2,598,960 possible five-card hands, but we were able to calculate the number from a fundamental understanding of probability. That is, if all outcomes are equally likely, the probability of any outcome is the number of ways it can occur divided by the total possible outcomes. The problem also gives the students practice with the simplification technique, as they can tackle smaller versions of the same problem.

Problem 12.6 The currently circulating one-cent piece (so-called penny) in the United States has a diameter of 0.75 inches. A penny is tossed on a large wooden floor that is composed of long slats that are 1.5 inches wide (see figure). What is the probability that a randomly tossed penny will land touching one of the lines between adjacent slats? If there is no easy access to US pennies, consider choosing an appropriate coin and line spacing.

―――――――――――――――――――○―――――――――――――――――

Strategies Utilized Consider all the possibilities. Draw a diagram. Simplify. Build a model.

> **Student Pitfall**
> This problem has a continuous rather than discrete probability distribution. That is, there is no way to calculate the total number of places that the coin can land.

> **Teacher Tip**
> A good way to introduce continuous probabilities is with a spinner, like one that might be used for board games. The number of places that the arrow of the spinner can "land" cannot be counted. However, if the color blue occupies a quadrant of the possible landing area, the students should be able to understand that the probability of the arrow pointing to blue after a random spin is 1/4.

Discussion 12.6 Once the students are somewhat accepting of the fact that they will not be able to enumerate the total number of places the coin can land, they should begin to use some of the problem-solving skills that were developed early in the course, for example, drawing a picture and building a model. This could be as simple as drawing long lines on a piece of paper that are 1.5 inches apart and moving a penny around between them.

While it is impossible to determine the total number of places the penny can land, the *ratio* of the number of places where the coin is overlapping a line to the total number of places that the coin can land *can* be determined. A good way for the students to make progress towards the solution is to each put a finger on a penny and slowly push it across the lines in the floor along a path that is at a right angle to the lines. This simplifies the problem by taking it from two dimensions down to only one.

> **Teacher Tip**
> Having a large number of students throwing pennies around is noisy and, depending on the level of exuberance, mildly risky. Younger or larger classes should be carefully monitored. Performing this experiment is *not*

(continued)

12 Probabilistic Reasoning

recommended for a fixed-desk classroom environment, especially one with tiered seating or wooden desks. However, it is a perfect opportunity for running a physical simulation.

Discussion 12.6 (cont) To make the answer even more visible, the students can treat the center of the penny as a point and shade in the areas of the paper that will result in the penny overlapping a floor line when the center of the penny is within the shaded area.

If the center of the penny is over an unshaded section of the paper, then the penny will not overlap a floor line. Once the paper is shaded in, it should become apparent that the chance of a randomly tossed penny overlapping a floor line is 50 %. The answer is one-half.

Debriefing 12.6 This is the first problem that requires the student to grasp the concept of a continuous probability distribution. That is, instead of a finite number of distinct outcomes, the coin can land at essentially an infinite number of positions. So, there is no way to calculate the number of places that the coin can land so it overlaps a line, nor is it possible to calculate the total number of places that the coin can land. However, it is possible to calculate the ratio of these two.

Problem 12.7 On a game show, a contestant has a chance to play the *Cash Wheel* game to win prizes. There are seven spaces on the wheel, and each is equally likely to be the result of a spin. There are three whammies and four prize numbers on the wheel, and the three whammies are next to each other. The contestant does not know where the three whammies are located.

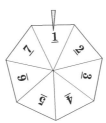

Here's how the game is played. The wheel is spun, and host reveals whether the resulting number is a whammy or a prize. If it is a whammy, the contestant wins nothing and the game is over. If it is a cash prize, the contestant gets to keep that prize and try for another, with a maximum of four prizes. The contestant has two options: to move the indicator one space clockwise and take the next space on the wheel (next highest number with the exception that 7 goes to 1) or to spin the wheel again. Assuming that the contestant keeps getting cash prizes, what is the optimal strategy for all three opportunities?

Strategies Utilized Understand the problem. Take inventory. Draw a figure. Enumerate the possibilities. Perform a gedanken.

Discussion 12.7 This question requires some investment of time to understand and frame the problem. Here's what the students might start with:
1. There are three whammies.
2. The contestant would like to avoid whammies and win four cash prizes.
3. The three whammies are next to each other on the wheel.
4. The seven possible positions for the three whammies are 1-2-3, 2-3-4, 3-4-5, 4-5-6, 5-6-7, 6-7-1, and 7-1-2.

> **Student Pitfall**
> Students who are inclined to plug numbers into equations without thinking are likely to come up with the probability of 3/6 that there is a whammy in the next spot after the first spin produces a cash prize. They reason, *"There are three whammies left out of six remaining spaces, so the probability that there is a whammy in the next space is 3/6."* This is incorrect.

Discussion 12.7 (cont) A thoughtful student might sketch a diagram and reason as follows:

"To decide whether to spin the wheel or to take the next clockwise spot, I must compare two probabilities; the probability that I get a whammy when spinning again and the probability of getting a whammy when taking the next clockwise spot on the wheel. So, I'll assume the wheel lands on a spot without a whammy. There are four spaces with a prize. How many of these four spaces have a whammy next?"

This is the key question. After avoiding a whammy with the first spin of the wheel, what is the probability that there is a whammy in the next space? Well, there are four spaces that do not have a whammy, and only one of those four has a whammy next. The probability that it landed on this one is 1 out of 4, or 1/4. This means that it is 3/4 that there is *not* a whammy next. If the wheel is spun again, the chance of getting a cash prize is 4/7. Since 3/4 is a higher probability than 4/7, the best strategy is to take the next space on the wheel rather than spin it a second time.

Assuming that this is also a prize, the contestant has another decision to make, spin or take the next space. At this point, the space the wheel is currently on is one of the three spaces that is at least two away from the whammies. The probability that the contestant is currently on the only one of these three that has a whammy next is 1 out of 3 or 1/3. Therefore, the probability it is *not* on this space is 2/3. This is still higher than 4/7, so the contestant would again opt to take the next space on the wheel.

Finally, if the contestant has collected three prizes, he has one more decision to make. He now knows he started at least three spaces away from the whammies and therefore started in one of two possible positions. The chance that the whammy is

12 Probabilistic Reasoning

next is 1/2, and of course, the chance that it is not next is also 1/2. Therefore, the optimal strategy is to spin the wheel because spinning the wheel produces a 4/7 chance of avoiding the whammy.

This problem may require a lot of time for the students to wrap their heads around. This is understandable. Be patient. To really see what is going on, the students should be able to perform a gedanken by asking the following questions: *"What if the first spin landed on 4 and it was not a whammy?" "What are the possible positions for the three whammies?" "What if the contestant opted for space 5 and it was not a whammy?" "What are the possible positions for the three whammies now?"*

Hopefully, the students will be able to come up with these questions by themselves; if not, try to gently push them in that direction. The table provides the possible positions of the whammies when the spinner landed on space 4 with the first spin and the contestant chose the next space on the wheel.

Space	Result	Possible positions for whammies
4	No whammy	1-2-3, 5-6-7, 6-7-1, 7-1-2
5	No whammy	1-2-3, 6-7-1, 7-1-2
6	No whammy	1-2-3, 7-1-2

As soon as it is known that space 4 is not a whammy, the possibilities of the whammies being in 2-3-4, 3-4-5, and 4-5-6 are eliminated. The only remaining possibilities are 1-2-3, 5-6-7, 6-7-1, and 7-1-2, and each of these four is equally likely. The only one of the four that will result in a whammy by choosing to take space five is 5-6-7. The probability that the whammies are in spaces 5-6-7 is one-fourth. Again, this is just the fundamental concept of probability.

If space 5 is also not a whammy, then only 1-2-3, 6-7-1, and 7-1-2 remain as the possible positions of the three whammies, and each of these three possibilities is equally likely. Only if the whammies are in 6-7-1 will taking the next space result in a whammy.

At this point, the students have an opportunity to calculate the probability of the contestant actually getting one, two, three, and four straight prizes with various strategies. Here is a table of the probability of getting two prizes with the two possible strategies.

Probabilities based on strategy		
Strategy	Spin–spin	Spin–next
Prob. of 2 prizes	16/49 (32.7 %)	3/7 (42.9 %)

The 16/49 comes from multiplying 4/7 by 4/7, and the 3/4 comes from multiplying 4/7 by 3/4.

The next table shows some probabilities for getting three prizes. Any of these are good test questions.

Probabilities based on strategy		
Strategy	Spin–spin–spin	Spin–next–next
Prob. of 3 prizes	64/343 (18.7 %)	2/7 (28.6 %)

There is also an opportunity to calculate the probability of getting three prizes using optimal strategy in a different way, and we recommend going through this with the students. In the table, we see that the probability of getting three prizes using optimal strategy is 2/7. Let's assume we are going to use optimal strategy and that the whammies are in spaces 5-6-7 – an assumption the students may have already made when solving the problem. What spaces can the spinner initially land on that will ensure the contestant will win three straight prizes using optimal strategy? Well, if the first spin lands on space 1 or 2, the contestant will avoid a whammy. The probability of this occurring is 2 out of 7, or 2/7 – an answer we got previously by multiplying the probability of three individual events as follows:

$$\frac{4}{7} \times \frac{3}{4} \times \frac{2}{3} = \frac{2}{7}$$

The three fractions represent the probability of avoiding a whammy with the first spin, the probability of avoiding a whammy by taking the next spot on the wheel, and the probability avoiding a whammy by taking the next spot after that.

Debriefing 12.7 Problems like these are often given at job interviews specifically for the purpose of separating the careful thinkers from those that simply plug numbers into equations without thinking. It is important to present your students with different problems that require new ideas and new thoughts. Don't let them fall into a mindless routine when problem-solving. The human mind requires new challenges to grow significantly.

Problem 12.8 There are three contestants remaining in the trivia challenge: Alec, Bob, and Charlie. The contest has an unusually fast-paced format that allows each contestant – in turn – to "challenge" any of the remaining contestants. The challenger is then presented with a trivia question to which he alone responds. If the challenger answers correctly, the contestant he challenged is eliminated. If the challenger does not answer correctly, the game continues with no one being eliminated. This procedure continues until one constant remains, and this contestant is declared the winner.

The game starts with Charlie making a challenge followed by Bob and then Alec. It is known that Alec will beat every person he challenges, Bob will eliminate 2/3 of the people he challenges, and Charlie will eliminate 1/3 of the people he challenges.

It can be assumed that each contestant will challenge the remaining contestant that is the biggest threat to them winning the contest.

The question is, what is the probability of each of the three contestants winning the event?

12 Probabilistic Reasoning

Strategies Utilized Understand the problem. Take inventory. Enumerate the possibilities. Recognize a pattern. Simplify. Perform a gedanken.

> **Teacher Tip**
> Have the students vote for the contestant they think is most likely to win. If they will volunteer to share their reasoning, it might generate some productive brainstorming. If you think that the students would like to tackle this problem individually or in groups, allow them to do so, but it might be a good idea to monitor their progress.

Discussion 12.8 A good first step is to frame the problem and to take inventory. A possible inventory might include the following:
1. Contestants take turns challenging one of the remaining contestants until there is a winner.
2. Alec wins all challenges, Bob wins 2/3 of his challenges, and Charlie wins 1/3 of his challenges.
3. Charlie gets the first challenge followed by Bob and Alec.
4. Contestants choose the opponent to challenge that will maximize the chance that they will win the contest.

The first step in solving this problem is to figure out what Charlie should do on his first challenge. This should generate some lively class discussion, and we recommend that this be allowed to come to a conclusion uninterrupted. Students should be able to perform a quick gedanken by first asking themselves, "*What would happen if Charlie challenged Bob and won?*" and then "*What would happen if Charlie challenged Alec and won?*" A thorough enumeration of the possibilities will lead to the interesting conclusion that Charlie's best move on their first challenge is to challenge either contestant and simply not answer. The students should eventually realize that if Charlie loses his first challenge, he is guaranteed the challenge of the lone remaining dummy at his next turn. With Charlie's optimal strategy on his first challenge known, we can start the quantitative part of the problem by determining the probabilities.

> **Teacher Tip**
> Once it is determined that Charlie's best chance to win is to "pass" on the first challenge, it might be a good idea to split the class into groups to determine how each of the three contestants can win. That is, what sequence of events will result in a particular contestant winning? Once these are enumerated, the probability of each can be determined.

Discussion 12.8 (cont) So, how can Alec win? The *only* way Alec can win is for Bob to fail when he challenges Alec, and then, after Alec challenges Bob and eliminates him, Charlie must fail when he challenges Alec. Shown below are the probabilities of each of these events occurring:

Result of challenge	Probability
Charlie fails on purpose	One
Bob fails against Alec	One-third
Alec eliminates Bob	One
Charlie fails against Alec	Two-thirds
Alec eliminates Charlie	One

The product of these five probabilities is 2/9, so the chance of Alec winning is 2 out of 9, which is not very good.

Now let's calculate the probability of Charlie winning. Unlike Alec, Charlie can win a number of different ways. In fact, this number is infinite. The simplest way Charlie can win is through the following sequence of events:

Event	Probability
Charlie fails on purpose	One
Bob eliminates Alec	Two-thirds
Charlie eliminates Bob	One-third

The probability of this sequence occurring is 2/9. Charlie can also win with this sequence:

Event	Probability
Charlie fails on purpose	One
Bob fails against Alec	One-third
Alec eliminates Bob	One
Charlie eliminates Alec	One-third

This is the only sequence in which Bob fails against Alec that can result in Charlie winning. Once Alec eliminates Bob, Charlie knows he will only have one chance to eliminate Alec simply because Alec never loses. The probability of this sequence of events occurring is 1/9.

Now let's consider yet another way Charlie can win:

Event	Probability
Charlie fails on purpose	One
Bob eliminates Alec	Two-thirds
Charlie fails against Bob	Two-thirds
Bob fails against Charlie	One-third
Charlie eliminates Bob	One-third

12 Probabilistic Reasoning

The probability of this occurring is 4/81. There can also be another round of misses like this:

Event	Probability
Charlie fails on purpose	One
Bob eliminates Alec	Two-thirds
Charlie fails against Bob	Two-thirds
Bob fails against Charlie	One-third
Charlie fails against Bob	Two-third
Bob fails against Charlie	One-third
Charlie eliminates Bob	One-third

The probability of this occurring is 8/729. Since the probability of another round of misses (Charlie fails and Bob fails) is 2/9, the likelihood of the Charlie winning the event after any number of turns is 2/9 times the probability of Charlie winning on his previous turn. So, the chance of Charlie winning when Bob eliminates Alec on his first turn is the following:

$$P = \frac{2}{9} + \left(\frac{2}{9}\right)^2 + \left(\frac{2}{9}\right)^3 + \left(\frac{2}{9}\right)^4 + \cdots$$

where the "$\cdots$" indicates that the sequence goes on to infinity. In this equation, the first 2/9 is the probability of the following: Charlie fails on purpose, Bob eliminates Alec, and Charlie eliminates Bob. Every additional power of 2/9 is from a round of "Charlie fails and Bob fails" between "Bob eliminates Alec" and "Charlie eliminates Bob."

> **Teacher Tip**
> The students are very likely to sum this series as they get the terms without seeing the pattern that is developing. If you ask them to write out each term separately, you might get this:
>
> $$P = 0.222222 + 0.0493827 + 0.0109739 + \cdots$$
>
> or perhaps this:
>
> $$P = \frac{2}{9} + \frac{4}{81} + \frac{8}{729} + \cdots$$
>
> If they do come up with this sum, ask them if they can find a pattern to predict the next term. You can ask, for example, if the probability of Charlie winning after eight rounds is P, what is the probability of Charlie winning after nine rounds? Or, simply, what is the probability of Charlie winning on his nth challenge?

Discussion 12.8 (cont) While the students can get a very good approximation by using their calculators to add the first few terms, this is a nice opportunity to calculate it exactly by summing the infinite series.

To sum the infinite series, we start by multiplying both sides of the equation by the geometric factor of 2/9. Expressed in this form, the students can *see* the pattern. Thus, we have

$$\frac{2}{9} \times P = \frac{2}{9} \times \left[\frac{2}{9} + \left(\frac{2}{9}\right)^2 + \left(\frac{2}{9}\right)^3 + \left(\frac{2}{9}\right)^4 + \cdots\right]$$

This equation can be rewritten as

$$\frac{2}{9}P = \left(\frac{2}{9}\right)^2 + \left(\frac{2}{9}\right)^3 + \left(\frac{2}{9}\right)^4 + \cdots$$

The two equations we have now are the original expression, repeated below,

$$P = \frac{2}{9} + \left(\frac{2}{9}\right)^2 + \left(\frac{2}{9}\right)^3 + \left(\frac{2}{9}\right)^4 + \cdots$$

and (2) the expression when both sides of the equation are multiplied by 2/9, also repeated below:

$$\frac{2}{9}P = \left(\frac{2}{9}\right)^2 + \left(\frac{2}{9}\right)^3 + \left(\frac{2}{9}\right)^4 + \cdots$$

Subtracting the second from the first gives us

$$\frac{7}{9}P = \frac{2}{9}$$

Note that all but one of the terms on the right-hand side cancel out.
Solving for P gives 2/7.

So, the chance that Charlie wins when Bob eliminates Alec in the first round is 2/7. We have also seen that Charlie has a 1/9 chance of winning when Bob fails against Alec in the first round. Thus, the probability that Charlie wins is the sum of 2/7 and 1/9. This is calculated as follows:

$$P = \frac{2}{7} + \frac{1}{9} = \frac{18}{63} + \frac{7}{63} = \frac{25}{63}$$

So, the probability that Charlie wins with optimal strategy is 25/63.

At this point, we can determine the probability of Bob winning by subtracting the probability that Charlie wins and the probability that Alec wins from one

12 Probabilistic Reasoning   213

because the probabilities of Alec, Bob, and Charlie winning must sum to one. However, it is a good exercise for the students to determine the probability of Bob winning independently.

So, how can Bob win? It should be clear that the only way Bob can win is if Alec never gets a turn.

The simplest way Bob can win is through the following sequence of events:

Event	Probability
Charlie fails on purpose	One
Bob eliminates Alec	Two-thirds
Charlie fails against Bob	Two-thirds
Bob eliminates Charlie	Two-thirds

The probability of this occurring is 8/27.

The next simplest way Bob can win just adds a round of misses, like this:

Event	Probability
Charlie fails on purpose	One
Bob eliminates Alec	Two-thirds
Charlie fails against Bob	Two-thirds
Bob fails against Charlie	One-third
Charlie fails against Bob	Two-thirds
Bob eliminates Charlie	Two-thirds

This is 2/9 of the previous probability of 8/27. Therefore, the sum of the number of ways Bob can win is

$$P = \frac{8}{27} + \left(\frac{8}{27} \times \frac{2}{9}\right) + \left(\frac{8}{27} \times \left(\frac{2}{9}\right)^2\right) + \left(\frac{8}{27} \times \left(\frac{2}{9}\right)^3\right) + \cdots$$

Factoring out the 8/27 gives

$$P = \frac{8}{27}\left[1 + \frac{2}{9} + \left(\frac{2}{9}\right)^2 + \left(\frac{2}{9}\right)^3 + \cdots\right]$$

To sum this, multiply both sides of this equation by the geometric factor of 2/9 and then subtract it from the original equation. This leaves

$$\frac{7}{9}P = \frac{8}{27}$$

Solving for P gives a probability of 8/21 that Bob wins. To compare all three probabilities, it is useful to use the common denominator of 63rds. We saw that the probability of Alec winning is 2/9, which is 14/63. The probability of Bob winning was 8/21, which is 24/63. So, Charlie is a slight favorite to win at 25/63, Bob is a close second at 24/63, and Alec is a distant third at only 14/63. These three indeed sum to 1, as they must. Make sure to congratulate all the students who predicted that Charlie was the favorite to win.

This table shows a more detailed analysis of the probabilities.

Probability of each contestant winning after Bob's first turn			
If Bob eliminates Alec (2/3)		If Bob fails against Alec (1/3)	
Charlie wins	18/63	Charlie wins	7/63
Bob wins	24/63	Bob wins	0
Alec wins	0	Alec wins	24/63

Here, it can be seen that Charlie is the only contestant that has a chance regardless of the result of Bob's first challenge against Alec. This is just enough to give Charlie the edge.

Debriefing 12.8 This two-part problem provides numerous opportunities to develop problem-solving skills. The qualitative part involves performing a gedanken and being thorough by considering all the possibilities. Next, it is important to determine Charlie's best strategy on his first turn. Then there is the quantitative part that involves pattern recognition, particularly recognizing the geometric factor of 2/9 in the infinite series and a careful enumeration of the possibilities. The problem yet again demonstrates the main concept of probabilistic reasoning and also provides an opportunity to develop skills in calculating probabilities.

Problem 12.9 Kate is going to a speed-date event to meet ten potential dinner dates. The evening starts with a half-hour session in which she "speed-dates" with up to ten bachelors for two minutes each. Her goal is to select the best one out of the ten. After each speed date, she can choose that bachelor to go out with that evening and pass on the remaining bachelors, or she can pass on that particular bachelor and move to the next one. She cannot choose any bachelor that she passed on earlier. Her strategy going into the event is to never choose any of the first three and then select the first one that is better than all of the first three thereafter. If none is better than the best of the first three, she is stuck with the last speed date. In other words, her strategy is to use the first three bachelors to "calibrate" the group. What is the probability that this strategy will result in her going out with her top choice of the ten?

Strategies Utilized Understand the problem. Take inventory. Enumerate the possibilities. Simplify. Perform a gedanken. Recognize a pattern.

Discussion 12.9 This problem requires a lot of thought, which is why it is included here. Students will need to spend a lot of time wrapping their heads around the problem, and this is exactly the type of training they need. They can discuss whether her strategy is a good one; they can estimate what is the likelihood of her ending up

12 Probabilistic Reasoning   215

with her #1 choice with this strategy. So it is possible to invest an entire class period just framing the problem.

A good first step towards the solution is to see if they can figure out *any* way that she will end up with the best of the ten.

> **Teacher Tip**
> If the students have trouble getting started, they can use the very useful problem-solving technique of tackling a simplified version of the same problem. For example, what if there were three speed dates and her strategy was to never pick the first one, and then pick the first bachelor thereafter that is better than the first one? What percent of the time would she end up with her first choice?

Discussion 12.9 (cont) The students should quickly realize that if her number one choice is among the first three speed dates, then she has no chance of picking him. The second thing they should realize is that if her second choice is among the first three and her first choice is not, she will always get her top choice. This is a great place to start, but before we do, let's frame the problem by naming the bachelors – and for convenience, we'll use the letters A to J and we'll name them in alphabetical order by Kate's ranking. So, if Kate ranked the ten bachelors in order of her preference, they would be Andy, Brad, Charles, David, Eddie, Fred, Gavin, Harold, Ian, and Jake. So, Andy is her top choice and Jake would be her last choice.

Let's focus on all the ways that Kate can end up with Andy using her strategy. One possibility is that Brad is among the first three and Andy is not. If this happens, Kate will always pick Andy.

The probability that Andy is not among the first three is 7/10. There are nine remaining spots for Brad, and he must go somewhere in the first three. The chance of this happening is 3 out of 9. The probability of both of these happening is

$$P = \frac{7}{10} \times \frac{3}{9} = \frac{21}{90} = \frac{7}{30}$$

This is already a pretty decent chance (7/30 = 23.3 %). But there are other ways she can get her top choice. What if Charles was among the first three and both Andy and Brad were among the last seven? As long as Andy was before Brad in the sequence, she would pick Andy. Now we have to calculate the probability of this happening.

There are initially ten slots available. The chance that Andy is not among the first three is 7 out of 10. The chance that Brad is not among the first three given that Andy is not among the first three is 6 out of 9, and the chance that Charles is among the first three is 3 out of 8. Finally, the chance that Andy is ahead of Brad in the dating order is 50 %. The chance that all of these things occur is

$$P = \frac{7}{10} \times \frac{6}{9} \times \frac{3}{8} \times \frac{1}{2} = \frac{126}{1,440} = \frac{7}{80}$$

There's still more work to be done. Another way she can get Andy is if Andy, Brad, and Charles are among the last seven *and* David is among the first three *and* Andy is before both Brad and Charles. The probability of this happening is

$$P = \frac{7}{10} \times \frac{6}{9} \times \frac{5}{8} \times \frac{3}{7} \times \frac{1}{3} = \frac{630}{15,120} = \frac{1}{24}$$

where the 1/3 is the chance that Andy is before both Brad and Charles among the last seven (one of them has to be first among the group of seven, and each one is equally likely to be first, so the chance that Andy is first is 1/3).

We can see a pattern developing here, and we can see the proverbial "light at the end of the tunnel." Let's march on towards it. The next way that she can select Andy is if Andy, Brad, Charles, and David are all in the last seven and Eddie is in the first three. The probability of this happening is

$$P = \frac{7}{10} \times \frac{6}{9} \times \frac{5}{8} \times \frac{4}{7} \times \frac{3}{6} \times \frac{1}{4} = \frac{2,520}{120,960} = \frac{1}{48}$$

where the 1/4 is the chance that Andy is before Brad, Charles, and David.

The pattern should be clear by now. We finish up by calculating the probability that Fred is the best of the first three, Gavin is the best of the first three, and Harold is the best of the first three. It is worthwhile to look at the last calculation, because we can start by placing Harold, Ian, and Jake in the first three places, which means that bachelors 1–7 must be distributed among the last seven places. After speed-dating for two minutes with bachelors Harold, Ian, and Jake, she'll take the next bachelor in line.

The chance of Andy being 4th when Harold, Ian, and Jake are among the first three is

$$P = \frac{3}{10} \times \frac{2}{9} \times \frac{1}{8} \times \frac{1}{7} = \frac{6}{5,040} = \frac{1}{840}$$

where the 1/7 is the chance that Andy is in the fourth position after positions 1–3 are filled. Here is a complete table of the probabilities:

Best of first three	Probability of getting Andy (fraction)	Probability of getting Andy (percentage)
1 (Andy)	0	0
2 (Brad)	7/30	23.333 %
3 (Charles)	7/80	8.750 %

(continued)

12 Probabilistic Reasoning

Best of first three	Probability of getting Andy (fraction)	Probability of getting Andy (percentage)
4 (David)	1/24	4.167 %
5 (Eddie)	1/48	2.083 %
6 (Fred)	1/100	1.000 %
7 (Gavin)	1/240	0.4167 %
8 (Harold)	1/840	0.119 %
Total	3,349/8,400	39.869 %

The answer here was calculated by determining the probabilities of Kate getting her top choice for every possible "best of the first three." However, as is often the case with probability problems, there is more than one way to get the correct answer. When we have presented this problem, some groups of students perform the calculation for each of the ten possible positions of Andy. That is, they complete the following table.

Andy's date position	Probability of getting Andy (fraction)	Probability of getting Andy (percentage)
First	0	0
Second	0	0
Third	0	0
Fourth	1/10	10.000 %
Fifth	3/40	7.500 %
Sixth	3/50	6.000 %
Seventh	3/60	5.000 %
Eighth	3/70	4.286 %
Ninth	3/80	3.750 %
Tenth	3/90	3.333 %
Total	3,349/8,400	39.869 %

The students should find it quite satisfying to fill in the blanks in this table and come up with the same probability that they did previously. If your students performed the calculation the second way, perhaps they would like to try to get the same answer using the first technique.

It is worthwhile to go through a couple of the rows in this table. To start, if Andy is among her first three speed dates, Kate has no chance of getting him. When Andy is in the fourth position, she is 100 % to get him. The chance of her getting Andy this way is 1/10 because it is one-tenth that he is in the fourth position.

Now let's go to the fifth position. In order for her to get Andy, there must be a bachelor in one the first three positions that is better than the bachelor in the fourth position. Another way of asking this question is, "*What is the probability that the fourth bachelor is the best of the first four?*" The answer is 1/4. Now we can ask, "What is the probability that the fourth bachelor is *not* the best of the first four?" The answer is 3/4. Therefore, the probability of selecting Andy in position 5 is the probability of him being in position 5 (which is 1/10) multiplied by the probability of Kate selecting Andy if he is in position 5 (which is 3/4). The result is 3/40.

Moving on to the sixth position, the relevant question is, what is the probability that any one of the three bachelors in positions 1–3 is better than both the bachelors in positions 4 and 5? Without good problem-solving skills, this problem seems very complicated: there are certainly a lot of probabilities to consider for the first five bachelors. However, a quick gedanken will reveal that only the position of the best bachelor among the first five matters. If the best of the five is among the first three, Kate will reject bachelors 4 and 5 and get to Andy in position 6. If the best of the first five is in either position 4 or 5, she will not get to Andy. So, the question is, what is the probability that the best of the first five bachelors is in position 1–3? Clearly, it is 3/5. Multiplying this by the probability that Andy is indeed in the sixth position gives us 3/50. The rest of the table is filled in with similar reasoning.

Debriefing 12.9 This problem has many layers – each of which demonstrates key stages in the problem-solving process. The first stage is to understand the problem, and this may take some time, but it is time well invested. The next stage is deciding how to tackle the problem. This also requires some careful thought and perhaps a simplification of the problem to gauge the utility of the technique. Once the method is chosen, the next step is to perform the calculations thoroughly, being careful to include all the possibilities. This problem offers an opportunity to check the answer by performing the calculations a different way. A couple of class periods invested on this problem will return handsome dividends in the students' future. If any of the students are considering a grand challenge problem (see Chap. 15 for many examples), they can consider what would happen if Kate had N dates to choose from. What strategy will maximize the chance of her getting the best one, and what is this chance?

Problem 12.10 Contract bridge is a card game in which thirteen cards from a standard, well-shuffled deck are dealt to each of four players. So, the entire 52-card deck is dealt, and each player gets 13 cards. What is the probability of each of the four players getting exactly one ace?

Strategies Utilized Simplify. Perform a gedanken. Construct a model. Recognize a pattern.

Discussion 12.10 This is a nice problem because the ratio of time spent thinking to time spent calculating is so high. Clearly, this is not a problem in which you can manually write out every possible set of the four hands and then count the ones that have one ace in each of the four hands. In fact, there are 635,013,559,600 different 13-card bridge hands. Getting students to calculate this value can be a useful way to get them to understand how quickly problems can grow in size.

The instructor may choose to start by polling the students' guesses, perhaps asking them to choose from all the 5 % increments (10 %, 15 %, 20 %, 25 %, etc.). It

12 Probabilistic Reasoning

is our experience that the students will guess too high – perhaps because the average number of aces in each hand is, indeed, one. In fact, we have never had a class in which the average of the guesses of the entire class was too low. The fact that the answer is not intuitive is part of what makes it an interesting problem.

> **Teacher Tip**
> It is very likely that the students will not choose an efficient way to do this problem. There are a number of ways to attempt this problem that are incorrect, and there are a number of ways that are correct but inefficient. That's OK. Resist the urge to tell them the most efficient way. An important part of becoming a good problem-solver is discovering what is incorrect or inefficient about an approach. In fact, we recommend that you encourage the students to first find any method, however inefficient, and then look for a way to simplify or improve it.

Discussion 12.10 (cont) There are a number of ways to get the answer to this problem. Students may start to wrap their heads around the problem by simplifying it. For example, taking all the aces and twos out of the deck and dealing out four hands of two cards each from these eight cards. This is a great start, and, as long as they are thinking, the students should be allowed to tackle the problem using their methods. We strongly believe in not interrupting the students when they are thinking hard, because that is the goal of the class: getting the students to think hard so they develop their brains.

A brute-force method to get the answer is to start by calculating the chance that the first hand gets exactly one ace. By performing a gedanken, you can see that there are 13 different ways to get exactly one ace and twelve non-aces when dealt a 13-card hand. The first card can be an ace and the remaining twelve cards non-aces, the second card can be an ace and the remaining twelve cards can be all non-aces, etc. The probability of getting the ace first is

$$P = \frac{4}{52} \times \frac{48}{51} \times \frac{47}{50} \times \frac{46}{49} \times \frac{45}{48} \times \frac{44}{47} \times \frac{43}{46} \times \frac{42}{45} \times \frac{41}{44} \times \frac{40}{43} \times \frac{39}{42} \times \frac{38}{41} \times \frac{37}{40} = \frac{703}{20,825}$$

The probability of getting the ace in exactly one of the 13 positions is 13 times this, which is 43.885 %.

> **Student Pitfall**
> If the student correctly calculates the probability of a particular player getting exactly one ace (0.43885), he/she might make the mistake of raising this to the fourth power to determine the probability of each of the four players on a

(continued)

single deal getting exactly one ace. This is not correct. Raising 0.4389 to the fourth power gives the probability of dealing out a 13-card hand that has a single ace *four times in a row* from a well-shuffled deck. The problem with the student's calculation is that the four hands in a single deal are not independent. That is, once the first player gets exactly one ace, the probability that the second player gets exactly one ace is higher than 43.88 %. AND, once the first two players are known to have exactly one ace, the probability of the third player getting one ace is even higher. Of course, once the first three players are known to have one ace, the fourth player is certain to have one ace. Why? Because all of the cards are dealt out and if the first three players only have one ace each, then the cards remaining to be dealt can only go to one place – the fourth player. Therefore, he/she must also have one ace.

Teacher Tip
A simpler version of this problem, as a warm-up or for younger classes, is to ask *"If we deal out all 52 cards to four players and three of the players have exactly one ace, what are the chances that the fourth player also has one ace?"* Students should realize that if each of the first three players has exactly one ace, the fourth player must have exactly one ace as well.

Discussion 12.10 (cont) Once the first player has exactly one ace, we can move to the next player. At this point, there are 39 cards left in the deck. The second player also needs one ace and twelve non-aces. The probability of getting the ace first followed by twelve non-aces is

$$P = \frac{3}{39} \times \frac{36}{38} \times \frac{35}{37} \times \frac{34}{36} \times \frac{33}{35} \times \frac{32}{34} \times \frac{31}{33} \times \frac{30}{32} \times \frac{29}{31} \times \frac{28}{30} \times \frac{27}{29} \times \frac{26}{28} \times \frac{25}{27} = \frac{25}{703}$$

The probability of getting the ace at any of the 13 positions is 13 times this, which is about 46.230 %.

Now we have dealt out half the deck and two aces are gone. We have 26 cards left to deal, and there are two aces among them. Let's deal out the next thirteen and, as before, assume that we get the ace first:

$$P = \frac{2}{26} \times \frac{24}{25} \times \frac{23}{24} \times \frac{22}{23} \times \frac{21}{22} \times \frac{20}{21} \times \frac{19}{20} \times \frac{18}{19} \times \frac{17}{18} \times \frac{16}{17} \times \frac{15}{16} \times \frac{14}{15} \times \frac{13}{14} = \frac{1}{25}$$

So, one-twenty-fifth of the time, the deal will consist of the ace first followed by 12 non-aces. But again, the ace could appear at any position, and each is equally likely, so the probability of getting exactly one ace is 13/25.

12 Probabilistic Reasoning

The remaining 13 cards must have one ace. In order for each player to get one ace, the first player must get exactly one ace, the second player must get exactly one ace, and the third player must get exactly one ace. Once the first three players get exactly one ace, the last hand must have exactly one ace. So, the probability of a bridge deal in which each of the four hands has exactly one ace is

$$P = \frac{703 \times 13}{20,825} \times \frac{25 \times 13}{703} \times \frac{13}{25} = \frac{13^3}{20,825} = 0.1055\ldots$$

This is 10.55 %. So, when a bridge hand is dealt, each player will get one ace only slightly more than one-tenth of the time.

There is, not surprisingly, a less complicated way to calculate the probability of each player getting exactly one ace. It involves modeling each of the four hands as a set of 13 slots that each holds one card. The first hand will be slots 1–13, the second hand will be slots 14–26, the third hand will be slots 27–39, and the fourth hand will be slots 40–52. Let's start by dealing out the four aces into these 52 available slots; after all, the aces are the only cards that matter.

It does not matter where the first ace goes. That is, it can go in any one of the 52 slots. The second ace must go in a different hand than the first ace. The probability of this happening is 39/51. That is, there are 39 slots available that are not in the hand that already contains the ace, and there are a total of 51 slots available (we started with 52 slots, but one is filled with the first ace). The third ace must not go in either of the two hands that already contain an ace. The probability of this happening is 26/50 (26 slots that will give of the result we want, out of 50 equally likely slots). Finally, the last ace must go in the only hand that does not yet contain an ace. The probability of this happening is 13/49 (13 slots that will give of the result we want, out of 49 equally likely slots). The probability of all four of these events happening is

$$P = \frac{52}{52} \times \frac{39}{51} \times \frac{26}{50} \times \frac{13}{49} = 0.1055\ldots$$

which is exactly the same probability calculated using the "brute-force" method.

Debriefing 12.10 This problem demonstrates how a probability problem can be tackled effectively and efficiently without a complete listing of all of the possibilities. It also demonstrates that the calculations can be made simpler by modeling the system creatively and focusing on what matters. The 48 non-aces are not the key issue in this problem; they're just "filler." In fact, it might be easier to model the problem with this: there are 52 players in the high school football team. Four of them are named John. The coach breaks the team into four practice squads of 13. What is the probability that there is one "John" on each squad?

Problem 12.11 An airline flight with 100 seats is full. All of the 100 passengers have tickets for their assigned seats. The first person in line to board the plane misplaces his boarding pass on the way down the jet bridge and decides to pick a seat randomly. All the remaining passengers sit in their assigned seats unless someone's seat is taken, and if it is, a different seat is chosen randomly. When the last passenger enters the plane, there is one seat remaining. What is the chance that it is that person's assigned seat?

Strategies Utilized Take inventory. Enumerate the possibilities. Reason backwards. Perform a gedanken. Increment and iterate. Simplify.

Discussion 12.11 An inventory of the facts we have to work with is as follows:
1. The first person sits randomly.
2. Every passenger that gets on the plane after the first person sits in their assigned seat unless it is taken. If it is taken, they choose a seat randomly.
3. When the last person gets on the plane, 99 seats are taken and one seat is available.

A quick poll of the students for an estimate of the probability that the 100th person's seat is still available is likely to draw a wide range of estimates from the class, most of which will be too low.

It might be a good idea to first have the students write their names and their guesses, perhaps using integer percent values, on a piece of scratch paper and collect these. Then divide the class into groups of three and allow the students within each group to come up with a consensus guess. The average of the group guesses is virtually always better than the average of the individual guesses. After the estimations are collected, the students can then try to solve the problem.

> **Teacher Tip**
> This is a problem that will challenge the student's problem-solving skills. Many will struggle because of the large number of people on the plane. However, this should compel the students to utilize the common problem-solving technique of attacking a much simpler version of the same problem. Of course, they should be able to come to this conclusion without any prompting from the instructor, so don't be in a hurry to give them any direction. If you have a classroom with fixed (or denotable) seating, you can run this as an experiment in your own class to see what happens!

Discussion 12.11 (cont) Often, trying to solve a much easier version of the same problem provides insights into the more complicated version. There is no reason not to try a *much* simpler version. So, let's start two seats instead of 100, and let's simplify by numbering the passengers the same as the assigned seats. That is, passenger 1 is assigned to seat 1, and passenger 2 is assigned to seat 2. Note that in practice the first passenger down the jet bridge is usually not assigned seat number 1 and assigned seats on passenger planes usually have a row number and a letter rather than just a number, but this is just another simplification that makes the problem more manageable.

With two passengers and two seats, passenger 1 will sit in either seat 1 or 2 with equal probability, so the chance the seat 2 is available for passenger 2 is 50 %.

Now let's try three seats. The question is, what is the probability that the third and last passenger on the plane sits in his assigned seat? To determine this, we calculate the probability of each of the ways that passenger 3 can get his assigned seat. There are two ways that this can happen. The first is that passenger 1 sits in seat 1. The chance of this happening is 1/3. Once passenger 1 sits in his assigned seat, passenger 3 is guaranteed his assigned seat.

The second way that passenger 3 can get his assigned seat is if passenger 1 sits in seat 2 and passenger 2 sits in seat 1. The probability of this happening is 1/3 times 1/2, the product of the two probabilities of the individual events. This is 1/6:

$$P = \frac{1}{3} \times \frac{1}{2} = \frac{1}{6}$$

So, the probability that seat 3 is available for passenger 3 if passenger 1 sits randomly is 1/3 + 1/6, which is one-half. The table below shows the possibilities.

Seat 1	Seat 2	Seat 3	Probability
Passenger 1	Passenger 2	Passenger 3	1/3
Passenger 2	Passenger 1	Passenger 3	1/6

Let's try four people.

If passenger 1 sits in seat 1, so will everyone else, and the chance of this happening is one-fourth because passenger 1 is choosing randomly from four seats.

If passenger 1 sits in seat 2 and passenger 2 sits in seat 1, both three and four will sit in their assigned seats. The probability of this sequence of events is

$$P = \frac{1}{4} \times \frac{1}{3} = \frac{1}{12}$$

If passenger 1 sits in seat 2 and passenger 2 sits in seat 3 and passenger 3 sits in seat 1, 4 will sit in his assigned seat. The probability of this sequence of events is

$$P = \frac{1}{4} \times \frac{1}{3} \times \frac{1}{2} = \frac{1}{24}$$

If passenger 1 sits in seat 3, then passenger 2 must sit in seat 2, and if passenger 3 sits in seat 1 rather than seat 4, then passenger 4 will sit in his assigned seat. The probability of this sequence of events is

$$P = \frac{1}{4} \times \frac{1}{1} \times \frac{1}{2} = \frac{1}{8}$$

These are the only four ways that passenger 4's seat will be available, and these are summarized in the table below:

Seat 1	Seat 2	Seat 3	Seat 4	Probability
Passenger 1	Passenger 2	Passenger 3	Passenger 4	1/4
Passenger 2	Passenger 1	Passenger 3	Passenger 4	1/12
Passenger 3	Passenger 1	Passenger 2	Passenger 4	1/24
Passenger 3	Passenger 2	Passenger 1	Passenger 4	1/8

These sum to 12/24, which is again one-half. This is interesting. At this point, it might be tempting to declare that the answer to the original problem must be one-half as we have "recognized the pattern!" However, at this point, it is only an educated guess (see Problem 7.6 for a good example of the perils of jumping to a conclusion too early). Let's see if we can come to this conclusion by another route.

Let's look at a possible scenario, considering the original problem, and perform a gedanken. What would happen if passenger 1 sits in, say, seat 3? This means that passengers 2, 3, ..., 36 sit in their assigned seat and passenger 37 has to take one of the unoccupied seats. The only seats he can choose that will immediately determine whether passenger 100 sits in his seat are seat 1 and seat 100. If he sits in seat 1, then passenger 100 is certain to get his seat, and if he sits in seat 100, then passenger 100 has no chance of getting his seat. Let's say he sits in seat 88. So, now passengers 38 through 87 sit in their assigned seat and passenger 88 has to sit randomly. As before, the only seats he can choose that will immediately determine whether passenger 100 sits in his seat are 1 and 100. This process continues until the last passenger gets on the plane.

At this point, we can use the problem-solving strategy *reason backwards*. So, instead of starting at the beginning, let's look at the situation when passenger 100 gets on the plane. We know that there is only one seat remaining. What seats can they be? When reasoning backwards, it might become clear that the only two seats that can possibly be available when passenger 100 gets on the plane are seat 1 and seat 100. If *any* seat between 1 and 100 were open, the passenger who is assigned that seat would have sat in it! Since every passenger who sits randomly has an equal chance of choosing seat 1 or seat 100, it must be 50/50 that seat 100 is taken when passenger 100 gets on the plane. The answer is 50 % – even if there are 300 seats on the plane.

Debriefing 12.11 This problem clearly demonstrates the value of thinking, pondering, and wondering. It is possible to grind out the problem with a lot of math, but it also can be solved with a flash of insight. Any calculations the students perform are not a waste of time; they provide a foundation in probabilistic thinking while developing the students' problem-solving intuition.

12 Probabilistic Reasoning   225

Problem 12.12 There are five chrysalides in a 6th grade science class. Three are those of the monarch butterfly and two are of the swallowtail butterfly. They will emerge in a random order. What is the probability that the third butterfly to emerge is a swallowtail?

Strategies Utilized Consider all the possibilities. Perform a gedanken. Recognize a pattern. Simplify.

Discussion 12.12 This problem would be easier for the students if there were only one swallowtail and four monarchs. With only one swallowtail, it is relatively easy to conclude that the swallowtail is equally likely to emerge any one of the five positions. That is, the chance of it emerging at any of the five positions is clearly one-fifth.

Similarly, it would be easier if the five chrysalides would all produce different butterflies, say, swallowtail, monarch, buckeye, queen, and a skipper, and the question was, "*What is the probability that the third butterfly to emerge is the swallowtail?*" Again, the answer is one-fifth.

With two swallowtail marbles and three monarchs, we can theorize that the chance of a swallowtail emerging at any of the five positions is the same. Since the chance of a swallowtail emerging first is 2/5, it seems reasonable to assume that the chance of a swallowtail emerging third should be 2/5 as well.

> **Teacher Tip**
> If you have the right students, you might want to try something even more obfuscating first. For example, start with 1 chrysalis, 40 monarchs, 30 swallowtails, 20 buckeyes, and 10 queens. When the butterflies emerge randomly one at a time, what's the chance that the 39th to emerge is a buckeye? When some students read this one, they may be nonplussed and adopt a defeatist attitude. Therefore, this problem provides the opportunity for the student to develop a trait that all good problem-solvers have – the calm confidence needed to start the problem. The chance that the 39th butterfly to emerge is a buckeye is the same as the chance that the first butterfly to emerge is a buckeye. This is one-fifth.

Discussion 12.12 (cont) Now let's confirm that the answer to the question as posed is 2/5 by calculating the probability that the third butterfly to emerge is a buckeye.

> **Teacher Tip**
> This is a good opportunity to reiterate the difference between independent events and dependent events. In this problem, the chance of a swallowtail emerging third is *dependent* upon the species of the first two butterflies that emerged.

Discussion 12.12 (cont) The probability of a swallowtail emerging first is 2/5 because two of the five butterflies are swallowtails.

There are two possible sequences in which the second butterfly to emerge is a swallowtail. The first sequence is monarch, swallowtail, and the second possibility is swallowtail, swallowtail.

The probability that the second butterfly to emerge is a swallowtail can be calculated by summing up the probabilities of these two events occurring. The probability that the first butterfly is a monarch and the second is a swallowtail is

$$P = \frac{3}{5} \times \frac{2}{4} = \frac{3}{10}$$

The probability that the first and the second butterflies to emerge are swallowtails is

$$P = \frac{2}{5} \times \frac{1}{4} = \frac{1}{10}$$

The sum of these is 4/10, which is 2/5, which is consistent with the theory that the probability must be 2/5 to draw a white marble in any of the five positions.

> **Teacher Tip**
> If the students are familiar with probability trees, this is a fantastic way to visualize the probability calculations, and we recommend that they be utilized. Anything that permits the student to better *understand* the calculation at a fundamental level is recommended.

Discussion 12.12 (cont) Now let's focus on the third butterfly to emerge. There are three ways to get a swallowtail in the third position: M-M-S, S-M-S, and M-S-S. The probability of M-M-S is

$$P = \frac{3}{5} \times \frac{2}{4} \times \frac{2}{3} = \frac{1}{5}$$

The probability of S-M-S is

$$P = \frac{2}{5} \times \frac{3}{4} \times \frac{1}{3} = \frac{1}{10}$$

The probability of M-S-S is

12 Probabilistic Reasoning

$$P = \frac{3}{5} \times \frac{2}{4} \times \frac{1}{3} = \frac{1}{10}$$

Again, these sum to 2/5, as they must. It would be time well spent for a young student to finish the calculation all the way to the last butterfly to emerge to confirm that the probability of a swallowtail to emerge at any of the five positions is 2/5.

> **Teacher Tip**
> Students can have difficulty telling the difference between independent events (such as throwing dice) and dependent events (such as butterflies emerging). This problem can be a useful way to illustrate the difference because the probability that the next butterfly to emerge is a swallowtail depends on what emerged previously. Getting students to take inventory after each emergence will demonstrate this principle.

Debriefing 12.12 This problem again demonstrates the usefulness of investing time to think about the best way to tackle a problem before starting to perform the calculations. The appreciation of the clever method is heightened when the student has utilized the brute-force method. For this reason, if the student is doing the problem inefficiently, don't interrupt. The student learns best when the teacher is silent, and the student is engaged in System 2 thought. Remember, the goal is not to quickly move on to the next problem.

Finally, it is worth mentioning that this problem is reminiscent of the "reverse raffle," in which the last ticket remaining is declared the winner. Any ticket is just as likely to get pulled out first as it is to get pulled out last – and at every position in between.

Problem 12.13 In the game Yahtzee,[6] players roll five six-sided dice to make various combinations, one of which is a large straight, which is a run of five consecutive numbers. These can only be 1-2-3-4-5 or 2-3-4-5-6. What is the probability that a large straight comes up on a single roll of five dice?

Strategies Utilized Enumerate the possibilities. Perform a gedanken. Simplify.

Discussion 12.13 Here is another problem that offers two distinct paths to the solution. The first is to enumerate the total number of ways the dice can form a large straight and divide it by the total number of ways the dice can land. Let's start by determining the number of ways a large straight can be formed. There are two possible large straights. These are 1-2-3-4-5 and 2-3-4-5-6. The large straight 1-2-3-4-5 can be rolled a number of different ways. Examples include 4-3-2-5-1 and 1-2-

---
[6] Yahtzee is a commercial dice game.

4-3-5. For comparison, there is only one way to roll all four: 4-4-4-4-4. If we calculate the number of different permutations of the numbers 1-2-3-4-5, we can determine the ratio of the probability of rolling a large straight to the probability of rolling all the same number, which is called a Yahtzee.

So, we have five numbers and are putting them in five positions. The 1 can go in any one of five positions, the 2 can go in any one of the remaining four positions, the 3 can go in any one of the three remaining positions, the 4 must go in any of the two remaining positions, and the 5 must go to the last position. Therefore, the number of possible permutations of the five different results is

$$N = 5 \times 4 \times 3 \times 2 \times 1 = 120$$

There are also 120 permutations of the large straight 2-3-4-5-6, which means that there are 240 ways the dice can form a large straight. There are 6^5 possible ways the dice can land, making the probability of rolling a large straight:

$$P = \frac{240}{7,776} = \frac{5}{162}$$

which is about 3.1 % or about once in 32 rolls.

We can now determine the relative likelihood of a large straight and a Yahtzee. Since there are 240 ways to roll a large straight and 6 ways to roll a Yahtzee, a large straight is 40 times more likely to be rolled than a Yahtzee. The probability of rolling a Yahtzee in a single roll of the five dice must be

$$P = \frac{6}{7,776} = \frac{1}{1,296}$$

because there are 7,776 different rolls and only six of them are Yahtzees.

A second way to get the answer is by performing a gedanken. The gedanken here is to mentally roll the dice one at a time and calculate the probability of getting a particular large straight at each step. As an example, let's select the large straight consisting of the numbers 1–5. The result of the first roll has to be one of the numbers 1–5. The chance of this is 5/6. The second roll can't be a six and it can't match the first number. The chance of this happening is 4/6. The third roll can't be a six and it can't match either of the first two numbers. The chance of this happening is 3/6. Continuing in this fashion, we see that the probability of rolling the numbers 1–5 with five dice is

$$P = \frac{5}{6} \times \frac{4}{6} \times \frac{3}{6} \times \frac{2}{6} \times \frac{1}{6} = \frac{120}{7,776}$$

The chance of rolling *either* large straight (1-2-3-4-5 or 2-3-4-5-6) is twice this, or 240/7,776, which is the same as we got with the first method.

12 Probabilistic Reasoning

> **Teacher Tip**
> Many popular games involve probabilities. If there is a popular card, board, or computer game with which your students are familiar, these can provide opportunities to challenge the students with good probabilistic reasoning problems.

Debriefing 12.13 This problem provides a good example of how the number of ways an event can happen increases the probability of it happening. This point can be further emphasized by asking the students, "*Which of the two sequences of heads and tails is more likely?*"

H-H-H-H-H-H-H-H-H-H

or

H-H-T-H-T-T-T-H-T-H

The answer is that they are equally likely. The reason why the probability that a toss of ten coins is much more likely to contain five heads and five tails than ten heads is that there is only one way the coins can all land on heads but there are a large number of ways five heads and five tails can be arranged among ten coins. Can your students calculate this? Can they derive the formula? This is actually a key principle in thermodynamics, and it intimately connected with the concept of entropy.

Problem 12.14 There are two identical opaque bags. One contains nine white marbles and one black marble, and the other contains ten white marbles. Your goal is to choose a white marble with your one selection. However, you are allowed to sample as many marbles as you want, without replacement, before declaring that the next marble selected will be your one selection. Of course, if you sample, you hope to select the black marble early in the sampling process, because both bags will then contain only white marbles when you make your selection. If you are unfortunate enough to sample all 19 white marbles, you will be forced to count the last marble, which must be black, as your selection.

In one experiment, the contestant decided to sample marbles. He selected eight marbles from the same bag, and all were white. Then he decided to make the next marble he selected the one that counted. Should he select from the bag with only two marbles left, or should he select from the bag that still has all ten marbles? Is there a better strategy for sampling the bags? If so, what is it?

Strategies Utilized Enumerate the possibilities. Perform a gedanken.

Discussion 12.14 The question we'll address first is which bag to select a marble from that will maximize the probability of selecting a white ball after removing eight white balls from bag #1. The fact that eight balls were removed from bag #1 means that it is probably the bag that contained ten white balls originally. We need to calculate this probability.

The chance of selecting eight white marbles from a bag that contains nine white marbles and one black marble is

$$P = \frac{9}{10} \times \frac{8}{9} \times \frac{7}{8} \times \frac{6}{7} \times \frac{5}{6} \times \frac{4}{5} \times \frac{3}{4} \times \frac{2}{3} = \frac{1}{5}$$

The chance of selecting eight white marbles from a bag that contains ten white marbles and no black marble is 8/8, which is 100 %.

So, we can perform a gedanken and consider, say, 6,000 trials in which we select eight marbles from one of the two bags. In the 3,000 trials in which the bag with the ten white marbles was selected, eight white marbles will be selected every time. In the 3,000 trials in which the bag with the nine white marbles was selected, eight white marbles will be selected in only 1/5 of them, or 600 times.

So, out of the 6,000 trials, 3,600 of them will be drawn that contained all white marbles. Of these 3,600, 3,000 were from the bag that contained ten white marbles. So, the probability is 5/6 that the eight white marbles were selected from the bag that contained ten white marbles.

With this information, we can determine the probability of pulling a white marble from the bag that only has two marbles left:

$$P = \left(\frac{5}{6} \times \frac{2}{2}\right) + \left(\frac{1}{6} \times \frac{1}{2}\right) = \frac{5}{6} + \frac{1}{12} = \frac{11}{12}$$

As before, it is useful to go over what these fractions represent. From left to right, the 5/6 is the probability that the bag from which the marble is chosen contained ten white marbles originally, the 2/2 is the probability of selecting a white marble from this bag if the bag originally contained ten white marbles, the 1/6 is the probability that the bag originally contained one black marble and nine white marbles, and the 1/2 is the probability of selecting a white marble if there is a black and a white marble left in the bag. The result is that the probability of selecting a white marble from the bag that eight white marbles were taken from is 11/12.

Now let's calculate the chance of selecting a white marble from the bag that still has ten marbles in it:

12 Probabilistic Reasoning

$$P = \left(\frac{1}{6} \times \frac{10}{10}\right) + \left(\frac{5}{6} \times \frac{9}{10}\right) = \frac{1}{6} + \frac{9}{12} = \frac{11}{12}$$

Here, the 1/6 is the probability that the bag with the ten marbles is the one that has ten white marbles in it and the 10/10 is the probability of getting a white marble from this bag if this is the case. The 5/6 is the probability that the bag with the ten marbles still in it is the bag that has the black marble, and the 9/10 is the probability of getting a white marble from this bag if this is the case.

It's the same. The probability of selecting a white marble from either bag is 11/12 after eight white marbles were removed from one of the bags. A gedanken should convince you that this must be the case; there are 12 marbles remaining, 11 white and one black, and each is equally likely to be chosen.

Now we can address the question of whether sampling marbles is a good idea. After all, if you do not sample any marbles, declaring that the first marble is the one that counts, your chance of selecting a white one is 19/20.

Let's base our sampling strategy on first example, in which eight white marbles were selected. The strategy will be to select up to eight marbles from the same bag. If the black one is selected, it is 100 % that the next marble will be white. If the black one is not selected, we have seen that it is 11/12 to get a white marble from either bag.

So, what is the probability that we get eight white marbles from the same bag? There are two possibilities. One is that we select the bag with all white marbles, and the second is that we select the bag with one black marble and pick eight straight white ones. The sum of these two probabilities is

$$P = \left(\frac{1}{2} \times \frac{8}{8}\right) + \left(\frac{1}{2} \times \frac{1}{5}\right) = \frac{6}{10}$$

Again, let's go over what each of these four fractions represents. The first one-half is the probability of selecting the bag with 10 white marbles. The 8/8 is the probability of selecting eight white marbles from this bag. The second one-half is the probability of selecting the bag with the black marble, and the 1/5 is the probability of selecting eight straight white marbles from this bag. So, it is 60 % that if eight marbles are drawn from the same bag, they will all be white. This means that this strategy will find the black marble 40 % of the time.

With these numbers, we can now calculate the probability that this strategy is successful. The calculation is shown below:

$$P = \left(\frac{4}{10} \times \frac{12}{12}\right) + \left(\frac{6}{10} \times \frac{11}{12}\right) = \frac{19}{20}$$

As before, let's go over what each of these four fractions represents. The 4/10 is the probability of drawing a black marble with eight selections. The 12/12 is the probability of selecting a white marble when the black one has been already selected, the 6/10 is the chance of not drawing the black ball with the first eight marbles, and the 11/12 is the probability of selecting a white marble after removing eight white marbles from one bag. The result is the *same* as not sampling any marbles at all.

In fact, there is *no* marble-sampling strategy that makes it more likely (or less likely) that you select a white marble with the selection that counts – including sampling the first 19 and choosing the last one. It is always 19/20 (95 %) to select a white marble.

Debriefing 12.14 It is very likely that students will have proposed numerous sampling strategies to avoid the black marble. Hopefully, it will make an impression on them when they discover that there is no sampling strategy that can improve or otherwise change the probability of selecting a white marble. It is always 95 %. A student's pattern recognition skills might help him/her make the connection between this problem and an "optimization" puzzle (Problem 11.6) – whatever the strategy of breaking the chocolate bar, the number of breaks is always the same.

Reference

1. Taleb N (1994) Fooled by randomness. Random House, New York

Logical Reasoning   13

> *A good puzzle, it's a fair thing. Nobody is lying. It's very clear, and the problem depends just on you.*
> – Erno Rubik

This chapter contains a set of problems that do not require any high-level mathematics or formal training in logic. They do not require any knowledge of vocabulary or culture. There are no "tricks." The problems just require a focused mind that is able to ask the appropriate "What if" questions and then follow the line of reasoning to the only result that makes *logical* sense.

At this point in the course, the students should not be intimidated by a problem that has no obvious pathway to the solution. They should patiently invest time trying to understand and frame the problem, they should perform gedankens in order to wrap their heads around the problem, and they should simplify, draw a diagram or build a model if appropriate.

Problem 13.1 Let's return to the Monty Hall problem (see Problem 5.5) one more time. But here we'll add the twist that there are five doors – one with a grand prize behind it and four with goats behind them. As a reminder, the procedure starts when the contestant selects one of the five doors. Monty then opens one of the other four doors that has a goat and asks if the contestant wants to trade the door he/she originally selected for one of the three other remaining unopened doors. After the contestant makes a choice, Monty will open another door with a goat behind it (remember, Monty knows where the grand prize is and will not reveal it) and again ask if the contestant wants to trade the door he/she currently has for one of the other two remaining unopened doors. After the contestant decides, Monty opens one more door with a goat behind it. At this point, there are three opened doors that each contained goats. The two unopened doors include the door the contestant had originally chosen and another unopened door. Finally, Monty gives the contestant another opportunity to switch doors. What should the contestant's three choices be, when offered the opportunity to switch, to maximize the chance of getting the grand prize? What is the chance of getting the grand prize utilizing this strategy?

Strategies Utilized Recognize a pattern. Perform a gedanken.

Discussion 13.1 It is likely that the students' initial reaction is to switch every time because in the earlier example, switching doors was a good strategy. However, after Monty Hall opens the first door with a goat behind it, the contestant is effectively picking randomly from four unopened doors – the one that was chosen originally and the three remaining unopened doors. After Monty opens the second door with a goat behind it, the contestant is effectively choosing from three unopened doors each of which has a probability of 1/3 of having the grand prize. Thus, the problem is reduced to the original Monty Hall problem with three doors. The switching strategy here will produce a 2/3 chance of success.

However, the contestant can do significantly better by not switching the first two times and switching the last time. This will only fail to get the grand prize when the contestant actually picked the door with the grand prize with his original choice. This strategy will be successful 4/5 or 80 % of the time.

To see if the students really understand what is going on, they can calculate the probability of success for all eight strategies by filling out the rightmost column in the table below.

Choice one	Choice two	Choice three	Fraction success
Keep	Keep	Keep	
Switch	Keep	Keep	
Keep	Switch	Keep	
Keep	Keep	Switch	
Keep	Switch	Switch	
Switch	Keep	Switch	
Switch	Switch	Keep	
Switch	Switch	Switch	

Debriefing 13.1 If this is presented later in the course, it should be a confidence builder, as it is not too challenging to come up with the optimal strategy. For this reason, this question might be a good candidate for a 10–15 minute quiz that the students work on individually. If you have the right students, you can ask them what is the probability of success with the optimal strategy if there are N doors and Monty shows the contestant $N - 2$ goats.

Problem 13.2 On a TV game show, a husband and a wife team has a chance to win a car. There are three doors from which to choose. One door has a car behind it, another has the key to the car, and the third contains the ever-present goat. To win the car, the husband must pick the door with the car and the wife must pick the door with the key. However, each gets *two* chances. The wife is in a sealed room when the husband makes his choices, and the three doors are closed when the wife comes on stage to make her two choices. If each guesses randomly, each has a 2/3 chance

13 Logical Reasoning

of succeeding, which means that they have a probability of 4/9 of winning the car (you might ask your students to calculate this). The question is, can the husband and wife utilize the short commercial break to decide on a door-picking strategy that will increase their chance of success? When using this strategy, what is the chance that they will win the car?

Strategies Utilized Perform a gedanken. Enumerate the possibilities.

Discussion 13.2 There is nothing that either the husband or the wife can do to increase their individual chance of success – it is always 2/3. The trick is to figure out a way to group the events when both the husband and the wife are successful.

> **Student Pitfall**
> In our experience, it is a big leap for the students to see that the number of second door checked can be based on what is behind the first door. They often get stuck trying various combinations of doors. For example, husband checks 1 and 2, wife checks 2 and 3, or they both check 1 first and then the husband checks 2 and the wife checks 3. They will often struggle for a long time without being able to have the key insight. If they can't get it, that's OK, let the problem go and perhaps come back to it later in the term.

> **Teacher Tip**
> This is a problem that can be easily acted out by students. All you need are three small boxes, a toy car, a key, and a goat (or other suitable booby prize). The two "contestants" can start in the hallway and come into the classroom to check two boxes. When actually modeling the problem, one of the students is more likely to have an *Aha!* moment.

Discussion 13.2 (cont) Let's say that the husband finds the key first and the wife finds the car first, each should use their second choice to select each other's first choice. That is, if the husband opens door 1 and finds the key, he wants the wife to select door 1 so she finds the key. Similarly, if the wife starts with door 2 and finds the car, she wants her husband to select door 2. To do this they can make a simple rule:

The husband checks door 1 first; if it is the key, he checks door 2 next.
The wife checks door 2 first; if it is the car, she checks door 1 next.
    To complete the set of rules, if the husband finds the goat behind door 1, he checks door number 3 and if the wife finds the goat behind door number 2, she checks door 3. Of course, if the husband finds the car behind door number 1 or if the wife finds the key behind door number 2, they don't need to check a second door.

There are six possible distributions for the goat, car, and key. These are shown in the table.

Door 1	Door 2	Door 3	Husband		Wife	
Goat	Key	Car	1, 3	Yes	2	Yes
Goat	Car	Key	1, 3	No	2, 1	No
Key	Car	Goat	1, 2	Yes	2, 1	Yes
Key	Goat	Car	1, 2	No	2, 3	No
Car	Goat	Key	1	Yes	2, 3	Yes
Car	Key	Goat	1	Yes	2	Yes

The columns headed Husband and Wife give the number of the doors that were checked and the *Yes* and *No* indicate whether the strategy was successful. Note that when the husband is successful, the wife is successful, and when the husband fails, the wife fails as well. Note further that each of them is successful 2/3 of the time individually and they are 2/3 successful as a team as well.

If the husband finds the car behind door 1, he knows his wife will find the key using their strategy. Similarly, if the wife finds the key behind door 2 with her first choice, she knows immediately that they have won the car because she knows that her husband will check doors 1 and 3.

Debriefing 13.2 This problem has some similarities to others presented in this book. Making the connection will challenge the students' pattern recognition skills. The better they are, the sooner they will make the connection. If they get to the end and no connection had been made, you can ask them, "*Does this problem seem like any others we have solved previously?*"

Problem 13.3 A young boy was the only one to see a criminal toss a gun down one of three sewers. The police want to retrieve the gun, but searching a sewer is an arduous and unpleasant task, so they would like to know which one contains the gun. The boy, however, is reluctant to say because the criminal threatened to harm him if he revealed where he threw the gun. One detective had an idea. He proposes, "*Let's ask the boy to identify one of the three sewers that does not contain the gun.*" This would eliminate the possibility that all three would have to be searched and make the possibility that they find it the first sewer equal to 50 %. Another policeman pipes in with, "*I have a better idea, I'll stand next to sewer 1 and ask the boy, of sewers 2 and 3, point to one that does not contain the gun.*"

Is this strategy really better? Why or why not?

Strategies Utilized Recognize a pattern. Perform a gedanken. Draw a diagram.

Discussion 13.3 This is a thinly disguised Monty Hall problem (see Problem 5.5). Hopefully the students will make the connection. When the boy identifies any one of the three sewers that does not contain the gun, as proposed by the first policeman, the probability that the gun in one of the other two sewers is 50 % (assuming the boy

13 Logical Reasoning

is telling the truth). When the second policeman stands next to sewer 1 and asks the boy to identify one of the other two sewers that does not contain the gun, the probability that the gun is in the sewer that the boy does not name is now 2/3.

> **Teacher Tip**
> To best utilize this problem, do not present it in the middle of a bunch of problems involving the Monty Hall problem. Perhaps this one can be saved for an exam – perhaps even the final exam. Of course, if you want the students to get the answer, it is best to present it immediately after the previous two problems, but, as mentioned before, the goal is not to solve as many problems as possible.

Discussion 13.3 (cont) The policeman standing next to sewer 1 is analogous to the contestant choosing door number 1 in the Monty Hall problem. The boy pointing to a sewer that does not contain the gun is analogous to Monty Hall opening a door that does not contain the prize. So, "switching sewers" from sewer 1 to the sewer that the boy did not point to is the correct move, with a 2/3 probability of success.

The key to understanding this one is to ask, "*What is the probability that the gun is in sewer 1?*" Since there are three sewers and each is equally likely the answer is 1/3. Now the boy points to either sewer two or three and reveals a sewer that does not contain the gun. The probability that gun is in sewer 1 is still 1/3. Therefore the probability that the gun is in the sewer that the boy did not point to is 2/3.

Debriefing 13.3 This is a problem that offers the students a great chance to make a connection between a problem previously presented. The difference between the two strategies proposed by the policemen is subtle but significant. It is worthwhile spending time on this to allow the students to grasp the significance. If they have not yet drawn a diagram to solve this one, perhaps it would be a good idea to ask them to enumerate the possibilities for *both* strategies. They might look like this:

Strategy #1

Gun location	Boy points to	Prob. of success
Sewer 1	Sewer 2 or 3	50 %
Sewer 2	Sewer 1 or 3	50 %
Sewer 3	Sewer 1 or 2	50 %

Strategy #2

Gun location	Boy points to	Prob. of success
Sewer 1	Sewer 2 or 3	0 %
Sewer 2	Sewer 3	100 %
Sewer 3	Sewer 2	100 %

Using the first strategy, the probability of finding the gun is 50 % in either of the two sewers that does not contain the gun. With the second strategy, the gun is found 100 % of the time when it is not in sewer 1. The probability that it is not in sewer 1 is 2/3. Therefore the probability that the gun is in the sewer that the boy does not point to is 2/3.

The next two problems are set in an environment of liars and truth-tellers – we have already explored such environment earlier in the text (see Problem 8.6).

Problem 13.4 There are two communities on an island. One always lies and the other always tells the truth. A visitor to the island would like some information so he would like to identify a truth-teller. He meets two of the locals; let's call them Alice and Bob. The visitor asks Alice, "*Are you a truth-teller?*" Alice understands the question, but responds in her own language by saying, "*tsxyk.*" Bob notices that visitor's puzzled look and clears things up by saying, "*Alice said yes, but she is a liar.*" Now:
(a) Which community is Alice from?
(b) Which community is Bob from?

Strategies Utilized Perform a gedanken.

Discussion 13.4 This problem, like so many others, yields to a good gedanken. The key question to ask involves the response to the question, "*Are you a truth-teller?*" How would a liar respond? How would a truth-teller respond? These are questions that the students should be able to ask themselves without any hinting. If you tell them to think about how both a liar and a truth-teller would respond to the question, "*Are you a truth-teller?*" you prevent them from developing their ability to problem-solve by performing a series of gedankens on their own. You also prevent them from having the joy of independently reaching a new level of understanding.

Once the key question is asked, the answer is only a quick series of logical steps away. The answer to the key question is that both a liar and a truth-teller would answer "*yes*" to the question, "*Are you a truth-teller?*" Therefore, "*tsxyk*" must mean "*yes.*" By itself, this provides no information. However, when Bob says, "*Alice says yes,*" we know that Bob is a truth-teller because Alice must have said "*yes.*" At this point, we still don't know whether Alice is a truth-teller or a liar. But Bob tells us this by saying, "*But she is a liar.*" Since we know Bob is a truth-teller, we know Alice is indeed a liar.

So, Alice is a liar and Bob is a truth-teller.

Debriefing 13.4 It is easy to use this problem as a barometer for the degree to which you have been a good cognitive trainer. At this point in the course, the students should not be looking at you for hints. They should know that the course is about the development of their thinking and reasoning skills, not the transference of knowledge. If you present this problem at the beginning of the course, it is likely they will not know how to solve the problem and, as a result, will immediately try to get some help. At this point in the course, they will still not know how to solve the problem immediately, but they will move forwards by understanding the problem, framing the problem, and performing gedankens until they have uncovered the solution. Many times, when we have presented this as a quiz, the students will unconsciously release an audible *Oooo*! or an *Aaaa*! when they solve the problem.

Problem 13.5 A Monday–Wednesday–Friday liar always lies on Monday, Wednesday, and Friday and always tells the truth on other days. A Tuesday–Thursday–Saturday liar always lies on Tuesday, Thursday, and Saturday and always tells the truth on other days. A couple is known to consist of a Monday–Wednesday–Friday liar and a Tuesday–Thursday–Saturday liar. One says, "*Yesterday was Monday.*" The other says, "*Yesterday I told the truth.*" What day is it?

Strategies Utilized Perform a gedanken. Draw a diagram. Enumerate the possibilities.

Discussion 13.5 This one has a lot of information and it is easy for the students to get confused. A good way to understand and frame the problem is to draw a diagram that reveals on which days each tells the truth and on which days they lie. This is shown in the table below.

	Mon	Tue	Wed	Thu	Fri	Sat	Sun
MWF liar	L	T	L	T	L	T	T
TTS liar	T	L	T	L	T	L	T

Another way to make things more clear is to make another table indicating the days on which the two intermittent liars can make each of the two statements.

	"*Yesterday was Monday*"	"*Yesterday I told the truth*"
MWF liar	Mon, Tue, Wed, Fri	Sun
TTS liar	Thu, Sat	Mon

When looking at the second table, the MWF liar can only say, "*Yesterday I told the truth*" on Sunday. The TTS liar can't say "*Yesterday was Monday*" on Sunday, so it can't be Sunday. The only remaining possibility is that it is Monday and the TTS liar said "*Yesterday I told the truth*," and the MWF liar said "*Yesterday was Monday.*"

It must be Monday.

Debriefing 13.5 The students may also solve this one by going through the days of the week one by one, and there is nothing wrong with this because there are only seven days. However, in this case it is more efficient to examine the statement, "*Yesterday I told the truth*" because it significantly narrows down the possibilities. Of course, there is nothing wrong with either approach. The key is to get them thinking and solving problems independently.

Problem 13.6 A jeweler makes a single strand of beads by threading onto a string in a single direction from a clasp a series of solid-colored beads. Each bead is green, orange, purple, red, or yellow. The resulting strand satisfies the following specifications:

- If a purple bead is adjacent to a yellow bead, any bead that immediately follows and any bead that immediately precedes that pair must be red.
- Any pair of beads adjacent to each other that are the same color as each other must be green.
- No orange bead can be adjacent to any red bead.
- Any portion of the strand containing eight consecutive beads must include at least one bead of each color.

Now, let's consider four multiple-choice questions[1]:

I. If the strand has exactly eight beads, which one of the following is an acceptable order, starting from the clasp, for the eight beads?
   (A) Green, red, purple, yellow, red, orange, green, purple
   (B) Orange, yellow, red, red, yellow, purple, red, green
   (C) Purple, yellow, red, green, green, orange, yellow, orange
   (D) Red, orange, red, yellow, purple, green, yellow, green
   (E) Red, yellow, purple, red, green, red, green, green

II. If an orange bead is the fourth bead from the clasp, which one of the following is a pair that could be the second and third beads, respectively?
   (A) Green, orange
   (B) Green, red
   (C) Purple, purple
   (D) Yellow, green
   (E) Yellow, purple

III. If on an eight-bead strand the second, third, and fourth beads from the clasp are red, green, and yellow, respectively, and the sixth and seventh beads are purple and red, respectively, then which one of the following must be true?
   (A) The first bead is purple.
   (B) The fifth bead is green.
   (C) The fifth bead is orange.
   (D) The eighth bead is orange.
   (E) The eighth bead is yellow.

IV. If on a six-bead strand the first and second beads from the clasp are purple and yellow, respectively, then the fifth and sixth beads CANNOT be
   (A) Green and orange, respectively
   (B) Orange and green, respectively
   (C) Orange and yellow, respectively

---

[1] This series of four questions is taken directly from an LSAT exam – the exam that students take in an attempt to get into law school. The questions have nothing to do with law; they are designed to test the candidate's ability to think logically and to solve problems – a skill that finds applications everywhere.

13 Logical Reasoning   241

   (D) Purple and orange, respectively
   (E) Yellow and purple, respectively

Strategies Utilized Perform a gedanken. Enumerate the possibilities.

Discussion 13.6 This is a good candidate for an individual quiz. The problems are not that challenging; they just require a clear mind and a depth of focus. To better explain the answers, the rules are numbered below:
1. If a purple bead is adjacent to a yellow bead, any bead that immediately follows and any bead that immediately precedes that pair must be red.
2. Any pair of beads adjacent to each other that are the same color as each other must be green.
3. No orange bead can be adjacent to any red bead.
4. Any portion of the strand containing eight consecutive beads must include at least one bead of each color.

   For Problem I, four of the five possibilities break one of the rules. The rule that is broken is next to the choice below in parentheses. The answer is C.
I. If the strand has exactly eight beads, which one of the following is an acceptable order, starting from the clasp, for the eight beads?
   (A) Green, red, purple, yellow, red, orange, green, purple (rule 3)
   (B) Orange, yellow, red, red, yellow, purple, red, green (rule 2)
   (C) Purple, yellow, red, green, green, orange, yellow, orange (OK)
   (D) Red, orange, red, yellow, purple, green, yellow, green (rule 1)
   (E) Red, yellow, purple, red, green, red, green, green (rule 4)

   For Problem II, four of the five possibilities break one of the rules. The rule that is broken is next to the choice below in parentheses. The answer is D.
II. If an orange bead is the fourth bead from the clasp, which one of the following is a pair that could be the second and third beads, respectively?
   (A) Green, orange (rule 2)
   (B) Green, red (rule 3)
   (C) Purple, purple (rule 2)
   (D) Yellow, green (OK)
   (E) Yellow, purple (rule 1)

   Problem III is a little more involved. Rule 4 states that any eight consecutive beads must contain one of each color. There is no orange, so orange must go on the bracelet somewhere, but it can't go in the first or eighth spot because it would be next to a red bead and this violates rule 3. Therefore, the answer is C; the fifth bead is orange.
III. If on an eight-bead strand the second, third, and fourth beads from the clasp are red, green, and yellow, respectively, and the sixth and seventh beads are purple and red, respectively, then which one of the following must be true?
   (A) The first bead is purple.
   (B) The fifth bead is green.
   (C) The fifth bead is orange.
   (D) The eighth bead is orange.
   (E) The eighth bead is yellow.

If the fifth and sixth beads are yellow and purple, the third and fourth beads have to be red according to rule 1. However, this violates rule 2 because two red beads can't be next to each other. The answer is E.

IV. If on a six-bead strand the first and second beads from the clasp are purple and yellow, respectively, then the fifth and sixth beads *CANNOT* be

(A) Green and orange, respectively
(B) Orange and green, respectively
(C) Orange and yellow, respectively
(D) Purple and orange, respectively
(E) Yellow and purple, respectively

Debriefing 13.6 It is very likely that your students – sometime in their education – are going to have to sit down for a few hours and take an exam. For this reason, they need not only mental strength but mental stamina. It might be a good idea to remind the students that this question is just one of many on the LSAT. Being able to maintain focus and concentration while efficiently solving problems is an important skill, not only in the students' formal education, but in their career and personal life as well. Tackling these types of problems will develop this skill.

In Chap. 11 (Sect. 11.2), we presented a few optimization problems of diverse nature (scheduling of travelers, designing a bridge, finding the best strategy). The following problem can be considered also as an optimization problem, but with a special twist: the number of variables (in this case, numbers) is not known.

Problem 13.7 Choose a set of positive integers that add to 50 that will produce the largest product when they are all multiplied together.

Strategies Utilized Perform a gedanken. Simplify. Enumerate the possibilities.

Discussion 13.7 Once the problem is understood, the solution is best accessed through a series of trials that will thoroughly explore the solution space. The extremes are 50 and zero, fifty 1s, and we can throw in 25 and 25. All of these add to 50. The 50×0 gives us zero, the fifty 1s make a product of 1, and the 25×25 gives us 625. Well, $25 \times 25 = 625$ is the clear winner here.

But the students shouldn't stop there. What about five 10s? They add to 50 and produce a product of 100,000. Now we're getting somewhere.

But we can replace a ten by five 2s. When five 2s are multiplied together, we get 32, which is more than ten.

If a student gets this far, they are very likely to conclude that the answer is twenty-five 2s. The product of which is $2^{25} = 33{,}554{,}432$.

There is, however, one more possibility to check. Let's simplify the problem by assuming that we are trying to find a set of integers that sum to six and have the highest product. Three 2s will have a product of eight, but two 3s have a product of

13 Logical Reasoning

nine. Therefore, 3s are better than 2s when trying to maximize the product of numbers that have a particular sum.

Sixteen threes will give us a sum of 48 and we'll add a two to get up to fifty. The answer we have now is $2 \times 3^{16} = 86,093,442$, which is significantly bigger than $2^{25} = 33,554,432$.

To be complete other possibilities should be considered – for example, 5×3^{15} or even 2×4^{12}. If you have the right students, you can ask them for a proof that 86,093,442 is the maximum.

> **Teacher Tips**
> If the students tackle this problem in groups, it is sometimes fun to announce the highest product as it slowly increases. Usually, the students are motivated to beat the current high with their product. If your students are proficient with numbers, it might not be a good idea to separate them into groups because the answer may come very quickly. This one might be a good candidate for an exam question.

Debriefing 13.7 While this problem deals with numbers and powers, it is not primarily a math problem. It is an optimization problem using numbers. If the students understand and frame the problem and then follow it up with an *increment and iterate* strategy, they should home in on the answer in short order. If there are students in the class that are not well-trained mathematicians, they should actually become more confident and comfortable when dealing with numbers after solving this one.

Problem 13.8 You have eight gold coins, one of which is counterfeit and weighs slightly less than each of the others. You have a balance but no other weights. The balance can be used to determine whether one set of coins balances another set. Describe a procedure to isolate the counterfeit coin in the minimum number of weighings.

Strategies Utilized Understand the problem. Perform a gedanken.

Discussion 13.8 Almost invariably, students will come up with three weighings. They start with four on each side and then compare two vs. two from the side that was light and then compare the two that are light from that weighing.

It seems like this is the most efficient way to isolate the counterfeit coin. However, what the students usually do not realize is that they get information about the coins that are not weighed.

> **Teacher Tip**
> While many problems are good to solve in groups, we recommend that the students work on this one individually – perhaps even as a quiz. This is a good opportunity for the students to learn the importance of not going with the first answer they get. If this is a quiz, some students will get three weighings as their answer and will want to hand it in early. If some students are done early, we suggest giving them something else to work on, or you may consider asking them, *"Are you convinced that your answer is optimal?"*

When the students are finished (and the quiz is collected), you can poll them by asking, *"Who isolated the counterfeit coin in three weighings?"* Then, *"Who isolated the counterfeit coin in two weighings?"*

Then we recommend giving the students the opportunity to solve it in only two weighings.

Isolating the counterfeit coin in two weighings can be accomplished by starting with three coins on either side of the balance. If it balances, compare the two coins that were not on the balance and the lighter one is counterfeit. If it does not balance, compare any two of the three coins on the light side. If they balance, the coin that was not weighed is counterfeit, and if it does not balance, the lighter coin is counterfeit.

Debriefing 13.8 Hopefully, the students will be surprised and impressed by the efficiency of the two-weighing solution. If they were confident that the minimum number of weighings was three, they did not thoroughly explore the sample space of the problem; they got an answer and assumed it was the best. School is the best place to make these types of mistakes – not when you are making important decisions involving large numbers of people or large amounts of capital. This is what Puzzle-based Learning is all about – developing problem-solving skills in a safe, friendly environment so the students are better prepared to make important decisions as contributing adults in the real world.

Problem 13.9 The proprietor of a rural farmer's market would like to be able to weigh out any integer amount of grain from 1 to 40 pounds in only one weighing using a two-pan balance and a set of standard weights. What is the minimum number of standard weights that will accomplish this and what are their values?

Strategies Utilized Understand the problem. Recognize a pattern. Perform a gedanken.

Discussion 13.9 Understanding the problem is an important first step. The students might test a few numbers to wrap their head around the problem by trying to fill an order for, say, 28 pounds of grain. Some possibilities include a 20-pound weight and an 8-pound weight on one side of the scale and the grain on the other.

13 Logical Reasoning

After trying several things, the students might adopt the pattern of coin values (1 cent, 5 cents, 10 cents, and 25 cents). For example, they could make four 1-pound weights, a 5-pound weight, a couple of 10-pound weights, and a 25-pound weight. This is a total of eight weights and it will do the job. However, it is not optimal.

One breakthrough that the students need to have to get the minimum number of weights needed is to realize that the weights can be placed on *both* sides of the scale. So, to weigh out one pound, a 3-pound weight can be on one side and a 2-pound weight and the grain can be on the other side.

Perhaps the best way to solve this is to start with a 1-pound weight. The next weight that is needed to prevent skipping a number is a 3-pound weight. These two standard weights will measure all the integer weights from one to four pounds.

To get five pounds, we can put the 1-pound and the 3-pound weight on one side of the scale and a 9-pound weight on the other. Since we can get integer weights from one to four pounds with the 1-pound and 3-pound weights, we can get all the integer weights up to nine pounds. So, a 1-, 3-, and 9-pound weight will measure all integer weights from 1 to 13 pounds in only one weighing. To measure 14 pounds of grain, we can use a 27-pound weight on one the other side of the scale and the three other weights with 14 pounds of grain on the other side. And we're done. The three smaller weights can be rearranged on the opposite side of the scale as the 27-pound weight to get all the integer weights from 14 to 26 and above 27; they can be stepped up similarly on the side with the 27-pound weight.

So, only *four* standard weights are needed and their values are 1, 3, 9, and 27 pounds.

If you are giving a test or exam, a reasonable five-part problem might be something like, using only 1-, 3-, 9-, and 27-pound weights and a balance, show how to measure the following amounts of grain in only one weighing:

(a) 2 pounds
(b) 8 pounds
(c) 11 pounds
(d) 17 pounds
(e) 32 pounds

Debriefing 13.9 This is a nice problem to give as part of a problem-solving competition because it has degrees of success. We have scored the results based upon the number of weights needed in five-point increments, starting at 5 points for a total of 8 weights needed, 10 points for seven weights, 15 points for six weights, etc. So, if a team can get all the integer weights from 1 to 40 with only five weights, it is awarded 20 points. We have given this problem at many competitions and no team has ever gotten the minimum of four weights in the allotted time.

Problem 13.10 The ten-digit "autobiographical" number is an integer between 1,000,000,000 and 9,999,999,999. It gives the number of times each digit appears in it in the following order: 0, 9, 8, 7, 6, 5, 4, 3, 2, and 1. The reason why it is called the autobiographical number is that it "tells its own story" by revealing how many of each digit appears in it. That is, the leftmost digit – the one in the billions column – gives the quantity of zeros in the number itself. The next digit, which is in the 100-millions column, gives the quantity of nines in the number itself. The digit after that gives the quantity of eights, etc. The rightmost digit, which is in the ones column, gives the quantity of 1s in the number. What is the *unique*, 10-digit, autobiographical number?

#0s	#9s	#8s	#7s	#6s	#5s	#4s	#3s	#2s	#1s

Strategies Utilized Understand the problem. Perform a gedanken. Increment and iterate. Simplify.

Discussion 13.10 This is a neat, multiple-step problem that requires an upfront investment for the student to understand exactly what is being asked. If the students do not yet understand that the teacher in a Puzzle-based Learning course is not an answer provider but a cognitive trainer, hands will shoot up after about a minute. Students will say, "*I don't understand what I'm supposed to do*," and they might ask, "*Can you explain the problem? I don't get it.*"

We recommend responding, "*All the information you need is there. If you don't understand it, read it again, this time more carefully and more slowly. If you still don't understand it, read it again.*" As we advised earlier, "Don't Blink." That is, get the students to think for themselves. You might consider advising them to guess a number and then check to see if it is right, but they should be doing this by themselves at this point in the course. It is natural for the students to test the teacher to see how much help they can get. As the students' cognitive trainer, your goal is to get the students to think for themselves.

If they are not making progress and are reaching the point of frustration, you might consider prompting them to tackle a simpler version of the problem by finding the four-digit autobiographical number by putting digits on these four blanks.

#0s	#3s	#2s	#1s

Here the answer is 1,012 (one 0, no 3s, one 2, and two 1s).

After the students understand what is being asked, the natural first try is 9,000,000,000. That is, they put a "9" above the zero's blank and then fill in the rest of the blanks with nine zeros. This is a great start and it demonstrates an understanding of the problem. However, it is not correct, because there is a zero above the nine's blank and there is one nine in the number.

13 Logical Reasoning 247

> **Teacher Tip**
> Many students want to be the first one to solve the problem, so they will raise their hand or shout, "*I got it!*" without checking their answer thoroughly. Since this is a very easy problem to check, just tell the students, "*To check your answer once you have filled in all the blanks, simply look at the digit above the zero's blank and count the number of zeros in your number. If they agree, look at the digit above the nine's blank and count the number of nines in the number. Continue down to the one's blank. If they are all correct, you have the right answer.*"

The next step is usually to put an 8 above the zero's blank and a 1 above the eight's blank. The problem now is that if a 1 is put above the one's blank, then there are two 1s in the number. Eventually, the student will iterate and increment down to the unique answer of 6,000,100,012.

Debriefing 13.10 This problem does not need any special math skills. It just needs persistence and logical reasoning. For example, "*If I put a one here, I have to put a two in the ones column, but then I have a two, so...*" We have presented this problem in classes ranging from fourth grade through graduate school and at teaching conferences and at problem-solving clubs. We have noticed no strong correlation between the time needed to solve the problem and the venue. Some fourth graders can bust it out in less than ten minutes, and some PhD students will struggle. Finally, in the framing process, a student might notice that the digits in the number must add up to ten because there are ten digits in the number.

Problem 13.11 Three college students, Alan, Bob, and Chris, get drunk at a party and fall asleep. Their friends take this opportunity, as friends do, to prank them. This particular prank involves writing with a permanent marker all over their faces.

The next morning, they all wake up simultaneously. Each, upon seeing the faces of the other two, breaks out in laughter. After a minute, Chris realizes, based on the continued laughter of Alan and Bob, that his face must be covered with marker as well.

Describe Chris' *logical* thought process that leads him to this conclusion.

Strategies Utilized Perform a gedanken.

Discussion 13.11 The gedanken that Chris performed is to wonder, "*What if my face did not have ink on it?*" A reasonable answer would be, "*Well, if my face did not have ink on it, Bob would be looking at only one face with ink on it; Alan's. So, if my face did not have ink on it, Bob would wonder what Alan is laughing at.*" That is, Chris would think, "*If my face did not have marker on it, Alan would think that Bob is not looking at any faces with marker.*" Since Bob is still laughing, I must be a victim of the prank as well.

Debriefing 13.11 This problem involves the valuable skill of wondering what other people are thinking. This skill is applicable in personal relationships and at business meetings. Many people are focused too much on themselves. Solving problems like this one can help them expand their view and begin to empathize with others.

Problem 13.12 The sum of the monkey's age and her mother's age is 12 years. In addition, the monkey's mother is twice as old as the monkey was when the monkey's mother was half as old as the monkey will be when the monkey is three times as old as the monkey's mother was when the monkey's mother was three times as old as the monkey. What are the ages of the monkeys?

Strategies Utilized Understand the problem. Simplify. Reason backwards.

Discussion 13.12 This problem will likely require a few readings for the students to wrap their heads around it. We like to present it in two parts, perhaps by using a slide show. The first part is, "The sum of the monkey's age and her mother's age is 12 years." When shown just this part of the problem, the students are feeling pretty confident because they have seen problems like this before. In fact, they might write down $x + y = 12$ on their paper. Their confidence is short lived, however, when they are shown the second part of the problem.

If students read through the problem a few times, they might simplify it by changing the names of the two monkeys – the alliteration of monkey and monkey's mother can be distracting. Perhaps this is better: "The sum of the kid and her mother's age is 12 years. In addition, the mother is twice as old as the kid was when the mother was half as old as the kid will be when the kid is three times as old as the mother was when the mother was three times as old as the kid. What are the ages of the two monkeys?"

While this one can be solved with a mathematical equation and it can be solved using the increment and iterate technique, it yields nicely to the reason backward technique.

The last sentence fragment is: "...when the mother was three times as old as the kid." Let's define the kid's age at this point in the past to be x. Thus, his mother's age at this time is $3x$. So, the difference between their ages is $2x$ and it always is $2x$. This is an important concept to grasp. The *factor* by which the mother is older than the kid changes with time, but the *difference* in their ages does not.

We can now replace the fragment, "...when the kid is three times as old as the mother was when the mother was three times as old as the kid," with the simpler statement, "...when the kid is $9x$." So now we have, "The sum of the kid's age and her mother's age is 12 years. In addition, the mother is twice as old as the kid was when the mother was half as old as the kid when the kid is $9x$." Notice how much less complex the problem is already.

13 Logical Reasoning

This quickly reduces to, "The sum of the kid's age and her mother's age is 12 years. In addition, the mother is twice as old as the kid was when the mother was $9x/2$."

Recalling that the mother is $2x$ older than the kid, we know that the kid is $5x/2$ when the mother is $9x/2$.

Now we have, "The sum of the kid's age and her mother's age is 12 years. In addition, the monkey's mother is twice as old as $5x/2$."

Which finally gets us, "The sum of the kid's age and the mother's age is 12 years. In addition, the monkey's mother is $5x$."

When the mother is $5x$, the kid is $3x$.

We can now, finally, use the fact that their ages sum to 12:

$$5x + 3x = 12$$

The solution to this equation is $x = 1.5$ which means that the monkey is 4.5 years old and the mother is 7.5 years old.

Debriefing 13.12 This problem may cause the students' eyes to glaze over upon their first reading. In one class, the problem was presented as a handout to solve individually. One student read the problem, promptly turned the paper facedown on his desk, slouched in his chair, and crossed his arms, thus signaling that he clearly was not going to "waste" any time thinking about the problem. While the problem does not have any obvious direct application, it does exercise the mind. It's completely analogous to running on a treadmill for one hour. The purpose of running on a treadmill is not to get anywhere; it is for physical exercise. The purpose of working on the monkey problem is not to have the answer; it is for mental exercise. Who cares what the answer is? It is 4.5 and 7.5. This is not interesting. This has no value. The purpose of working on the problem is to develop mental strength and mental stamina.

For a harder version of this problem (where there is also a rope, pulley, weight, and the monkey's brother), see Chapter XIII of *How to Solve It: Modern Heuristics* by Z. Michalewicz and D.B. Fogel (Springer, 2004).

Problem 13.13 You are in a team-of-three intellectual competition. There are ten events. In one of the events, you are isolated from your two teammates, and a black or white hat, chosen randomly, is placed on your head. Then you are briefly taken into a room where your other two team members are – each with a hat on his/her head. You can see the "colors" of their hats but you do not know the color of your own hat. After this brief encounter, you are now separated from your teammates again and told the nature of the challenge. Each of the three members of the team has a hat that was chosen randomly from black and white via a coin toss. At this point, each team member – still isolated – is asked if they want to guess the color of their hat, choosing one of three responses. They are:

(a) I decline to guess.
(b) I guess my hat is black.
(c) I guess my hat is white.

Your team will receive full marks for this event only if at least one person guesses right *and* everyone who guesses is correct. That is, if no one on your team chooses to guess, your team receives zero points and if anyone guesses wrong, your team receives zero points. Of course, your team members, like you, are very intelligent and will all work out the optimal strategy independently. What is this correct strategy and what is the probability of your team succeeding in this event if each person uses the optimal strategy?

Strategies Utilized Understand the problem. Enumeration. Perform a gedanken. Draw a diagram.

Discussion 13.13 If there was a designated team captain before the event began, the captain might take the initiative and choose to guess and the other members might decline to guess. This will give the team a 50 % chance of success. However, a strategy can be developed that is *better* than this.

A quick inventory of the facts of the problem should reveal that anyone that guesses has a 50 % chance of being correct. There is no strategy that can change this. The hat colors were chosen independently of each other so the other two hat colors that each team member sees can't possibility provide any information about the color of the hat on their head.

A path to the discovery of this strategy starts with a framing of the problem and a thorough enumeration of the possibilities. Certainly, you would like to avoid the possibility of each team member declining to guess. Also, it seems reasonable to try to avoid the situation in which there are multiple guesses.

Without any brilliant ideas, a team member might further frame the problem by drawing a diagram of the possibilities, as shown below:

Eight possible distributions of hat colors		
Member 1	Member 2	Member 3
Black	Black	Black
White	Black	Black
Black	White	Black
Black	Black	White
White	White	Black
White	Black	White
Black	White	White
White	White	White

Each of these eight possibilities is equally likely and each has a probability of 1/8 of occurring.

13 Logical Reasoning

With this diagram, the student can perform a gedanken and ask, *"What if each team member declined to guess if they saw one of each color hat and guessed the other color if they saw two of the same color?"* Another table shows the result:

Hat colors			Guesses			
Member 1	Member 2	Member 3	Member 1	Member 2	Member 3	Correct?
Black	Black	Black	White	White	White	No, no, no
White	Black	Black	White	None	None	Yes
Black	White	Black	None	White	None	Yes
Black	Black	White	None	None	White	Yes
White	White	Black	None	None	Black	Yes
White	Black	White	None	White	None	Yes
Black	White	White	Black	None	White	Yes
White	White	White	Black	Black	Black	No, no, no

This strategy is successful a whopping 75 % of the time. Anytime the hats are two of one color and one of the other, the strategy is successful. The students should realize that the strategy did not change the possibility of any individual guess being correct. The final column in the table reveals that there were six correct answers and six incorrect answers. The strategy only grouped the six incorrect answers to maximize the likelihood of the team being successful.

Debriefing 13.13 If the students understand that there is no way to make any individual guess better than 50 %, it will be very hard for them to believe that there is a strategy that will produce better than a 50 % success rate and that the strategy need not be discussed beforehand. The fact that such a strategy exists will impress upon them that first impressions are not always reliable and that a thorough enumeration of the possibilities and a series of clever gedankens can lead to surprising results. If students are interested in probing this matter further, you can try it with 7 or 15 hats and teach them about error-correcting Hamming codes. This would also be a good project for the advanced student. We have done the seven-hat problem with university students, and the students were successful at working out who is supposed to decline to guess, who is supposed to guess, and what color they are supposed to guess. With seven randomly chosen black or white hats and seven clever students, they can succeed 7/8 of the time – more often than with three hats!

Problem 13.14 Tell the students that the next time the class meets, you will have five hats to place on five students. The hats are chosen from these five colors: white, red, yellow, green, and purple (of course, use any color hats that you have access to, and if there are no hats, use colored stickers on their foreheads). They can be *any* combination of these colors. There might be two reds, two yellows, and a green; there might be one of each color; or they might all be red. The five students can see the color of the other hats but not his/her own hat color. Each of the five students must guess the color of their hat. All that is required is *at least one* of the five to get the right answer. If everyone guesses wrong, the group fails the exercise. During the time before the next class meeting, the students can devise a guessing strategy based on the other hats that they see. What strategy will *guarantee* that someone gets the right answer?

Here, unlike the previous problem, the members of the team know the goal of the exercise before they see the hat colors on their teammates and thus can plan a guessing strategy in advance.

Strategies Utilized Take inventory. Perform a gedanken. Simplify.

Discussion 13.14 An inventory of the facts might include:
1. There are five possible hat colors and five students.
2. A hat will be placed on each student.
3. There may be multiple hats of the same color.
4. At least one student must guess right.
5. The students can discuss a guessing strategy before the event.

The students will likely invest time framing the problem by calculating the probability of at least one student getting it right if they all guess randomly. They might even calculate the number of possible combinations of hat colors or the probability that all the hat colors are different if they are chosen randomly. They might calculate the probability of everyone guessing wrong if they all guessed red irrespective of what hat colors they see. They might even try to use psychology to determine if you are the type of person to have multiple hats of the same color or have one of each color.

Once the problem is understood, the shortest path to the answer is through simplification. With five students and five hat colors, this is a very challenging problem. With two students and two hat colors, it is more reasonable to solve.

Let's consider two students and two hat colors: red and white. There are four possible combinations, RR, RW, WR, and WW. The goal is to ensure that at least one student guesses right for all of these combinations. What strategy is available?

It seems reasonable to conclude that the guess must be based on the color of the hat on the other student. Any guessing strategy that does not depend on the hat color of the other student will have an average success rate of 75 %, with success being at least one student guessing correctly. If your students do not come to this conclusion, perhaps you can test their probability skills by asking, "what is the probability that at least one person is correct if both guess randomly?"

The four possibilities for the two hat colors (RR, RW, WR, and WW) can be grouped into two states – the hats are the same color and the hats are different colors. Of course, neither person knows which is the case, but if one person assumes that the hats are the same and the other assumes that they are different, one of them must be correct.

Let's assume that student 1 is assigned to assume that the hats are the same color and student 2 is assigned to assume that the hats are different colors. We present the results in the following table for all possible hat color distributions:

13 Logical Reasoning

Student 1	Student 2	1's guess	2's guess	Success?
Red	Red	**Red**	White	Yes
Red	White	White	**White**	Yes
White	Red	Red	**Red**	Yes
White	White	**White**	Red	Yes

Correct guesses are shown in bold. With this strategy, it is guaranteed that one student will get it right and the other will be wrong. So, we have a strategy that works for two hat colors and two people. The problem now is extending this to three people and three hat colors (red, white, and pink).

What if one student sees two different hat colors? What if one student sees two red hats? What if one student sees two white hats? The transition from two hats to three hats is a significant one. What would happen if all three hats were pink? Would everyone guess pink? What if there were two whites and a red? A quick gedanken will reveal, as with only two people and two hat colors, that each student must have a different strategy.

The key is to have every person make a different assumption about the "state" of all the hats and have the three assumptions cover all the possibilities. With two people, the different assumptions were: the hats were the same color and the hats were different colors. With three possible hat colors, something else is needed to define the state of the hats.

> **Teacher Tip**
> The answer is not trivial and we strongly recommend not revealing it to the students. Trying to solve a problem like this is exactly the exercise that will develop problem-solving skills. Yes, it is challenging, but this is what is needed.
>
> Earlier in this book we have made the analogy between developing physical strength and developing mental strength. Imagine a football coach going into the weight room and finding a player struggling to bench press 200 pounds. It would not make sense for the coach to remark, *"You know, it's a lot easier if you take a couple of those weights off the bar."* When developing physical strength, it is important to perform challenging exercises. The same is true when developing mental strength. Just as the football coach would not help the player lift the weight, the teacher should not help the student solve this problem. If the problem goes unsolved, that's OK.

So, the goal is to get each one of the students to make a different assumption about the state of the hats in such a manner that one of them must be correct. The way to do this must include every possible hat distribution. If you are challenging the students with this one, they should have no problem enumerating the 27 possible distributions of the hats among the three students. These are given in the table:

Number	Student 1	Student 2	Student 3
1	Red	Red	Red
2	Red	Red	Pink
3	Red	Pink	Red
4	Pink	Red	Red
5	Red	Pink	Pink
6	Pink	Red	Pink
7	Pink	Pink	Red
8	Red	Red	White
9	Red	White	Red
10	White	Red	Red
11	Pink	Pink	Pink
12	Red	White	Pink
13	White	Red	Pink
14	White	Pink	Red
15	Red	Pink	White
16	Pink	Red	White
17	Pink	White	Red
18	Red	White	White
19	White	Red	White
20	White	White	Red
21	White	Pink	Pink
22	Pink	White	Pink
23	Pink	Pink	White
24	White	White	Pink
25	White	Pink	White
26	Pink	White	White
27	White	White	White

The key now is figuring out a way to "assign" each of the 27 possibilities in a manner that ensures one of the students guesses correctly. To get to the solution, the students should explore the sample space by trying numerous possibilities. Even though they probably will be flawed, the character of the flaws will allow them to progress towards a solution. In a brainstorming session, students might have the following proposals:

Proposal 1:
They all guess red.

This only works if there is at least one red hat. The probability of there being no red hats is $(2/3)^3$ which is 8/27. Therefore this strategy will work 19/27 of the time. By looking at the table of possibilities, it is easy to find the eight distributions in which this strategy will not work.

Proposal 2:
Student 1 guesses red.

13 Logical Reasoning

Student 2 guesses pink.
Student 3 guesses white.

This strategy will not work because there are distributions in which 1 is not red, 2 is not pink, and 3 is not white. The probability that 1 is not red is 2/3, the probability that 2 is not pink is 2/3, and the probability that 3 is not white is 2/3. Therefore, the probability that this strategy is successful is 19/27.

Proposal 3:
Student 1 assumes that all three hats are the same color.
Student 2 assumes that the hats are one of each color.
Student 3 assumes that the hats are two of one color and one of the other.

It doesn't appear that this will work because student 3 has two possibilities if he sees two different color hats. That is, if student 3 sees a red and a white, does he guess red or white?

When there were only two students and two hat colors, the solution was for one student to assume that the hats were the same color and for the other student to assume that the hats were different colors. With three or more students, the solution is not clear.

Let's define a new parameter to characterize the distribution of the three hats – "redness." Each of the 27 possible hat distributions will be assigned a redness value. The color white will be assigned a redness of zero, the color pink will be assigned a redness of 1, and the color red will be assigned a redness of 2.

With this strategy, the redness values for each of the 27 possible distributions of hat color will be assigned as follows:

Number	Student 1	Student 2	Student 3	Redness
1	Red	Red	Red	6
2	Red	Red	Pink	5
3	Red	Pink	Red	5
4	Pink	Red	Red	5
5	Red	Pink	Pink	4
6	Pink	Red	Pink	4
7	Pink	Pink	Red	4
8	Red	Red	White	4
9	Red	White	Red	4
10	White	Red	Red	4
11	Pink	Pink	Pink	3
12	Red	White	Pink	3
13	White	Red	Pink	3
14	White	Pink	Red	3
15	Red	Pink	White	3
16	Pink	Red	White	3
17	Pink	White	Red	3
18	Red	White	White	2
19	White	Red	White	2

(continued)

Number	Student 1	Student 2	Student 3	Redness
20	White	White	Red	2
21	White	Pink	Pink	2
22	Pink	White	Pink	2
23	Pink	Pink	White	2
24	White	White	Pink	1
25	White	Pink	White	1
26	Pink	White	White	1
27	White	White	White	0

We can see that the hat distributions can be characterized by a redness value from 0 to 6. It is also apparent that if any student is looking at, say, two pink hats, they are certain that the redness of the three hats is either 2 (they have a white hat), 3 (they have a pink hat), or 4 (they have a red hat). That is, if any student is looking at two pink hats, the redness value of the three hats can't be 0, 1, 5, or 6. In fact, any combination of two hats results in three possible values of redness for all three hats.

All that remains is to assign one of the three possible redness values to each of the three students. Here is one solution:

Student 1 assumes that the total redness of the three hats is either 0, 3, or 6.
Student 2 assumes that the total redness of the three hats is either 1 or 4.
Student 3 assumes that the total redness of the three hats is either 2 or 5.

Since only one redness value can exist and each student assumes a different possible value of the redness, exactly one student must be correct for each of the 27 hat distributions. This is shown in the diagram below.

Student 1	Student 2	Student 3	Guess 1	Guess 2	Guess 3
Red	Red	Red	**Red**	White	Pink
Red	Red	Pink	White	Pink	**Pink**
Red	Pink	Red	White	White	**Red**
Pink	Red	Red	Red	Pink	**Red**
Red	Pink	Pink	Pink	**Pink**	Red
Pink	Red	Pink	White	**Red**	Red
Pink	Pink	Red	White	**Pink**	White
Red	Red	White	Pink	**Red**	Pink
Red	White	Red	Pink	**White**	White
White	Red	Red	Red	**Red**	White
Pink	Pink	Pink	**Pink**	Red	White
Red	White	Pink	**Red**	Pink	White
White	Red	Pink	**White**	White	White
White	Pink	Red	**White**	Red	Pink
Red	Pink	White	**Red**	Red	Red
Pink	Red	White	**Pink**	White	Red
Pink	White	Red	**Pink**	Pink	Pink
Red	White	White	White	Red	**White**

(continued)

13 Logical Reasoning

Student 1	Student 2	Student 3	Guess 1	Guess 2	Guess 3
White	Red	White	Pink	Pink	**White**
White	White	Red	Pink	Red	**Red**
White	Pink	Pink	Pink	White	**Pink**
Pink	White	Pink	Red	Red	**Pink**
Pink	Pink	White	Red	White	**White**
White	White	Pink	Red	**White**	Red
White	Pink	White	Red	**Pink**	Pink
Pink	White	White	White	**White**	Pink
White	White	White	**White**	Pink	Red

By applying pattern recognition skills, it is evident why this works.

Student 1 is assigned to assume that the hats are either all the same color or one of each color. This covers nine of the 27 possibilities.

Student 2 is assigned to cover the three color combinations of two pinks and a red, two reds and a white, and two whites and a pink. This covers nine of the 27 possibilities.

Student 3 is assigned to cover the three color combinations of two pinks and a white, two reds and a pink, and two whites and a red. This covers nine of the 27 possibilities.

Since one of the 27 possibilities must be used and all 27 are covered by either student 1, 2, or 3, one and only one of them must guess his/her hat color correctly.

> **Teacher Tip**
> If the students are interested, you might want to take this opportunity to explain modulo arithmetic to them.

Discussion 13.14 (cont) Let's look at a particular example. What if the hats are distributed as follows?

Student 1	Student 2	Student 3
Red	White	White

Student 1 sees a white and a white and counts this as a redness of zero. Student 1 is assigned the total numbers of 0, 3, and 6 and guesses "White" to keep the redness value at zero. Student 2 sees a red and a white, which has a redness value of two. Student 2 is assigned the redness value of 1 and 4, so student 2 guesses "Red" to get the redness value to four. Finally, student 3 sees a red and a white, which has a redness value of two. Student 3 is assigned the redness values of 2 and 5, so student 3 guesses "White" to keep the redness value at 2. So, student 3 is the only one that is correct.

There are 27 different possibilities for the distribution of hat colors and the students should be allowed to check as many of these as they like before they are convinced that the strategy works.

Of course, we still did not tackle the problem as stated. However, going from 3 students to 5 students is a small step. Assign "violetness" numbers 1–5 to the hat colors of white, red, yellow, green, and purple and assign a series of violetness values to each participant that are spaced five apart. Student 1 is assigned the violetness totals of 0, 5, 10, 15, and 20. Student 2 gets 1, 6, 11, and 16. Student 3 gets 2, 7, 12, and 17, and so on.

It would be a good exercise for the students to go through a couple of these. If convenient, split the class into groups of three and assign each group a different hat color combination.

> **Teacher Tip**
> In the next class period following this exercise, actually do it with five students and five hats after letting the group assign numbers to both the hat colors and the participants. When we have done this, the students are *so* impressed that they can get it right they want to do it more than once – and the students in the audience want to try it as well. For this reason, it is best to have a few possible hat combinations available. Once, the students pulled this off for a parents' night demonstration. The parents were brought up on the stage to randomly put hats on each of five students – and they impressed the audience by getting exactly one guess right, as promised. Finally, it is important to use hats that are of obvious color (i.e., avoid using a hat that some people would call gray and others would call blue).

Debriefing 13.14 At first glance, it appears that this problem is impossible because the hats that can be seen by the participants contain *no* information about the color of the hat each participant is wearing. The procedure does not increase the likelihood of anyone guessing right. It only assures exactly one correct guess. That is, it prevents zero correct guesses and it prevents multiple correct guesses. This is a great example of the power of the simplification technique, as the solution to the problem involving five hats is much harder to see without going through the two- and three-hat solutions.

Geometric Reasoning

> *He has ability, but he is not a geometer, which, as you know,
> is a great defect.* — Blaise Pascal

The word geometry comes from the ancient Greek *geo*, meaning earth, and *metron*, meaning measurement. So, geometry originally meant measuring the earth. Today geometry has expanded to include the study of two- and three-dimensional shapes as well as how multiple three-dimensional shapes are connected and how multiple two-dimensional shapes will *tessellate*. Tessellation is covering a surface with shapes so that there are no gaps or overlapping, usually where we only have one shape. A mosaic floor is an example of a tessellated surface. (If you're interested in reading more about tessellations, you can look at the works of Roger Penrose.)

There are two reasons why a chapter on geometry problems is included in this book. The first is that the field of geometry offers a bounty of simply stated yet very challenging problems. The second is that problem-solving often involves spatial reasoning. The ability to perform qualitative and quantitative work on two-dimensional shapes and three-dimensional objects is important in a variety of fields. People can have a lot of trouble seeing how shapes fit together, even when they can physically manipulate them, and this provides a rich vein of physical and mental puzzles for the teacher to use.

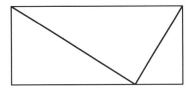

The problems presented here often have multiple paths to the solution, some of which are tortuous while others are clever. Many formal proofs in geometry can be reduced to a simple clever diagram. These are called *look-see* proofs. For example, in the diagram, which has a larger area: the large triangle or the sum of the two small triangles? It is not at all obvious just by looking at the diagram as presented, but the addition of a single line makes it clear that the area of the large triangle must be the same as the sum of the areas of the two smaller triangles, by symmetry.

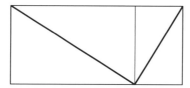

In this chapter we present a wide range of problems with a wide range of difficulty. All, however, are designed to develop the students' abilities to effectively attack new problems.

Students can become frustrated if they can't find a solution to a geometric problem, especially when they are handling a physical representation of the problem. Quite often this is because they are not really taking inventory and keeping track of what they have already tried. However, we have to accept that any record-keeping system for moves and attempts in a three-dimensional system is going to be quite complex – much as solving a Rubik's cube effectively requires us to start with an unambiguous representation of the way that the cube is set up and how we can change the position of the cube elements to move closer to a solution. When students aren't keeping track of what they are doing, they can repeat moves (either exactly or through accidental symmetry) and feel that they are getting nowhere. Persistence and an understanding of System 2 thinking can be crucial to develop when introducing these sorts of problems.

The first set of problems in this chapter will involve the square polyominoes. You would have seen dominoes as part of a game before – mathematically, a domino is a rectangle made by joining two squares edge to edge. A polyomino is any two-dimensional shape that can be made with any number of squares that are connected along at least one edge. There is only one possible shape for a two-square polyomino – the 2-by-1 domino. There are two possible shapes for a triomino (three squares) and five possible shapes for a tetromino (from the Greek prefix *tetra-*, meaning four). It is a nice challenge for the students to find all five independently by sketching them on a piece of paper. This actually makes a good warm-up for the first problem in this chapter, as it will get the students thinking about reflective and rotational symmetry. The five tetrominoes are shown here:

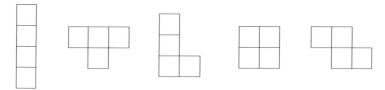

For reference, these five tetrominoes have been given names. From left to right, they are I, T, L, O, and Z.

14 Geometric Reasoning

Problem 14.1 Sketch all the possible different pentominoes (from the Greek prefix *penta*, meaning "five").

Strategies Utilized Enumerate the possibilities. Draw a diagram.

Discussion 14.1 One useful strategy is to start with the set of tetrominoes and add single squares at various positions on this set to produce the entire set of twelve pentominoes. The complete set of twelve pentominoes is given below:

For reference, these have been given names. They are I, L, Y, N, W, S, X, U, T, F, P, and V from top left to bottom right.

> **Teacher Tip**
> Start off this problem without telling the students how many pentominoes there are. This will give them practice being thorough and organized when solving a problem that requires a complete enumeration of all the possibilities, much like the bracelet with black and white beads (see Problem 5.7). There will likely be a wide range of times needed to find the complete set. We recommend not helping those students who are struggling to find the last one or two. This will help them to practice thinking under pressure and to develop mental stamina as well as to learn not to give up. This will also increase the feeling of accomplishment when they finally get the complete set. To keep the students busy who are finished early, have other problems ready for them. (For example, there are 12 ways to connect six equilateral triangles along an edge. These shapes have been named hexiamonds. Sketch all 12 hexiamonds.) Remember, the primary purpose of this exercise is not to be able to find the pentominoes. It is to develop the students' problem-solving skills. Helping the students find the pentominoes is similar to running their three miles for them when they are trying to earn a position on a track team.

Debriefing 14.1 Students will typically get six to seven of the total relatively quickly, but the last two may take as long as the first ten. There are a number of questions you can ask them about their process when they have found all twelve pieces. Ask them what techniques they used – especially when they got stuck. Typically, students will gladly volunteer which piece they found last. If the students started with three squares in a row and then tried to place two more squares around them, they will have trouble getting the "W" pentomino, as it is the only one that does not have three squares in a row.

Problem 14.2 A complete pentomino set has 60 squares, twelve pieces each with five squares. It is a significant challenge to make rectangles that are 6-by-10, 5-by-12, 4-by-15, and 3-by-20 with a set of pentominoes. A complete tetromino set has a total of 20 squares, five pieces each with four squares. Construct a rectangle using all five tetrominoes, or develop an argument that it is not possible.

Strategies Utilized Enumerate the possibilities. Draw a diagram. Build a model. Perform a gedanken.

Discussion 14.2 This problem can be solved with an *Aha!* moment, that is, a flash of insight. This moment will come more quickly if the students build a model of the five pieces and try to produce a rectangle with them. They should first discover the "problem piece" and then discover why that particular piece is a problem.

> **Teacher Tip**
> This is a somewhat challenging proof. So, perhaps it is best utilized at something like a meeting of the math club or for extra credit rather than regular class time.

Discussion 14.2 (cont) It might help if the students performed this problem on a grid and if the pentomino grids are available, they could be used for this as well.

The key to the proof involves the fact that four of the five tetrominoes will cover two black squares and two red squares when placed over a standard checkerboard grid. One of the pieces covers three of one color and one of the other. Any rectangle with even number of squares always covers an equal number of black squares and red squares. So the presence of the "T" tetromino ensures that a complete set of five tetrominoes will always cover nine of one color and eleven of the other, making it impossible to produce any rectangle.

14 Geometric Reasoning

Teacher Tip
If students have any difficulties following these explanations, we can try a simpler problem. There is a regular chessboard with two opposite corners removed, as shown in the diagram. There are 31 dominoes. Each rectangular domino can cover two squares of the chessboard. Your challenge is to cover the chessboard with these dominoes by placing them horizontally or vertically.

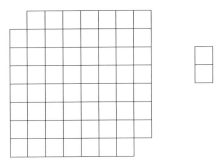

We can spend hours trying to arrange the dominoes on the board, but a simple observation can terminate this process very quickly. Note that if we color the squares of the board black and white in the usual manner, the board will have 30 squares of one color and 32 squares of another. The squares at opposite corners must be of the same color.

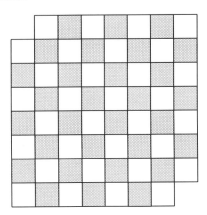

Since each domino always covers one white square and one black square, regardless of its placement on the board (and this is the invariant in this problem), it's impossible to cover this particular chessboard with these dominoes![1]

---

[1] There is a three-dimensional instance of this puzzle, where the task is to fill a $6 \times 6 \times 6$ box with bricks. Each brick is $1 \times 2 \times 4$. Can you do it?

Debriefing 14.2 This is a clever proof and, hopefully, the students appreciate its elegance and simplicity. If they do, they will be motivated to perform a gedanken when problem-solving.

The next problem requires sets of pentominoes (perhaps one set for each group of three students). These can be cut out of durable cardboard, but we recommend purchasing commercially available plastic sets which have one-inch squares or perhaps having the woodshop construct pentomino sets as a project.

Problem 14.3 Enclose the largest area with a set of 12 pentominoes. This problem can be called the Pentomino Pasture. The fence of the pasture is formed by the 12 pentominoes, and the fence must be at least one square unit thick all the way around (see the figure for an example).

Strategies Utilized Understand the problem. Enumerate the possibilities. Increment and iterate. Draw a diagram. Perform a gedanken.

> **Teacher Tip**
> To better define the problem, it is helpful to have a grid upon which the students can build their pasture. For a pentomino set with one-inch squares, we recommend a 16-by-16 grid o one-inch squares with the middle 100 squares grayed out. This will make counting the area enclosed much easier and the students can easily align the pentomino pieces with the grid. It is relatively simple to construct a square grid with a computer and then have a printer enlarge the grid so that the squares are one inch on a side. It is also useful to have a border of 3 inches all the way around the 16"-by-16" grid. A typical Pentomino Pasture grid is shown below with the pentomino set arranged to enclosed 120 square units:

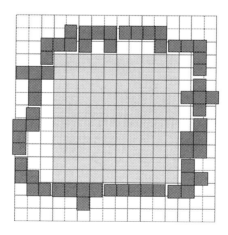

14 Geometric Reasoning   265

Discussion 14.3 This is a nice problem for a number of reasons. First of all, it is a challenge just to circumscribe the middle 100 squares. Second, students do not need to memorize formulas to intelligently tackle this problem. Third, this problem tends to level the playing field as far as previous training in mathematics is concerned; often, the students who are not "straight A" perform the best at this exercise. Fourth, the maximum of 128 is very difficult to obtain. When the students are trying to maximize the area, keep track of the current high value and write it on the board as a target for others to aspire to. If a team beats the current high, first inspect it to ensure that it meets the qualifications and then erase the old high and put the new one in its place. If any team gets to 125 in a normal class period, they should be congratulated. We have been presenting the Pentomino Pasture exercise in classes, workshops, problem-solving competitions, and on university "open days" for several years now. Once every couple of years, an individual or team circumscribes the maximum of 128 squares. Their accomplishment is immortalized by tracing the positions of the pentominoes on the grid and then signing and dating it with a marker. These grids are "retired" and then placed in a large artist's portfolio case that we call the Pentomino Pasture Hall of Fame.

> **Teacher Tip**
> Often, a person or group wants to continue to work on the pasture after the class period is over. Some teachers leave a pentomino set available for students to work on before school, after school, and/or during lunch periods. Or, if appropriate, roll up a grid and allow the student to take a set of pentominoes home. As a teacher, it is very satisfying to get a text message at 2:00 am with a picture of the pentominoes circumscribing 128 square inches. It's an accomplishment that the student(s) will remember for a long time.

Debriefing 14.3 It is interesting to discuss the students' strategies after they are finished tackling the Pentomino Pasture Problem. It is best to do this when they have their best Pentomino Pasture in front of them. To get the discussion going, ask them what the worst piece was. They should agree that it is the x-piece, which is also known as the plus sign. Next, you can ask them which of the pentominoes can be used to cover the diagonally opposite corners of a 3-by-3 grid. Once they identify the three that can (the S, V, and W pieces), ask them where they are best used on the fence. They should agree that they belong in the corners.

This is a classic example of an iterate and increment problem and it involves pattern recognition and optimization skills as well.

Problem 14.4 Arrange a complete set of the twelve pentominoes in groups of three to form these four shapes:

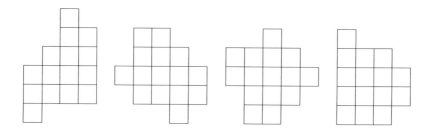

Strategies Utilized Enumerate the possibilities. Recognize a pattern.

Discussion 14.4 There are a lot of different ways that the pentominoes can be combined. In fact, the students will find that there is more than one way to make each of the four shapes. However, to make all four simultaneously with one complete set is a significant challenge, as the problem involves two steps, grouping the pentominoes correctly and then constructing the shapes from each group. A solution is shown below:

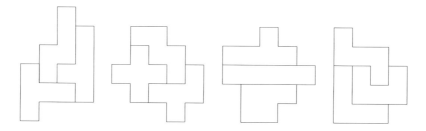

> **Teacher Tip**
> This problem also offers an opportunity for the students to devise their own pentomino-grouping puzzle. For example, two groups can compete with each other by secretly arranging their set of 12 pentominoes into four groups with three pieces each and then tracing the outlines of the four shapes. The outlines are then swapped and each team tries to produce the four shapes given to them with a set of pentominoes. Depending on the length of the class period and the size of the class, you might consider holding a single-elimination tournament to declare a champion.

Debriefing 14.4 There are *a lot* more things you can do with a set of pentominoes and the Internet is a great resource. It is a significant intellectual challenge to stare at a set of shapes and figure out how they would combine. Actually working with the shapes in a group will develop consensus-building skills and the ability to work efficiently as a team.

14 Geometric Reasoning

Problem 14.5 Assume that you are a long distance away from the US Pentagon building, at some randomly selected position, and are looking at it through a high-powered telescope. What is the probability that you see only two of the five walls?

(Unsurprisingly, the US Pentagon is a building constructed in the shape of a regular pentagon, that is, a building with five sides of the same length, all joined together at the same angle.)

Strategies Utilized Understand the problem. Draw a diagram. Build a model. Perform a gedanken.

Discussion 14.5 The first thing to do when trying to solve any problem is to try to understand what the problem is asking. For this problem it is important to understand what is meant by a "long distance away." In this context, a long distance implies that the distance from the pentagon is much bigger than the pentagon itself.

Next, the solver can perform a gedanken and ask *"Is it ever possible to see three sides of the pentagon when viewing it from the side?"* *"How close do you have to be to the pentagon to see only one side?"* and *"If I start next to a wall facing it and walk backwards, how far do I have to go before another side comes into view?"*

These questions can be answered more easily by building a model or drawing a diagram.

> **Student Pitfall**
> The students may see no obvious way to build a simple physical model of a pentagon. But there are a lot of ways to do this with the materials in an ordinary classroom. Don't tell them what to do, just allow them the freedom to explore. Building a model can be as simple as balancing five whiteboard markers vertically on their ends to represent the five corners of a pentagon on the top of a desk and then walking around it at various distances. Another modeling strategy we have seen consists of balancing playing cards or flashcards to form the walls of the pentagon. The angles don't have to be exact and the model doesn't have to be perfect. It is just to aid the students to better wrap their head around the problem so they can attack it more efficiently.

Discussion 14.5 (cont) With the problem understood, one path to the answer is by drawing the pentagon from the top and extending the five sides in both directions. It is best to use a relatively small pentagon as in the figure:

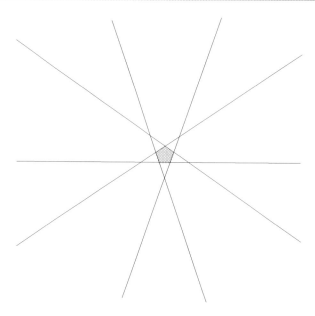

Once the figure is drawn, it is useful to identify the areas in which one, two, and three sides of the pentagon can be seen. Only one side of the pentagon can be seen from within the five small triangles that form a five-pointed star with the pentagon. Beyond this distance, there are ten areas and the number of pentagon sides that can be seen within these areas alternates from 2 to 3 going around the pentagon.

The apex angle of these ten areas is the same, as they are all formed by the tip of a five-pointed star. With a little geometry, it can be shown that the angle of the tip of a five-pointed star is 36 degrees, although knowledge of this angle is not necessary to get the solution.

So, very far from the pentagon, there are five areas in which two sides can be seen and each of these spans an angle of 36 degrees and there are five areas in which three sides can be seen and each of these areas spans an angle of 36 degrees as well. Therefore, if you are very far from the pentagon, you are just as likely to see only two sides as you are to see three sides. More precisely, the farther you are, the closer the answer is to 50 %.

Debriefing 14.5 This problem demonstrates the usefulness of investing time developing problem-solving skills. People who do not have a lot of problem-solving experience will likely struggle with the problem because they don't know where to start. They will be looking for a rule or a protocol rather than effectively and efficiently making progress by building a model, drawing a diagram, and performing a gedanken.

If you have students who find this problem interesting, you can suggest they co-plot the probability of seeing exactly one, exactly two, and exactly three sides of the pentagon as a function of the radial distance from its center.

14 Geometric Reasoning

Problem 14.6 Knowledge of the Pythagorean theorem allows a wealth of interesting problems to be solved. To start, however, the students have to know the Pythagorean theorem. The best way to "know" something is not by being told but by discovering it for oneself. This problem gives the students an opportunity to make that discovery. Consider the four identical right triangles in the figure arranged to form two squares. Let's define the lengths of the sides of the triangle as A, B, and C, where C is the hypotenuse. The outer surfaces of the triangle form a large square and the inner surfaces of the triangle form a smaller square. Use this figure to demonstrate that $A^2 + B^2 = C^2$.

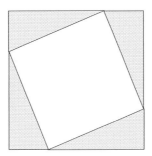

Strategies Utilized Understand the problem. Recognize a pattern.

Discussion 14.6 The length of a side of the large square is $(A + B)$. The length of a side of the interior square is C. The area of the large square can be expressed in two different ways. One is simply $(A + B)^2$ and the other is by summing up the area of the five shapes of which it is composed. There are four triangles, each of area $AB/2$ and the smaller square of area C^2. By setting these equal and performing some algebraic manipulation, it can be shown that $A^2 + B^2 = C^2$.

Debriefing 14.6 Not many people can derive the Pythagorean theorem. Students will appreciate the theorem and be able to use it to solve problems if they understand where it comes from. This problem makes a good warm-up for any problem that involves the Pythagorean theorem, as does the next one.

Problem 14.7 It certainly can be argued that the Pythagorean theorem is fundamentally about squares rather than triangles. The theorem states that the area of a square on the hypotenuse of a right triangle is equal to the sum of the area of the squares on the two legs. Consider the figure below of three squares placed to form a triangle. The challenge is to demonstrate that the area of the larger square on the bottom is equal to the sum of the areas of the two smaller squares.

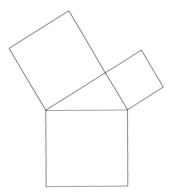

Strategies Utilized Understand the problem. Perform a gedanken. Build a model.

> **Teacher Tip**
> Before you present the students with this problem, show them the figure and ask them if the three squares were made from gold sheet, would they prefer to have the big square or the two smaller squares?

Discussion 14.7 There are a couple of ways to present this problem to the students. One is to hand out a piece of paper with a larger version of this figure with a pair of scissors and challenge the students to cut the two smaller squares into pieces that will fit together to form the larger square. Ask the students to try to minimize the number of pieces they produce. A second method is to provide instructions on how to perform the dissection. Perhaps it is a good idea to do them both, one after the other.

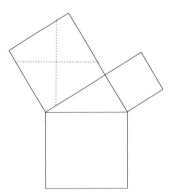

A clever procedure that results in a good 5-piece puzzle starts with the student drawing horizontal and vertical lines (parallel to the sides of the largest square) on the middle-sized square through its center. If the students ask how they can find the

14 Geometric Reasoning

center of the square, suggest that they figure it out for themselves. (They can simply draw both diagonals and find their point of intersection.) These four pieces and the entirety of the smaller square can be placed on top of the largest square to show that the area of the two smaller squares is the same as the largest square; hence, $A^2 + B^2 = C^2$.

> **Teacher Tip**
> If you have access to a woodshop, it is relatively easy to construct a wooden model of this dissection. Use about a 5-by-12-by-13 right triangle. It is a surprising challenge to put all five pieces together to make the largest square.

Debriefing 14.7 These type of "look-see" proofs will develop in the student a better understanding of the Pythagorean theorem, and hence, the student will be able to apply it in situations in which there is no obvious triangle.

As a historical note, this dissection was discovered by Henry Perigal (1801–1898), a British stockbroker and amateur mathematician. He was so proud of it that he had it printed on his business cards and it was engraved on his tombstone.

Problem 14.8 A circle is inscribed in a square as shown. There is a 1-inch by 2-inch rectangle in the corner that just touches the circle. What is the radius of the circle?

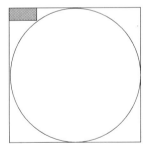

Strategies Utilized Simplify. Draw a diagram.

Discussion 14.8 The hallmark of a good problem is that there is no obvious way to start it. Students usually stare at this one for quite some time before making any progress. When confronted with a problem like this, one very helpful technique is to sketch some useful lines on the figure. Certainly, it is not obvious at first glance that this problem involves the use of the Pythagorean theorem. The key step is to sketch a triangle whose hypotenuse is a radius of the circle as shown in the second figure.

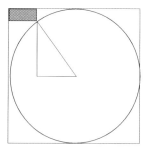

Since the long side of the rectangle is 2 inches, the short side of the triangle is $R - 2$ inches, where R is the radius of the circle. Similarly, the vertical side of the triangle is $R - 1$ inch and the hypotenuse is simply R.

Applying the Pythagorean theorem, we get

$$(R - 1 \text{ inch})^2 + (R - 2 \text{ inches})^2 = R^2$$

Doing some algebra gets us

$$R^2 - (5 \text{ inches})R + 6 \text{ inches}^2 = 0$$

This quadratic equation can be factored into

$$(R - 5 \text{ inches})(R - 1 \text{ inch}) = 0$$

The solution is $R = 5$ inches.

Debriefing 14.8 This problem is an excellent indicator of the students' problem-solving ability. An inexperienced problem-solver will not be able to make progress, but an experienced problem-solver will aggressively attack the problem by adding lines to the figure that will lead to the solution. Many times the lines added will not be useful, but there's nothing wrong with this. Problem-solving virtually always involves a few steps in an inefficient direction – it is these missteps that result in the development of problem-solving experience and intuition.

Problem 14.9 A man is standing on the shore of a calm ocean, with his eyes 2 meters above the surface of the water. What is the distance to the horizon? The radius of the earth is 6,371 km. (We model the earth as a sphere).

Strategies Utilized Understand the problem. Draw a diagram.

Discussion 14.9 The first step here should be to clearly understand what is meant by "distance to the horizon." Imagine a small float that is slowly moved away from the shore. Eventually, it will be behind the curved surface of the water. That is, the person standing at the shore could not see the float – even with a powerful telescope – because the water is blocking it. The question asks the distance the float is from shore when it disappears.

> **Teacher Tip**
> This is a good opportunity to challenge the students' ability to estimate. Ask them to make an educated guess of the distance to the horizon when a person's eyes are two meters above the water's surface. If you don't think it will prevent them from volunteering their guesses, record a few of the answers on the board and refer to them after the problem is solved.

Discussion 14.9 (cont) Any progress on this problem starts by drawing a figure, despite the fact that the scale will not be correct. The first figure below shows the person, the line of sight of the person, and the earth – clearly not to scale. The question asks for the distance between the person and the point at which the line of sight contacts the surface of the earth.

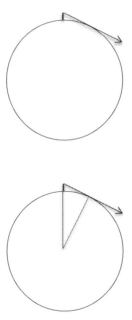

At first glance, many students will not see the path to the solution. However, a good problem-solver will sketch in two important lines – a radius to the person and a radius to the person's horizon. These reveal a right triangle that has a hypotenuse that is the sum of the radius of the earth and the height of the person's eyes above the surface. The other two sides of the triangle are the distance to the horizon and the radius of the earth. The Pythagorean theorem states:

$$(6{,}371{,}000 \text{ m})^2 + d_{\text{horizon}}^2 = (6{,}371{,}000 \text{ m} + 2 \text{ m})^2$$

Solving for the distance to the horizon reveals that it is about five kilometers away. In our experience, the students usually guess a larger distance. For distances close to the earth, the above equation can be reduced to

$$d_{\text{horizon}} = \sqrt{2R_e h}$$

where R_e is the radius of the earth and h is the height. Your students may want to plot this and compare the distance to the horizon for several heights, for example, a child, an ant, or an airplane cruising at 35,000 ft.

> **Student Pitfall**
> Students may object that the distance to the horizon is measured along the surface of the earth and that the calculation performed is for the straight distance from the person's eyes to the horizon. If they do, this is a good opportunity to have them perform the calculation of both distances – the arc length and the tangent line – so they can see that the difference is negligible.

Debriefing 14.9 This problem is another one that requires some thought before applying the relevant equation, which in this case is the Pythagorean theorem. The problem also provides an insight into the relative size of the planet on which we live. As a historical note, strong evidence for the curvature of the earth could be seen when viewing a distant tall ship through a telescope, as the hull of the ship was below the horizon and thus could not be seen. However, the sails of the ship were clearly evident. It must have been a strange sight indeed! When the ship returned to port, the sailors were puzzled by the concern of those on land that the ship's hull was underwater during the voyage.

Problem 14.10 Due to a design flaw, a manufacturing company produced one million defective circuit modules. The modules are housed in 5 cm by 5 cm by 24 cm sealed boxes. Engineers have determined that they can be repaired by adding an electrical connection between the centers of diagonally opposite short edges as indicated on the figure.

This connection will be made with a wire (or a printed circuit) that travels along the outside of the box. What is the minimum length of the wire that will connect these two points?

14 Geometric Reasoning

Strategies Utilized Understand the problem. Simplify. Draw a diagram. Build a model. Perform a gedanken.

> **Teacher Tip**
> It is nice to show the result of this problem with some string and a box. A block of wood models the sealed box quite nicely.

Discussion 14.10 At first glance, it is hard to see how the answer can be less than 29 centimeters, going straight across the top of the box from left to right and then straight down the square side at the right. But, that's what makes it an interesting problem. The connecting wire can actually be significantly shorter than 29 cm.

One way to tackle this problem is to simplify it. It is much easier to think in two dimensions than three dimensions, so let's unfold the box. There is more than one way to do this, but the most effective is shown in the diagram. With the box unfolded in this manner, the shortest "straight line" path that connects the two points in question becomes apparent. Its length can be calculated with the Pythagorean theorem. The height of the triangle shown in the figure is 24 cm and the base is 10 cm. The hypotenuse is only 26 cm. So, the connecting wire only has to be 26 cm long.

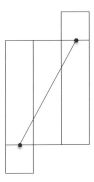

Another way to tackle this problem is with a three-dimensional model. The problem can be modeled with some string and a box. It is actually quite remarkable how quickly the students can find the shortest distance by placing the string between the two points in various ways and pulling it tight.

Debriefing 14.10 This problem demonstrates the importance of properly framing the problem, that is, understanding the problem, pondering the problem, drawing a diagram, and simplifying the problem. This investment can provide a direct path to the solution while eliminating dead ends and obstacles. It is hard to imagine a student coming up with the answer by just looking at the three-dimensional box.

Problem 14.11 A quilter has 20 sections of cloth, each of which is a right triangle. Any two of them will form a rectangle that is 10 inches by 20 inches. How can these twenty pieces be put together to form a square quilt?

Strategies Utilized Understand the problem. Build a model. Perform a gedanken.

> **Teacher Tip**
> We had the woodshop produce several hundred small wooden triangles to model this problem. The legs of the triangles are 1″ by 2″ and they are about 1/8″ thick. This is big enough so that the students can manipulate them easily, but small enough that a large bag of them is not cumbersome to transport. We also spray painted them several different colors so that the final product somewhat resembles a quilt.

Discussion 14.11 Students invariably will make numerous 4″ by 5″ rectangles with the 20 triangles. This is almost a square. There comes a point in the problem-solving process where the students will realize that what they are doing is not working. This point is where the good problem-solvers will separate themselves from the others. They will shift from System 1 thinking to System 2 thinking. They will reframe the problem and retake inventory. They will perform a gedanken by asking themselves several questions. Some of these might include the following: *"What will be the total area of the finished quilt?" "What is the area of one triangle?"* and *"What must be the length of a side of the quilt?"* A person, who is less experienced, is likely to keep rearranging the pieces without success before giving up.

Once the gedanken is performed, the path to the solution is clear. Each triangle has an area of one square inch. The total area of any rectangle made with 20 of these triangles must be 20 square inches. If the shape is a square, the side length must be the square root of 20 inches^2. The square root of twenty square inches can be written as $2\sqrt{5}$ inches. The last question is, "How can I get a length of $2\sqrt{5}$ inches from a bunch of 1″ by 2″ triangles?" If the students can apply the Pythagorean theorem, the answer is "from two hypotenuses of the triangles," because each hypotenuse is $\sqrt{5}$ inches. Once the students start constructing the quilt with two hypotenuses along each edge of the square, the solution will come quickly. Without the gedanken, it is virtually hopeless.

14 Geometric Reasoning   277

Debriefing 14.11 It is very common to try to tackle a problem with straightforward methods. This is not a bad idea because straightforward methods usually work. However, there are times when these methods don't work. When this happens, it is time to think, reframe the problem, and retake inventory. This problem would be much easier if you gave the students an actual framed square that was $2\sqrt{5}$ inches on a side into which the students had to pack the 20 squares. In fact, some students have solved this problem by calculating the size of the square that will be produced, drawing it on a piece of paper and then packing the triangles into that square.

Problem 14.12 A carpenter has produced a tabletop in two pieces that can be rearranged to form both a 120″ by 120″ table and a 160″ by 90″ table. What are the shapes of the two pieces?

Strategies Utilized Understand the problem. Simplify. Draw a diagram. Build a model.

Discussion 14.12 This problem is a significant challenge. An experienced problem-solver might start by drawing a 12 cm by 12 cm square on a piece of paper to model the 120″ by 120″ table and then sketch – in pencil – a few lines that might provide insights into the shape of the two pieces. Similarly, a good initial attack consists of drawing a 9 cm by 16 cm rectangle to model the 90″ by 160″ table and try to figure out how to cut it into two pieces that will form a square.

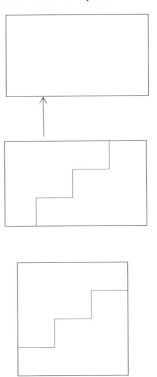

The key *Aha!* moment is to realize that the transition of the tabletop from one shape to the other involves a change of 40″ in one dimension and 30″ in the other. Therefore, it seems like a good idea to start with the 160″ length and cut straight in 40″ from the end, as shown in the first figure. This cut must be 30″ deep if it is to provide the extra 30″ needed on the 90″ edge.

Continuing in this fashion, it can be seen that the two pieces must be shaped as shown in the second figure.

To transform the tabletop from a 160″ by 90″ to a 120″ by 120″, one of the two pieces is shifted by 40″ in one direction and 30″ in the other, thus producing the square table.

This description of getting the answer belies the difficulty that most students have with this problem. Be patient. Don't help them if they get stuck – encourage them to keep thinking. Consider assigning it as a homework problem if class time runs out. It is better to leave it unsolved than to tell them the answer.

Debriefing 14.12 The students should see the need to work with the 30s and 40s that dominate the dimensionality of the shapes. The 160 consists of four 40s and it must be reduced to three 40s to get to 120. The 90 consists of three 30s and another 30 must be added to it to get to 120. If a student starts by sketching a 40-by-30 grid on either shape, the stair-step answer should not be far behind. This is yet another example of adding lines to a diagram as a framework upon which the solution can be found.

Problem 14.13 A treasured rug that is 4 feet by 5 feet was damaged in its center by an electrical generator that overheated. The two square feet that were affected were cut out leaving a hole as shown in the figure. Describe how the rug can be cut into two pieces that will make a complete 3-by-6 rectangle that might be used in a hallway.

Strategies Utilized Understand the problem. Simplify. Recognize a pattern. Draw a diagram. Build a model.

> **Teacher Tip**
> This problem can be modeled with paper, cardstock, cloth, or even with an actual rug. We had one student triumphantly bring in a floor mat from an old car that he cut to the proper dimensions with shears when the problem was left unsolved after a class period.

14 Geometric Reasoning

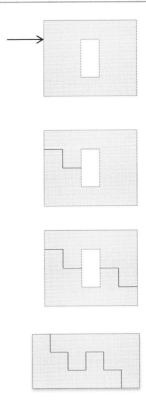

Discussion 14.13 The first thing we can try here is to cut in one foot from the 4 foot length because this dimension is going from 4 feet to 3 feet. Since the other dimension is growing by a foot, this cut should be one foot deep. Continuing the stair-step cut, we quickly reach the hole. To fill the hole, we need a piece that is one foot wide and two feet long. The first cut has produced exactly the piece we need. By shifting the top over one foot and down one foot, the hole will be filled. All that remains is to reproduce the same cut on the other side that will allow this shift. This is shown on the next figure.

Debriefing 14.13 The students' pattern recognition skills should be utilized here. The previous problem provided a nice example of how – in special cases – the stair-step cut and shift can produce one rectangle from another. Here, the original rectangle has a hole in the middle and this makes it a more challenging problem.

Problem 14.14 A fish-shape (see figure) has a round bottom and a neck that has the same curvature. If the diameter of the round bottom is 10 inches, calculate the area of the shape.

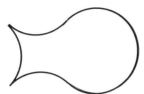

Strategies Utilized Recognize a pattern. Increment and iterate. Perform a gedanken. Draw a diagram.

Discussion 14.14 This is a problem that requires problem-solving experience. The key to an elegant and uncomplicated solution involves adding more lines to the figure to make the answer apparent simply by looking at the figure.

A good start is to expand the figure by completing the circles on the four arcs in the first diagram and then adding squares around each one. So, now we have a figure that consists of four squares and four circles.

With this figure it should be clear that each of the four pieces that make up the tail of the fish can be placed around the circle to make it a square. So the area of the fish-shape is the same as the area of the square, which is 100 square inches.

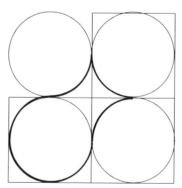

14 Geometric Reasoning

Debriefing 14.14 At first glance, this problem looks like a calculus problem. Students might look for a formula for the area of this shape. They might try to use the area of a circle in their calculations. By framing the problem completely and drawing some useful lines, the answer becomes easy to see. This is the essence of good problem-solving – making the answer easy to see by thinking about the problem in a more efficient and effective manner.

Problem 14.15 Four corridors in an animal shelter meet at a right angle as shown in the diagram. Each corridor houses caged animals separated based on their temperament and disposition. It is desired that any animal housed in one corridor not be able to see through to any of the other three corridors. To accomplish this, some screens will be set up in the square connecting region that will block all lines of sight from one corridor to another. What is the minimum total screen length necessary?

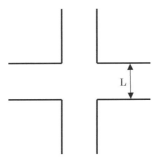

Strategies Utilized Understand the problem. Increment and iterate. Perform a gedanken. Draw a diagram.

Discussion 14.15 The students will probably need some time to wrap their heads around the goal of this problem. It should be understood that the screens can be along the edges of the square or anywhere inside the square.

The first and most obvious solution is simply four screens along the four edges of the square. This is a total length of $4L$ where L is the width of the corridor.

Now we can iterate and use incremental reasoning to gradually improve on this initial attempt. Performing a quick gedanken, it might become clear that only three sides of the square are needed to block all lines of sight from one corridor to the next. This is a total length of $3L$.

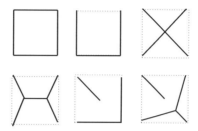

A little more thought might result in a combination of both diagonals. Applying the Pythagorean theorem, we get that the total length of the two screens is $2\sqrt{2}L = 2.828L$. This design looks very efficient and it would be easy to think that it is optimal.

Further thought might improve on this. What about introducing a branch in the middle of the X as shown in the next figure down? With the intersections of the three screens a distance of $L/4$ in from the right and left hallways, application of the Pythagorean theorem reveals that the total length of these screens is an even-better $(1 + \sqrt{3})L = 2.732L$. This really looks like the best possible screen arrangement. In fact, if there were four cities at the corners of a square, the minimum total length of roads connecting them all would follow this design and be 2.732 times the distance between adjacent cities.

Nonetheless, we can improve on this design as well. We can combine a couple of the previous designs and come up with two screens along adjacent edges with a half-diagonal opposite of them. Applying the Pythagorean theorem yet again, we get that the total screen length is $2L + L/\sqrt{2} = 2.707L$. How can it possibly be better than this? Well, there is one final "tweak" that will improve in this design.

By dragging the intersection of the two sides at the lower right towards the center and adding a fourth trailing screen, we can reduce the total length. An optimization technique such as calculus will reveal that the length of the fourth screen in the lower right corner is $(3 - \sqrt{3})L/6$.

Still more application of the Pythagorean theorem reveals that the total length of these four screens is equal to $\sqrt{2} + \frac{3}{\sqrt{6}} L$, which is only $2.639L$.

At the time of this publication, this is believed to be the minimum screen length to make a square region opaque. But there is no formal proof.

Teacher Tip
If the students are interested, they can plot the total length of the screens as the intersection point moves in towards the center from the lower right as a function of the distance the intersection point is from the lower right corner. When this distance is zero, the total screen length should be $2.707L$, and when it moves all the way to the center of the square, it is the X configuration and the total screen length is $2.828L$. The minimum total length of $2.639L$ should be clearly evident on the plot.

14 Geometric Reasoning

For more related information, search Internet for *Fermat point*[2] and for *Steiner's problem*.

Debriefing 14.15 This example demonstrates the potential importance of not stopping at the first solution. By iterating the solutions, we improved the answer multiple times. This puzzle also demonstrates the utility of mathematical skills such as algebra and calculus. If the students are interested, consider showing them the calculus needed to determine the configuration at which the total screen length is a minimum.

Problem 14.16 Chris the Carpenter has a piece of wood left over from a project that has the shape and dimensions as shown in the figure. He would like to make a square tabletop from this shape by cutting it into as few pieces as possible and then gluing the pieces together. What is the minimum number of pieces necessary? Ignore the kerf (the width of the cut from the blade).

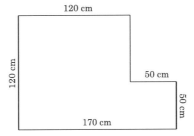

Strategies Utilized Recognize a pattern. Perform a gedanken. Draw a diagram.

Discussion 14.16 After drawing a diagram and some thought, the students will very likely find a dissection similar to the one shown here, which requires a total of six pieces.

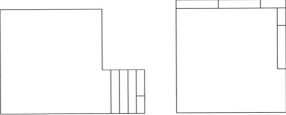

At this point you can ask if they can do it in fewer pieces. How about a total of three pieces? Performing the dissection in a total of three pieces requires another

---

[2] *Fermat point* of a triangle, also called the Torricelli point, is a point such that the total distance from the three vertices of the triangle to the point is the minimum possible. It is so named because this problem is first raised by Fermat in a private letter to Evangelista Torricelli, who solved it.

level of sophistication. It requires the engagement of System 2 thought and it requires a careful and thorough framing of the problem.

> **Teacher Tip**
> Avoid telling the students the dissection can be performed in just three pieces until they have found the six-piece solution. This will allow them to experience some level of success and it will also fixate them on trying to solve the problem by making orthogonal cuts. Their inability to lower the total number of pieces using only orthogonal cuts should develop their appreciation of the reframing work needed before attempting a single cut on the piece of scrap.

Discussion 14.16 (cont) After trying various orthogonal cuts that will cut the piece of scrap into three pieces that will form a square, the students should come to the conclusion that it is a hopeless challenge.

This stage of problem-solving is a very common one. It marks the transition from simplistic thought to a "higher-order" thought. This point separates the more experienced problem-solvers from the less experienced ones. Both groups tried some simple things but failed. The difference is that the less experienced ones will have no other move, whereas the experienced problem-solvers go back to the beginning and invest some time reframing the problem and retaking inventory.

A key first step is the determination of the final size of the table. The total area of the scrap piece consists of a 120-by-120 square and a 50-by-50 square. The total area of the piece is therefore 14,400 cm^2 + 2,500 cm^2, which is 16,900 cm^2. The final square table must have a side length of the square root of 16,900 cm^2, which is 130 cm.

Students who are using their System 2 might recognize that this calculation is simply the Pythagorean theorem: $A^2 + B^2 = C^2$. The A and the B are the side lengths of the two smaller squares and C represents the side length of the table we are going to construct.

At this point, it is clear that there are no 130 cm precut lengths anywhere on the piece of scrap lumber; we have a couple of 50 cm lengths, a couple of 120 cm lengths, a 70 cm length, and a 170 cm length. In the six-piece dissection, the 130 cm lengths were made from (starting with the bottom and moving clockwise) a 130-cm solid piece, 120 cm + 10 cm, 50 cm + 50 cm + 30 cm, and 10 cm + 20 cm + 50 cm + 50 cm.

At this point, an experienced problem-solver might perform a gedanken by asking, *"Where can I get four 130 cm lengths for the edges of the new table?"*

From the Pythagorean theorem, we know that we can get a 130 cm length from the hypotenuse of a right triangle that has legs of length 50 and 120 cm.

Therefore, the possibility of making the edge of the new table from the hypotenuse of a 50-cm by 120-cm right triangle should be considered.

14 Geometric Reasoning

Now, with our new inventory of the problem complete, we can consider how the scrap can be cut to get the lengths we need. Starting at the problematic small square in the lower right, it seems as if the only reasonable cut is as shown.

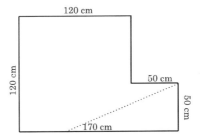

Once this cut is made, the only remaining cut that will produce both the right-angled corner and the 130 cm side length that we need for the new square table is shown in the next figure.

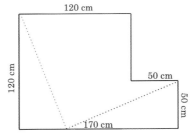

On this figure, the four 130 cm lengths that we need for the perimeter of the new table are evident. Each cut produces two – one on each side of the cut. The only thing that remains is to position the two triangles so that the hypotenuse of each forms the edge of the new table – as shown.

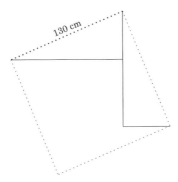

It should be pretty clear to the students that the reframing work was crucial to solving the problem. It would be very difficult to try to cut the piece of scrap into three pieces that would make a square without knowing either the size of the square that would be produced or the fact that the 130 cm needed is the hypotenuse of a right triangle with legs 120 and 50 cm.

Debriefing 14.16 When encountering any problem, it is natural to try simple solutions. It is also natural to stop when a solution is found. Often, this is not the best strategy. When a business is trying to solve a problem, the leaders try to come up with many possible solutions that can be debated rather than go with the first idea that is proposed.

# Grand Challenges **15**

> *Problems worthy of attack prove their worth by fighting back.*
> – Piet Hein

Here is a collection of challenging problems that require multiple problem-solving strategies and a solid foundation in understanding and framing the problem. These can be used as grand challenges that advanced students work on outside of the classroom or as a "special bonus" if the class is doing well.

These problems not only require System 2 thinking, they require mental stamina and the utilization of multiple problem-solving strategies.

At the risk of reiterating this point too often, there is no reason to help the advanced students solve these problems. The point of giving students a problem is to help them become independent problem-solvers. If they can't get it, you can respond, *"It's OK that you can't get it, it is a hard problem, do you want an easier one?"*

In smaller classes (20–30 students), these problems are great to assign when – for some reason – you can't attend the class. Rather than cancel the class and rob the students of a mental workout, have an assistant hand out or otherwise present one of these problems. Left on their own, the students should self-organize and leaders will emerge – and some may just leave. The students that leave will miss out on the opportunity to develop their teamwork and interpersonal skills as well as grow their social capital.

Before presenting any problem to the students, it is a good idea to try to solve it yourself – before looking at the answer. This way, you can empathize with the student's struggles.

Here we go....

Problem 15.1 A crafty crab has constructed seven blind holes in the configuration shown. Every day at noon it changes from one hole to an adjacent hole, and every night at midnight a raccoon comes and looks in a single hole in the hopes of finding the crab. After five consecutive days of failure, the raccoon takes a day off in an attempt to discover a hole-checking protocol that will *guarantee* it will catch the crab. In other words, the raccoon is going to do some System 2 thinking to discover a sequence of hole-checking that will guarantee it catches the crab even if the crab knows the hole-checking protocol. What sequence of hole-checking will inevitably trap the crab?

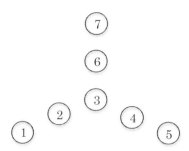

Strategies Utilized Understand the problem. Take inventory. Simplify. Enumerate the possibilities. Draw a diagram.

Discussion 15.1 This is a formidable problem that must start with the thorough understanding of what is being asked. Taking inventory gives us the following rules based on the numbers used in the diagram:
1. A crab never spends two consecutive nights in the same hole.
2. If the crab is in hole 1, it can only move to hole 2.
3. If the crab is in hole 2, it can move to hole 1 or hole 3.
4. If the crab is in hole 3, it can move to hole 2, hole 4, or hole 6.
5. The hole-checking protocol is a series of hole-checks that will assure that the crab is caught when the sequence is complete.

Perhaps the best way to tackle this problem is to simplify it by reducing the number of holes. The trivial simplification is to reduce the number of holes to one. Here the crab is caught on the first night. The next simplification is two holes, which we'll label 1 and 2. The hole-checking protocol that will work is 1-1, that is, checking hole number one on consecutive days. A sequence of 2-2 also works. Next we'll tackle three holes in a straight line, numbered 1, 2, and 3, respectively. The most efficient protocol here is 2-2. This guarantees that the crab will be found in two nights. Let's try four holes in a line, 1, 2, 3, and 4.

At first glance this might appear impossible, but let's simplify yet again. Let's assume that the crab started in an even-numbered hole, either hole 2 or hole 4. The sequence that assures success is simply 2-3, because if the crab started in 2, it is caught immediately, and if the crab started in 4, it must be in hole 3 the next day. So if 2-3 does not catch the crab, it must have started in an odd-numbered hole. Since two days have passed, it must be back in an odd-numbered hole again. To catch a crab that started in an odd-numbered hole, the sequence 3-2 can be used. Therefore, the sequence, 2-3-3-2 will guaranteed that the crab would be found in a maximum of four days. Note how the simplification of the problem helps to understand and frame the problem. After solving simplified versions of the same problem, the students can experience success and begin to see what the solution might look like.

> **Teacher Tip**
> The students will come up with many hole-checking sequences that are wrong and will want you, as the perceived authority, to verify their answer. We recommend promulgating a rule that you only check an answer if it was checked against another student twice, and if you check a sequence and it is wrong, you will not check another sequence from that group for, say, 15 minutes. Checking if a sequence will trap the crab can be as simple as one person playing the role of the crab putting his/her index finger on the hole that are "in" and another playing the role of raccoon putting his/her index finger on the hole that is being checked. Both should be aware of the proposed hole-checking sequence, with the person playing the role of the raccoon blindly following it and the person playing the role of the crab trying to avoid capture.

Discussion 15.1 (cont) Now let's try four holes in the triangular configuration. Here it should be clear that 3-3 is enough because if the crab did not start in hole 3, it will be there the next night.

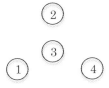

At this point, the students should have their heads wrapped completely around the problem and can be left to attempt to solve it by themselves – either in class or at home. Either they do or they don't.

For the seven-hole problem as stated, a ten-step sequence that guarantees the crab would be caught is 2-3-4-3-6-2-3-4-3-6. The first five steps in the sequence will catch the crab if it started in an even-numbered hole. If these five steps do not catch the crab, it must have started in an odd-numbered hole. After five steps, a crab that started in an odd-numbered hole must be in an even-numbered hole, so repeating the five-step sequence will trap the crab.

Debriefing 15.1 In our experience, students find this problem very challenging. They tend to produce sequences in which the same hole is checked two or three times in a row. It would be an interesting study to compare the results from two groups of students – one that was presented the problem as is and another that was given the hint, "*Assume that the crab started in an even-numbered hole.*" Being a good problem-solver means being able to make these types of simplifications without being prompted to do so. The ability to simplify a difficult problem in order to gain insights into the nature of the problem and the shape of its solution space is crucial to being a good problem-solver.

Problem 15.2 Imagine that you have twelve cannonballs, all but one of equal weight. You do not know whether the odd cannonball is heavier or lighter than the others, only that it is different. You are given a balance scale and told to try to find the odd cannonball in the minimum number of weighings.

Strategies Utilized Understand the problem. Take inventory. Simplify. Perform a gedanken.

> **Teacher Tip**
> There will very likely be a lot of wrong/incomplete "solutions" here. Students will think that they have the solution because it works for a particular set of weighing outcomes. So, you will find yourself troubleshooting the students' answers. We suggest having students test their solutions on other students. This will help both the solution proposers and the troubleshooters better understand the problem. What happens on many occasions is that the student proposing the solution will realize that it is not right in the process of making the proposal, rendering the need for the troubleshooter superfluous.

Discussion 15.2 The key to solving these types of optimization problems is to be ambivalent about the result of the weighing. It is similar to guessing a number from 1 to 1,000 when given a series of higher or lower answers after each guess. If you guess 500 first, you don't care whether the answer is higher or lower. On the other hand, if you guess 100 first, you are hoping that the answer is lower. So, when you put the first set of cannonballs on the scale, you should not be hoping for a particular result. That is, you are trying to get as much information as possible from each weighing, irrespective of the result.

If you balance 6 cannonballs vs. 6 cannonballs, you know it will not balance, so you get very little new information (what new information do you get?). If you balance 5 vs. 5, you are hoping that it balances. If it does not balance, you do not get very much information.

If you start by comparing any two cannonballs with each other, you are hoping that it does not balance. It seems like the two possibilities to consider are starting with three on a side of the balance and starting with four on each side of the balance.

If we start with 4 vs. 4 and it balances, we have isolated the defective one among the remaining 4. If it does not balance, we have isolated the cannonball among the eight that we compared, but we also know which pile of four is lighter and which pile of four is heavier *and* we have found four good ones (the four we did not weigh).

Let's analyze each of the two outcomes with a 4 vs. 4 first weighing and see what happens. To better frame the problem, let's number the cannonballs 1–12 and weigh 1–4 vs. 5–8 with the first weighing.

If the 4 vs. 4 first weighing balances, then we take three good ones (1, 2, and 3) and three of the four we did not weigh (9, 10, and 11). If this balances, we know cannonball 12 is defective. Weigh 12 against a good one to uncover whether it is heavier or lighter. If the 3 vs. 3 weighing does not balance, the defective cannonball is isolated among 9, 10, and 11, and we know whether it is heavier or lighter based on the result of the second weighing. Take any two cannonballs from that group of three (e.g., 9 and 10) and compare them on the balance. If they balance, the one that was not weighed is the defective one and the results of the second weighing revealed whether it is heavier or lighter. If they do not balance, you know the defective one is 9 or 10 and you know whether it is lighter or heavier based on the result of the second weighing.

Now let's consider the procedure when the 4 vs. 4 weighing does not balance and let's assume that the 1, 2, 3, 4 side is light. The result of this weighing reveals that 9, 10, 11, and 12 are good and either one of the group from 1, 2, 3, and 4 is light or one of the group from 5, 6, 7, and 8 is heavy. If we weigh 1, 2, 3 vs. 9, 10, 11, we will be OK if it does not balance. Then we weigh 1 vs. 2 and if they do not balance, we know the lighter one is defective and light. If they balance we know 3 is light. However, if 1, 2, 3 balances 9, 10, 11, we are stuck because we now have the possibility that 4 is light or 5, 6, 7, or 8 is heavy and will not be able to get the answer with one weighing.

Something different has to be done with the second weighing; something that gives as much information as possible. At this point the students have to take inventory and consider the possibilities. After a while they may come up with this:

After the first weighing we know that either the set 1, 2, 3, 4 contains a light cannonball or the set 5, 6, 7, 8 contains a heavy cannonball.

Now we take 7 and 8 off the scale and switch 2 and 3 to the other side and switch 5 to the other side. So, 1, 4, and 5 are being compared with 2, 3, and 6.

There are three possibilities: they balance; the 1, 4, 5 side is lighter; or the 1, 4, 5 side is heavier.

If they balance, either 7 or 8 is heavy. Compare either 7 or 8 with a known good one for the third weighing and you have the answer.

If the 1, 4, 5 side is light, either 1 or 4 is light *OR* 6 is heavy. Compare 1 vs. 4 with the third weighing and you have the answer (if they balance, 6 is heavy).

If the 1, 4, 5 side is heavy, either 5 is heavy or 2 or 3 is light. Compare 2 vs. 3 and you have your answer (if 2 and 3 balance, 5 is heavy).

It is also possible to solve this problem by using one of the known good cannonballs in the second weighing, and students have uncovered this solution as well.

Debriefing 15.2 Solving this problem is a tremendously iterative process. The students will usually have to try many combinations before they get the answer. As with many challenging problems, there are numerous dead ends and many times the solver has to stop and retake inventory. This problem should give the students plenty of practice critically analyzing their proposed solutions as well as thinking of clever ways to perform the second weighing.

Problem 15.3 There are nine hats. Two are blue and seven are red. The teacher takes seven students and puts them in a single file line in front of the class. He puts hats randomly on each student and hides the remaining two hats. The students can see all the hats in front of them, but can't see their own hat nor any of the hats behind them.

That is, the arrangement of hat colors is

R	R	R	R	B	R	R
#1	#2	#3	#4	#5	#6	#7

with the students facing to the right. As it happens, only the third person from the front gets a blue hat (student #5).

This means that a red hat and a blue hat are hidden. The teacher will then poll the students starting from the back of the line (from student #1, who can see all the hats but his/her own), asking them if they know the color of their hat by logical deduction. The students will respond either "yes" or "no." Note that the students are not revealing the color of their hat, only whether they know by logical thought what color it must be.

All the students hear every yes-or-no answer from behind them and can use this information to try to logically determine the color of their own hat. If all are good, logical thinkers, what will the answers be starting from the back of the line (the left in the diagram)?

Strategies Utilized Take inventory. Simplify. Build a model.

Discussion 15.3 This is a great problem to just watch the students solve – especially if given later in the term. Once the students are comfortable in the classroom environment and with each other, they can really flourish.

> **Teacher Tip**
> Consider just presenting this problem and getting up and leaving. Don't assign groups; don't tell them what strategies to use. Just say, I'm going to run some errands; I'll be back in a half hour and when I come back I want the consensus on a sequence of seven yes–no answers to be written on the board. If the students know they are going to be left on their own, it is amazing how resourceful they will be. US General George Patton said, *"If you tell people where to go but not how to get there, you'll be amazed at the results."*

15 Grand Challenges

Discussion 15.3 (cont) On one occasion, the teacher walked back into the room to find seven students seated in a line in the front of the class each with Post-it notes on their upper back. The color of the Post-it note represented the color of the hat they were wearing. What a great model of the problem! One student acted as the pollster, starting from the back and asking each student whether they knew what color hat they were wearing. Here are the results with explanation.

#1 can't tell because he does not see two blue hats. He says NO.

#2 knows that #1 is not looking at two blue hats because that is the only way #1 could know the color of his hat. Since #2 is looking at a blue hat on number 5, he knows he is not blue, and therefore must be red. He says YES.

#3 reasons exactly like #2 and says YES. However, all those in front of #3 already know that #3 must say YES – even if they had their eyes closed the entire time. Once the students in positions #3 through #7 hear NO and YES from #1 and #2, they know that both #1 and #2 are looking at exactly one blue hat. If #3 sees no blue hats, he knows he is blue and says YES. If #3 sees one blue hat, then he knows he is red and also says YES.

#4 gets no new information from #3's YES, but he can see a blue hat on #5 and hence knows he must be red and says YES. So, #4 knows why #3 said yes – #3 knows he must be red.

#5 knows that #4 must see a blue hat. If #4 saw all red he could not determine the color of his hat because he does not know why #3 said YES. Since #5 can't see a blue hat, he knows he is blue and says YES.

#6 is looking at a red hat on #7 and hears NO, YES, YES, YES, YES from behind. This sequence is consistent with the hats being

R	R	R	R	B	R	R
#1	#2	#3	#4	#5	#6	#7

or

R	R	R	R	R	B	R
#1	#2	#3	#4	#5	#6	#7

So, #6 can't tell and says NO. In other words, #6 doesn't know why #5 said YES. #5 could say YES knowing he is red, and #5 could say YES knowing he is blue.

#7 can't see anyone's hat and hears: NO, YES, YES, YES, YES, NO from behind.

From the first two, he knows that there is exactly one blue hat on someone from #3 through #7. He gets no new information from #3, but when #4 says YES, he knows #4 is looking at a blue hat. When #5 says YES he gets no new information because #5 *must* says YES because of the answers he heard from #1 through #4. As soon as #4 says YES, #7 knows the sequence must be one of these three possibilities:

| R | R | R | R | B | R | R |
|#1|#2|#3|#4|#5|#6|#7|

or

| R | R | R | R | R | B | R |
|#1|#2|#3|#4|#5|#6|#7|

or

| R | R | R | R | R | R | B |
|#1|#2|#3|#4|#5|#6|#7|

#5 would respond YES in any of the three situations. However, when he hears a NO from #6, he realizes that #6 can't tell between

| R | R | R | R | B | R | R |
|#1|#2|#3|#4|#5|#6|#7|

or

| R | R | R | R | R | B | R |
|#1|#2|#3|#4|#5|#6|#7|

In both cases, #7 has a red hat and therefore says YES.

So the sequence of seven answers is NO, YES, YES, YES, YES, NO, YES.

> **Teacher Tip**
> The positions that cause the most trouble are #5 and #6. There is very likely to be a disagreement here. If you decided to present this to the students and leave the room, there might not be a consensus regarding the responses from #5 and #6 upon your return. This is an excellent opportunity to let the students talk among themselves to reach a consensus, with you stepping in only when the students are no longer being productive.

Debriefing 15.3 This problem can certainly last more than one class. There is ample room for debate and students often come to the next class with input about the problem from their family, friends, and coworkers. This is all good. When students problem-solve with family members, it can promote mutual respect and an appreciation for each other's talents and contributions.

The problem also teaches the students not to make decisions early and to utilize all the information that is available. This will prevent them from making the mistake of declaring that "*Well, I'm sure that #7 can't possibly know*" at the beginning of the problem. There will come a time in the student's life when other people are presenting their ideas – it could be a spouse, friends, or coworkers. It is important that the students develop the skill to listen to what others are saying and either refute their argument or change their own ideas. For many people the opposite of talking is not listening, it is waiting. They have their ideas and are waiting to present them, irrespective of what the other person is saying. They have already "uploaded" their argument into their brain and are politely waiting to fire away.

Problem 15.4 Imagine a stock that has a value of $100 on January 1. On the 15th of each month, a fair coin is tossed and the value of the stock is increased by $1 if the coin lands heads up and the value of the stock is decreased by $1 if the coin lands tails up. After one year (twelve tosses), the value of the stock will range from $88 to $112 and have an average value of $100.

Now let's consider an investor that would like to limit his/her loss to $5. He/she buys one share of the stock for $100 and applies a stop order at $95. If at anytime during the year the stock price drops to $95, he/she sells the stock for $95 and accepts the $5 loss. The question is: How does the stop order affect the *average* outcome of the investment? That is, will the stop order result in an average outcome that is different from $100? If so, will this average be higher or lower?

Strategies Utilized Understand the problem. Enumerate the possibilities. Reason backwards. Perform a gedanken.

Discussion 15.4 This is a problem that provides a fertile field for intelligent discussion. Interesting questions include: What percent of the time will the stop order be applied? In which month is the value of the stock most likely to hit $95 for the first time? Is the final value of the stock more or less likely to end on $100 with the stop order in place? Will a year consisting of 4 heads and 8 tails have more than 50 % stops?

> **Student Pitfall**
> The students will make two mistakes here and both can be classified as gambler's fallacy, which is that the result of previous independent events affects the result of future events. Students can assume that the reason that the stock got to $95 was that it was a bad investment and it should be sold. This is not the case. The reason it got to $95 is that the investor got unlucky. There is no reason to believe that the stock will have a tendency to go either up (reasoning that heads are "due" and that the value should return to the mean) or down (reasoning that the investment was "bad" and you should get out as soon as possible) in the future based on the results the previous months.

Discussion 15.4 (cont) It is possible to spend multiple classes on this problem and we suggest delving into the problem as deeply as the students desire.

The brute force attack on this problem is to calculate the average value of the stock at the end of the year with the stop order in place. In fact, students with advanced mathematical training will usually utilize this training to calculate this average. Students without advanced mathematical training are in a position to get the answer more quickly, because they might perform a gedanken to arrive at the answer.

We are going to start with the brute force method.

There are $2^{12} = 4,096$ possible sequences of heads and tails for the entire year, and one way to calculate the average outcome is to sum up the result of each of them and then divide by 4,096. With the stop order in place, this will include a number of outcomes in which the stop order was applied. When the stop order is applied, the final value of the stock is $95.

Another way to attack the problem is to model it with a computer. As discussed in Sect. 11.1, this requires some programming skills (which is another very valuable problem-solving tool). A computer model could model the stock price, say, a billion times with the stop order in place, and then calculate the average outcome. If there are some programmers in the class, they might be interested in coding a Monte Carlo simulation of the stock price.

To "do the math" we could start by calculating the total number of possible sequences of heads and tails. Some relevant questions to ask are "*How many of the 4,096 possibilities will be stopped because the stock reached a value of $95?*" and "*How many of the 4,096 possibilities will result in, say, eight heads and four tails?*"

Let's start by determining the number of ways to get four heads and eight tails with twelve tosses. We can start by thinking about it like this: we have to select four of the twelve months that will be tails (we can ignore the heads). The first tails can go in any one of twelve months; the second tails must go in one of the remaining eleven, the third into one of the remaining ten, and the fourth into one of the remaining nine. The number of different ways this can happen is

$$N = 12 \times 11 \times 10 \times 9 = 11,880$$

However, we are counting the actual arrangements more than once. That is, the 11,880 contain a large number of duplicates and we have to eliminate these. Consider the following situation in which tails came up only in the months of January, March, April, and November, shown in the table below.

Month	Jan	Feb	Mar	Apr	May	Jun	Jul	Aug	Sep	Oct	Nov	Dec
Flip	T	H	T	T	H	H	H	H	H	H	T	H

Now, if the first tails were placed in Apr, the second in Mar, the third in Nov, and the fourth in Jan, the final result would look exactly the same as if the tails were placed in the slots in the order of Jan, Mar, Nov, Apr, but our counting methods counted each of these separately. The degree of overcounting is simply the number of permutations of four months. The number is small enough that we can list them all:

Jan, Mar, Apr, Nov	Mar, Jan, Apr, Nov	Apr, Jan, Mar, Nov	Nov, Apr, Jan, Mar
Jan, Mar, Nov, Apr	Mar, Jan, Nov, Apr	Apr, Jan, Nov, Mar	Nov, Apr, Mar, Jan
Jan, Apr, Mar, Nov	Mar, Apr, Jan, Nov	Apr, Mar, Jan, Nov	Nov, Jan, Apr, Mar
Jan, Apr, Nov, Mar	Mar, Apr, Nov, Jan	Apr, Mar, Nov, Jan	Nov, Jan, Mar, Apr
Jan, Nov, Mar, Apr	Mar, Nov, Apr, Jan	Apr, Nov, Jan, Mar	Nov, Mar, Apr, Jan
Jan, Nov, Apr, Mar	Mar, Nov, Jan, Apr	Apr, Nov, Mar, Jan	Nov, Mar, Jan, Apr

There are 24. It is easy enough to do the calculation. There are four possible months for the first tails, three possible months for the second tails, and two possible months for the third tails, and the last tail must go in the only remaining month. This is

$$N = 4 \times 3 \times 2 \times 1 = 24$$

So we counted every possible arrangement of four tails 24 separate times. To correct for this multiple counting, we have to divide the 11,880 by 24.

$$N = \frac{11,880}{24} = 495$$

So, there are 495 possible sequences of 12 flips that contain four tails and eight heads.

> **Teacher Tip**
> The numbers are big here and many students will have trouble wrapping their heads around the fact that there are 495 ways to get four tails and eight heads in twelve tosses of a coin. To help them understand how many there are, ask the student to write, say, ten or even twenty possible sequences.

Discussion 15.4 (cont) If you have done enough probability calculations, you and/or the students might want to derive the formula for the number of ways to arrange x identical things in y spaces. The formula is

$$N = \frac{y!}{(y-x)!x!}$$

For the example of four tails in twelve flips, we get

$$N = \frac{12!}{(12-4)!4!} = 495$$

Since there are a total of 4,096 possible outcomes and 495 of them consist of four tails and eight heads, the probability that a toss of twelve coins will result in eight heads and four tails is

$$P = \frac{495}{4,095} = 0.1208\ldots$$

which is about 12 %.

Using this formula we can determine the total number of ways each combination of heads and tails can occur. Since every one of the 4,096 heads–tails sequences is equally likely, we can use the number of different ways each can occur as a measure of its likelihood of occurring. The complete results are shown in the table below.

# of heads	# of tails	# occurrences
12	0	1
11	1	12
10	2	66
9	3	220
8	4	495
7	5	792
6	6	924
5	7	792
4	8	495
3	9	220
2	10	66
1	11	12
0	12	1
	Total	4,096

When at least eight tosses are heads, the stop order will never be used because the value of the stock can never get to $95. Also, the stop order will always be used in years in which at most three heads are tossed. We have to do some work with the tosses that result in four through seven heads because in some cases the stop order will be applied and in others it will not.

Let's do an example with the result of 7 heads and 5 tails. There is only one way that this can be stopped and that is when the first five tosses are tails and, of course, the last seven must be heads. So, of the 792 combinations of 7 heads and 5 tails, only one will lead to the stop order being applied.

How about 6 heads and 6 tails? The only way that a 6 heads and 6 tails result can be stopped at $95 is if the first five tosses are tails or if six of the first seven tosses

15 Grand Challenges   299

are tails and the head is in position 1–5. There are 12 ways that this can happen, as shown below.

	Month											
#	1	2	3	4	5	6	7	8	9	10	11	12
1	T	T	T	T	T	T	H	H	H	H	H	H
2	T	T	T	T	T	H	T	H	H	H	H	H
3	T	T	T	T	T	H	H	T	H	H	H	H
4	T	T	T	T	T	H	H	H	T	H	H	H
5	T	T	T	T	T	H	H	H	H	T	H	H
6	T	T	T	T	T	H	H	H	H	H	T	H
7	T	T	T	T	T	H	H	H	H	H	H	T
8	H	T	T	T	T	T	T	H	H	H	H	H
9	T	H	T	T	T	T	T	H	H	H	H	H
10	T	T	H	T	T	T	T	H	H	H	H	H
11	T	T	T	H	T	T	T	H	H	H	H	H
12	T	T	T	T	H	T	T	H	H	H	H	H

In this table we can see that the first seven possibilities include all the ways that one tail and six heads can be arranged among the last seven tosses. Similarly, the last five possibilities include all the ways one head can be arranged among the first five positions. So, of the 924 possible ways to get 6 heads and 6 tails, only 12 of them will result in the stop order being applied.

The table below presents all of the calculations and a complete summary of the results.

	WITH stop order					WITHOUT stop order	
	NOT stopped		Stopped				
H-T	Occurrences	Result	Occurrences	Result		Occurrences	Value
12-0	1	$112	–	–		1	$112
11-1	12	$110	–	–		12	$110
10-2	66	$108	–	–		66	$108
9-3	220	$106	–	–		220	$106
8-4	495	$104	–	–		495	$104
7-5	791	$102	1	$95		792	$102
6-6	912	$100	12	$95		924	$100
5-7	726	$98	66	$95		792	$98
4-8	275	$96	220	$95		495	$96
3-9	–	–	220	$95		220	$94
2-10	–	–	66	$95		66	$92
1-11	–	–	12	$95		12	$90
0-12	–	–	1	$95		1	$88
Totals	3,498	$352,790	598	$56,810		4,096	$409,600

So, the stop order will be applied in 598 of the 4,096 possible sequences, which is about one-seventh of the time.

Now we can calculate the average outcome with the stop order applied. To do this we sum the final value of the stock for all 4,096 possibilities. This is $352,790 plus $56,810, which is $409,600. When this sum is divided by the total number of occurrences, 4,096, the result is $100.

It is interesting to compare the final distributions with and without the stop order. These are shown in the two graphs.

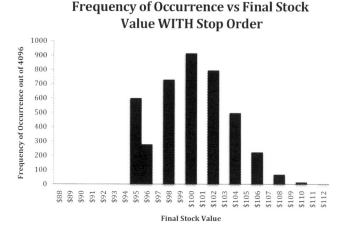

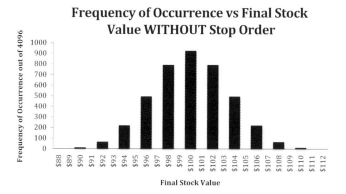

In both cases the sum of the final value of the stock for all 4,096 possible sequences of heads and tails is $4,096,600. When you divide by the number of occurrences, the average value of the stock is $100 in both cases.

A gedanken may convince the student that this *must* be the case. The value of this stock is a random event and not only is there no way to have a positive expectation value when investing in a 50/50 event, there is also no way to have a negative expectation value.

This exercise illustrates a number of the problem-solving principles in Part II of this book. The strategy *reason backwards* can certainly offer insights. Let's start at the end of the year and work backwards. Everyone should agree that the average value of the stock at the end of the year is $100. In fact, the average value of the stock after any number of months is $100. Let's consider the average value of the stock *after* it reaches $95 and is sold. What is the average value of the stock at the end of the year that was sold for $95? The answer must be $95 – after all, its value is determined by flipping a coin. Since the average *change* in the value of the stock after it is sold must be zero, selling the stock can have no effect on the average outcome. Therefore the average value with the stop order in place must be the same as without.

Another principle that is useful here is "Simplify!" To start, we can consider an investment period of two months and a stop order at $99. If the toss for the first month is a head, the value will go to $101 and the final result will be either $102 or $100 with an average of $101. If the flip after the first month is a tail, the stop order will be employed and the final value of the stock is $99. The average of $101 and $99 is $100 – as it must be.

Debriefing 15.4 It is a basic principle of gambling that there is no betting strategy that will result in a positive expectation value on series of coin flips. Those familiar with this concept should be able to immediately see that the stop order can provide no difference in the *average* outcome of the investment. However, it is important to note that experienced investors do not base their decisions on the *average* result and this is why stop orders are used by investors.

Problem 15.5 A teacher shows the class nine hats. Five are red and four are white. Five volunteers are chosen from the class and these five are seated in front of the class. The teacher then announces that he/she will place a hat on each of their heads and hide the four hats that are not used.

The placement is:

Student number	#1	#2	#3	#4	#5
Hat color	R	R	R	R	R

That is, everyone gets a red hat and the four white hats are hidden. The students do not know the colors of their own hats, and they do not know the color of the four hats that are hidden. However, each student can see the color of the hats on the other four students.

The five students are polled from left to right to determine whether they know the color of their hats. The students answer *yes* or *no*. The students can hear all the answers and may use the answers to deduce the color of their own hats. What will the responses be if the students all are perfect logicians?

Strategies Utilized Perform a gedanken. Simplify. Take inventory.

Discussion 15.5 The answers will be

Student number	#1	#2	#3	#4	#5
Hat color	R	R	R	R	R
Answer	NO	NO	NO	NO	YES

> **Teacher Tip**
> If a student has trouble making progress on this problem, you might consider recommending that they simplify the problem and then perform a gedanken. Simplification can involve reducing the number of students and/or the total number of white hats.

Discussion 15.5 (cont) Here is the solution to the problem as stated.

#1 says NO because he is not looking at four white hats. Before #1 answers, #2 knows #1 can't tell the color of his hat. This means that #2 gets no new information from #1's answer. So, #2 can't tell either and says NO.

Let's consider #3's thought process if he assumes that his hat were white. If his hat were white, #2 would have to consider the possibility that #1 was looking at two white hats (on #2 and #3). However, even if #1 were looking at two white hats, he couldn't tell. So, this means that #3 can't tell either. That is, what #3 sees and what he hears from #1 and #2 is consistent with his hat being either white or red.

Now let's consider #4. She hears NO, NO, NO and sees all red hats. #4 considers what would have happened if she were white. If it were white, then #1, #2, and #3 would be looking at one white (on #4). If this were the case, then #3 would have to consider the possibility that #2 thought that #1 was looking at three white hats (on #2, #3, and #4). However, even if #4 were white and #2 thought #1 was looking at three white hats, the answers still would be NO, NO, NO, so the answers NO, NO, NO are consistent with #4 having a white hat and #4 having a red hat. Therefore, #4 can't tell.

Now let's look at #5. #5 has to consider what would happen if he had a white hat.

If he had a white hat, #4 would have to consider the possibility that #1, #2, and #3 are all looking at two white hats (on #4 and #5). If this is the case, when #4 hears NO, NO, NO, he knows he must be red. If both #4 and #5 are white, #3 would know he was red as soon as #2 says NO.

15 Grand Challenges

Here's why.
If the situation is

Student number	#1	#2	#3	#4	#5
Hat color	R	R	R	W	W
Answer	NO	NO	?		

then #3 would say YES because he knows his hat can't be white once he hears #2's NO. #3 sees a white on #4 and #5. As soon as #1 says NO, #3 knows that #2 knows that #1 is not looking at four white hats. If #3 were white, then #2 would be looking at white hats on #3, #4, and #5 and would know he can't be white, because #1 is not looking at four white hats. Indeed, it is a challenging problem.

So, when #4 can't tell, #5 knows he must be red. If #5 were white and all the rest red, the answers would be

Student number	#1	#2	#3	#4	#5
Hat color	R	R	R	R	W
Answer	NO	NO	NO	YES	

So as soon as #4 says NO, #5 knows he must be red.

Debriefing 15.5 The students will very likely not be able to make any progress by working on the problem as given. This is where their problem-solving skills come in. They can change the number of white hats and they can change the number of students. A trivial problem is three hats, two red and one white, and two students. Red hats are put on the two students and the white hat is hidden. If the first one can't tell, what does the second one say? Slowly ramping this one up from the trivial to the actual problem is a great way not only to solve it but also gradually to frame the problem.

Problem 15.6 A rectangular block of wood has dimensions 8 cm × 8 cm × 27 cm. Cut the block into the minimum number of pieces that can be rearranged to form a cube. Ignore the width of the kerfs (the waste from the width of the blade).

Strategies Utilized Recognize a pattern. Take inventory. Simplify. Reason backwards.

Discussion 15.6 It is a significant challenge to get to the minimum of four pieces. A good first step towards a solution is to frame the problem by calculating the size of the cube that can be made from 8 cm × 8 cm × 27 cm rectangular box. This can be done with brute force using the equation

$$8 \text{ cm} \times 8 \text{ cm} \times 27 \text{ cm} = 1,728 \text{ cm}^3 = x^3$$

Solving for x gives 12 cm. Another way to see that the side of the cube must be 12 cm is to realize that the sides (8 cm, 8 cm, and 27 cm) are all cubes of integers and multiplying their cubed roots together gives the length of a side of the cube, 12 cm.

At this point perhaps it is best to make a cut and see what develops. The most straightforward first cut seems to be a straight cut 12 cm down the 27 cm length to produce a block of 8 cm × 8 cm × 12 cm long and a block of 8 cm × 8 cm × 15 cm long. This first block can be stood on its 8 cm by 8 cm face and placed in the corner of a 12 cm × 12 cm × 12 cm frame.

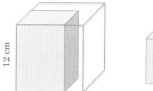

Next, we can cut two 4 cm × 8 cm × 12 cm blocks from the remaining 8 cm × 8 cm × 15 cm piece, leaving a 8 cm × 8 cm × 3 cm piece. These two larger blocks can be stood on their 4 cm × 8 cm edge and placed along the sides of the 8 cm × 8 cm × 12 cm block that is already in the frame. Now we have this:

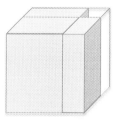 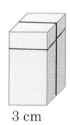

The block we have remaining is a 3-cm length of the original 8 cm × 8 cm × 27 cm block, and the space we have to fill is 4 cm × 4 cm × 12 cm. The most efficient way to do this is to cut the 8 cm × 8 cm × 3 cm block into four 4 cm × 4 cm × 3 cm blocks and stack them in the space on their 4 cm × 4 cm faces. This is a total of seven pieces – a good start. It is difficult to figure out a way to improve on this by producing only rectangular boxes.

If the students stop to use their System 2 thought process, they might make the connection between this problem and some of the 2-dimensional problems tackled in Chap. 14 (Geometric Reasoning) – specifically the stair-step technique to change the dimensions of a rectangle.

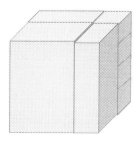

To start this process, we first cut the 27-cm length into an 18-cm length and a 9-cm length. The larger piece, shown in the figure below, can be stairstepped into a 12 cm × 12 cm × 8 cm with the cut shown in the figure:

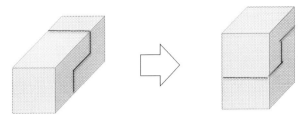

The 9 cm × 8 cm × 8 cm block that remains needs to be transitioned into a 4 cm × 12 cm × 12 cm block to complete the cube. To start we cut the block in half to produce two 4 cm × 8 cm × 9 cm pieces. The 8 cm × 9 cm face of each of these can be step-shifted into a 6 cm × 12 cm face as follows:

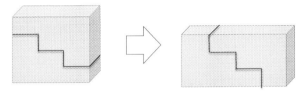

These two 4 cm × 6 cm × 12 cm can be stacked to produce the requisite 12 cm × 12 cm × 4 cm piece to stack right next to the 12 cm × 12 cm × 8 cm piece to produce the 12 cm × 12 cm × 12 cm cube.

There are six pieces, one fewer than previously. Can we do better?

Time to take inventory again. What other techniques are available? Let's try a problem-solving technique that is often very useful when the final state is known. Let's work backwards from the final state.

So, we'll start with a 12 cm × 12 cm × 12 cm cube and try to get an 8 cm × 8 cm × 27 cm box. Continuing with the stair-stepping theme, let's consider the possible rectangular boxes we could make from 12 cm × 12 cm × 12 cm cube. A good candidate is a 4 cm by 6 cm step size as shown.

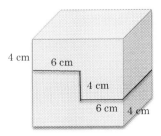

If we shift the top piece to the right and down, this produces an 8 cm × 12 cm × 18 cm rectangular box. Now let's see if we can get an 8 cm × 12 cm × 18 cm box into an

8 cm × 8 cm × 27 cm box. All that is needed is a repeat of the stair-step procedure just used after a 90-degree turn of the box, with the step size of height 4 cm and width of 9 cm. This gets the height down from 12 to 8 cm and the length up from 18 to 27 cm.

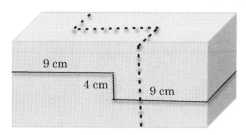

The dotted line is the previous cut that produced the 12 cm × 8 cm × 18 cm box from the 12 cm × 12 cm × 12 cm cube.

The four pieces produced are unusual looking. There are two right- and left-handed pairs; one pair each has a volume of 528 cm^3 and the other pair each has a volume of 336 cm^3.

Debriefing 15.6 This problem was solved by recognizing a pattern that was used to solve Problem 14.12 and applying it twice from the desired result (a 12 cm × 12 cm × 12 cm cube) to get back to the initial piece (an 8 cm × 8 cm × 27 cm block). It is a nice project actually to produce these four pieces from wood. It is a good idea to first try to make 3-D models of the pieces, perhaps with clay, paper, or even a potato. With the four pieces it is a significant challenge to produce all three different rectangular boxes, and it makes a great desk paperweight!

Problem 15.7 In Chap. 9 we presented the following problem (Problem 9.6).

Imagine a village of one-eyed aliens who are very logical, but have an unusual cultural tradition regarding the color of their eyes – which is known to be either brown or blue. If any members of the community can logically deduce their eye colors, they have to leave the village forever after making the announcement that they have figured out the color of their eye (at the daily meeting, which is mandatory for everyone in the village). The village has no contact with the outside world and it has no mirrors or reflecting surfaces. Let's say that there are 5 blue-eyed aliens and 25 brown-eyed aliens in this village. One day, a visiting anthropologist addresses them at their daily meeting and says, "*My word, your blue and brown eyes are beautiful!*" There was a collective gasp among the aliens because she mentioned eye color. It would be natural to think that no harm was done as a result of this announcement because every member of the community already knew this – any blue-eyed alien could see 4 blue eyes and 25 brown eyes, and any brown-eyed alien could see 5 blue eyes and 24 brown eyes.

The earlier problem asked why the village will be empty within a week. Here we ask, "What *new* information from the anthropologist doomed the community?" This question is significantly more challenging.

Strategies Utilized Understand the problem. Take inventory. Simplify. Enumerate the possibilities. Draw a diagram.

Discussion 15.7 As before, it is very useful to attack a simpler version of the problem. We'll start with only two blue-eyed aliens and the rest are brown eyed. Let's name the two aliens with a blue eye Alice and Bob. Before the anthropologist makes her announcement at the daily meeting, they are both aware, as is everyone else, that there is *at least* one blue-eyed alien in the community. However, Alice does not know that Bob knows this. As soon as the anthropologist makes her announcement, Alice knows that Bob knows there is at least one blue-eyed alien in the community. So, before anthropologist makes her announcement, Alice would guess that Bob doesn't see anyone with a blue eye. This is hard enough to wrap your head around with just two blue-eyed aliens – it is virtually impossible with five.

Let's take the next step and consider a total of three blue-eyed aliens: Alice, Bob, and Charlie. What do you think Alice would say if she was asked, "*What would Bob answer if you asked him how many blue eyes does Charlie think are in the community?*" The answer is zero. Alice thinks that Bob thinks that Charlie thinks that there are *no* blue-eyed aliens. As soon as the anthropologist makes her announcement, Alice now knows that Bob knows that Charlie knows there is at least one blue-eyed alien in the community. This is new information and this new information dooms the community. It is left to the reader to perform the analysis with five blue-eyed members: Alice, Bob, Charlie, Daniel, and Eddie.

It is worthwhile to note that nothing would have happened to the community if the anthropologist told every community member *individually* that there are both blue-eyed and brown-eyed members – as long as every member could not be sure that she gave others in the community the same information.

Debriefing 15.7 This problem, on its face, may appear to have little direct application. However, the ability to think about what other people are thinking is a tremendous skill to have, not only in business, but in your personal relationships as well. And, if you can think about what Alice thinks Bob thinks of you, then you have a rare and valuable skill.

Problem 15.8 Logicians Alice, Bob, and Charlie each wear a hat with a positive integer on it. They can see the numbers on the other two hats but not their own. Each logician knows that the number on one particular hat is the sum of the numbers on the other two hats and that all the numbers are positive integers. They are asked in turn if they can identify their numbers. In the first round Alice, Bob, and Charlie all in turn say they don't know. In the second round Alice is first to go and announces her number must be 50. What numbers are on Bob's and Charlie's hats?

Strategies Utilized Take inventory. Perform a gedanken. Enumerate the possibilities.

Discussion 15.8 When taking inventory, the students should realize two key things. One of these is that each person will have at most two possibilities for his or her number by looking at the other two. If Bob sees 15 and 25, he knows that the number on his hat is either 10 or 40. The second key fact is that Bob's and Charlie's answers in the first round eliminated one of Alice's two possibilities.

There are a number of questions a student can ask when performing a gedanken to solve this problem. One is, "*Can Alice ever tell what her number is on her first turn?*" The answer is yes; she will be able to tell if she sees two of the same number. If she sees 25 and 25, her number must be 50, because 0 is not a positive integer. This may seem trivial, but it is actually a key element of the solution.

Another gedanken is to guess at Bob's and Charlie's numbers and see if that can offer any insights. Knowing that Alice is 50, a reasonable start might be 49 and 1. So, Alice looks at 49 and 1 and thinks her number must be 48 or 50. Bob looks at 50 and 1 and thinks his number must be 49 or 51. Charlie looks at 50 and 49 and concludes that his number must be 99 or 1.

In this case, however, Bob's and Charlie's "no" answers do not eliminate the possibility that Alice's number is 48. What is needed is a set of numbers on Bob and Charlie that will make one of Alice's two possibilities impossible.

It is probably a good idea for the students to just keep guessing at the numbers and reason out what would happen – simply as a way to explore the problem space. After all, it is very easy to try some numbers to see what happens.

> **Teacher Tip**
> There are a couple of easier problems that may lead the students to the solution to this one. If the students have tried and failed to solve this one, ask them what would their answers be if Alice had 20, Bob had 30, and Charlie had 10. The students should be able to eventually realize that as soon as Alice can't tell, Bob knows he is 30. The next level of difficulty is when Alice has 20, Bob has 30, and Charlie has 50. Bob knows immediately that he must be 30 or 70 and Charlie knows immediately that he must be 10 or 50. As soon as Bob doesn't know, combined with the knowledge that Alice can't tell, then Charlie knows he can't be 10 and must be 50. If the students can't wrap their head around this one, put them in Charlie's position (literally, with some numbered hats, or even-numbered Post-it notes stuck to their forehead) and perform a gedanken with a 10 on Charlie. So, the assumed situation is
>
> | Alice | Bob | Charlie |
> |-------|-----|---------|
> | 20 | 30 | 10* |

(continued)

15 Grand Challenges

> Charlie is performing a gedanken asking, *"What if I had a ten?"*
> If Charlie were 10, Bob would know right away that he is 10 or 30. Since Alice doesn't know her number, Bob can't have 10 and must have 30. But Bob couldn't tell and therefore Charlie can't have 10 and must be 50. The actual problem is the next level of difficulty.

Discussion 15.8 (cont) Now let's return to the problem as stated and see what would happen if Bob has 20 and Charlie had 30.

Alice	Bob	Charlie
50	20	30

Before anyone answers:
Alice looks at 20 and 30 and knows she must be 10 or 50.
Bob looks at 50 and 30 and knows he must be either 20 or 80.
Charlie looks at 50 and 20 and knows he must be 30 or 70.

All three can't tell in the first round. In the second round, Alice is still looking at 20 and 30 and is thinking whether she can be 10 or 50.

At this point, Alice performs a gedanken and thinks, *"What if my number were 10?"*

Then Charlie would be looking at a 10 and a 20. Looking at a 10 and a 20 gives him two possibilities 10 and 30.

If he had 10, Bob would have known that he had 20 in the first round, because Bob would be looking at two 10s.

Since Bob didn't know and Charlie didn't know with the knowledge that Bob didn't know, Alice knows she can't have 10 and therefore must have 50.

> **Student Pitfall**
> It is not hard for the students to guess that the numbers on the other two hats are 20 and 30. This is something they might try early on in the process simply as a way to better understand the problem. If they do, it is unlikely that they will realize that it is the correct answer and dismiss it. Once dismissed, they will have trouble returning to it.

The best way to demonstrate this is to put a 10 on Alice and see what would happen. Again, if it is at all convenient, consider doing this literally with students and numbered hats (or the equivalent).

Alice	Bob	Charlie
10	20	30

Alice can't tell and Bob can't tell. However, if this were the case, Charlie would know he must be either 10 or 30 from the beginning. If Charlie were 10, Bob would be looking at two 10s and know he must be 20. Since Bob couldn't tell, Charlie can't be 10, so he must be 30. But, Charlie couldn't tell, so Alice concludes that she can't be 10 and must be 50.

Debriefing 15.8 This multilayer problem is a significant challenge for students to grasp. In this case, it is OK to help them wrap their heads around the problem by presenting two less complicated problems in an attempt to get them to solve the problem as presented. We recommend providing no hints to the problem as presented other than the two less complicated examples. If the students still can't solve the problem, that's OK. Having the solution is not important. The numbers 20 and 30 provide no value. Getting the answers without putting in the work develops mental stamina as much as buying a treadmill and not using it develops physical stamina. If you don't tell them the answer, there is the possibility they'll work it out sometime in the future.

Problem 15.9 A teacher has four cards – each has an integer greater than 1 written on it. He calls two students to the front of the room, Patrick and Samantha, and hands each of them a card and tells them to look at it without showing anyone else. He places the other two cards facedown on the desk without showing them to anyone. The teacher announces that the number on Patrick's card is the product of the numbers on the two cards facedown on the desk and that the number on Samantha's card is the sum of the numbers on the cards facedown on the desk.

The teacher then asks Patrick, "*Do you know Samantha's number?*" But before Patrick can answer, Samantha blurts out, "*There is no way Patrick can figure out my number just by looking at his number.*"

The teacher turns to Samantha and says, "*You are correct. You're very smart. Now wait your turn to answer.*"

The teacher then asks Samantha, "*Do you know Patrick's number?*" But before Samantha can answer, Patrick interrupts by saying, "*Samantha's number must be 17.*"

What are the two numbers on the cards that are facedown on the table?

Strategies Utilized Understand the problem. Take inventory. Simplify. Perform a gedanken. Enumerate the possibilities. Draw a diagram.

Discussion 15.9 This problem offers the students a tremendous opportunity to solve a very challenging problem that looks completely impenetrable at first look. Solving it will make a lasting and valuable impression on a young student.

Let's frame the problem space by performing a gedanken or two before we start trying to solve the problem. We know that Samantha has 17, but to establish a good baseline, let's explore numerous possibilities for Samantha. What if Samantha had

15 Grand Challenges

the number 7? If Samantha had 7, she would know that Patrick has either 10, which is 5×2, or 12, which is 4×3. If Patrick had ten, however, he would know that Samantha is looking at 7 because the only two positive integers greater than one that multiply to ten are 2 and 5. Therefore Samantha can't have 7 because she could not be sure that Patrick couldn't tell what her number is just by looking at his number.

Remember that Samantha's number must be such that every possible number for Patrick does not have a unique product of two integers that are bigger than one.

If Samantha had 8, she would know that Patrick had 12 (6×2), 15 (5×3), or 16 (4×4). But, if Patrick had 15, he would know Samantha must have 8. Therefore Samantha can't have 8 because she could not be sure that Patrick couldn't tell what her number is just by looking at his number.

If Samantha had 9, she would know that the possible numbers for Patrick are 14 (7×2), 18 (6×3), or 20 (5×4). But, if Patrick had 14, he would know that Samantha had 8 because the only two positive integers greater than one that multiply to give 14 are 2 and 7.

If Samantha had 10, she would know that the possible numbers for Patrick are 16 (8×2), 21 (7×3), 24 (6×4), and 25 (5×5). But, if Patrick had 21, he would know that Samantha had 10 because the only two positive integers greater than one that multiply to give 21 are 3 and 7. Patrick would also know that Samantha had a 10 if Patrick had a 25.

Now, if Samantha has an 11, she would know that Patrick had an 18 (9×2), a 24 (8×3), a 28 (7×4), or a 30 (6×5). Each one of these products has two different factors. 18 can be made by multiplying 2×9 or 6×3. 24 can be made by multiplying 12×2, 8×3 or 6×4. 28 can be made by multiplying 14×2 or 4×7. 30 can be made by multiplying 6×5 or 3×10. So, if Samantha has 11, she would know that there is no way for Patrick to know her number. If Patrick was looking at an 18, the sum of the two numbers on the table could be 11 or 9. If Patrick was looking at a 24, the sum of the two numbers on the table could be 14, 11, or 10. If Patrick was looking at a 28, the sum of the two numbers on the table could be 11 or 16. If Patrick was looking at a 30, the sum of the two numbers on the table could be 11 or 13. Now, some progress is being made.

Here is an abbreviated table of possibilities for a range of numbers for Samantha.

Sam	Possible numbers for Patrick						
7	10	12					
8	12	15	16				
9	14	18	20				
10	16	21	24	25			
11	18	24	28	30			
12	20	27	32	35	36		
13	22	30	36	40	42		
14	24	33	40	45	48	49	
15	26	36	44	50	54	56	
16	28	39	48	55	60	63	64

(continued)

Sam	Possible numbers for Patrick										
17	30	42	52	60	66	70	72				
18	32	45	56	65	72	77	80	81			
19	34	48	60	70	78	84	88	90			
20	36	51	64	75	84	91	96	99	100		
21	38	54	68	80	90	98	104	108	110		
22	40	57	72	85	96	105	112	117	120	121	
23	42	60	76	90	102	112	120	126	130	132	
24	44	63	80	95	108	119	128	135	140	143	144
25	46	66	84	100	114	126	136	144	150	154	156
26	48	69	88	105	120	133	144	153	160	165	168	169
27	50	72	92	110	126	140	152	162	170	176	180	182
28	52	75	96	115	132	147	160	171	180	187	192	195	196
29	54	78	100	120	138	154	168	180	190	198	204	208	210
30	56	81	104	125	144	161	176	189	200	209	216	221	224
31	58	84	108	130	150	168	184	198	210	220	228	234	238
32	60	87	112	135	156	175	192	207	220	231	240	247	252
33	62	90	116	140	162	182	200	216	230	242	252	260	266
34	64	93	120	145	168	189	208	225	240	253	264	273	280
35	66	96	124	150	174	196	216	234	250	264	276	286	294
36	68	99	128	155	180	203	224	243	260	275	288	299	308
37	70	102	132	160	186	210	232	252	270	286	300	312	322

The next highest number that Samantha can have that produces ambiguous products for every possible number for Patrick is 17. This is followed by 23, 27, 35, and 37. The table shows Patrick's possibilities for a range of numbers for Samantha.

Let's look at the "17" row to try to uncover how Patrick knew that Samantha was 17. The first thing that we know is that Samantha knows that Patrick can't tell what number he has, so Patrick knows that Samantha's possible numbers are 11, 17, 23, 27, 35, 37..... (It will be a significant challenge for any student to come up with a rule that generates all these numbers. The rule is that $n - 2$ is not prime.) The question now is what number can Patrick have so that he is certain that Samantha has 17?

Let's go over all seven of Patrick's possibilities one by one, all of which can be found in the "17" row of the table. What if Patrick has 30? Well, with 30, the sum could be 11 (6×5) or 17 (15×2), and 11 and 17 are both possibilities Patrick must consider.

Remember that at this point we are playing the role of Patrick, trying to figure out what Samantha's number could be knowing that Samantha knew that he could not figure out his number by looking at her number.

If Patrick had 42, the two numbers he must consider for Samantha are 17 (14×3) and 23 (21×2). The two positive integers that are greater than one and multiply to 42 are 17 (14×3) and 23 (21×2), and both 17 and 23 are possibilities Patrick must consider. Therefore, if Patrick were looking at a 42, he wouldn't know whether Samantha's number is 17 or 23.

What about 52? The sum of two positive integers that are greater than one and multiply to 52 are 17 (13×4) and 28 (26×2). However, the possibility of 28 is eliminated because if Samantha held 28, she would not be able to say that Patrick could not tell what number she had by looking at his number. If Samantha had 28, Patrick could have 187, which is the product of 11 and 17. If Patrick is looking at 187, he is certain that Samantha has 28 because the only two numbers greater than one that multiply to give 187 are 11 and 17. Similarly, Patrick could hold 115 and know that Samantha was 28 because 23 and 5 are the only two numbers greater than one that multiply to give 115.

Therefore, if Samantha had 28, she could not say with certainty that Patrick could not tell her number by simply by looking at his. So, the solution is that Patrick must have 52 because 52 is the only number that Patrick can have to be *sure* that Samantha's number is 17 knowing that Samantha knew he couldn't tell what she had by looking at his number. Therefore the numbers on the two cards on the table must be 13 and 4. However, let's continue for completeness.

The next possibility is 60. The sum of two positive integers that are greater than one and multiply to 60 are 17 (12×5) and 23 (20×3), and both 17 and 23 are possibilities Patrick must consider. So, if Patrick was looking at a 60, he couldn't tell if Samantha was 17 or 23.

Next is 66. The sum of two positive integers that are greater than one and multiply to 66 are 17 (11×6) and 35 (33×2). Since both of these are possibilities, Patrick can't say for sure that Samantha's number is 17; it might be 35.

What about 70? The sum of two positive integers that are greater than one and multiply to 70 are 17 (10×7) and 37 (35×2). Since both of these sums are possibilities, Patrick can't say for sure that Samantha's number is 17 – it might be 37.

Finally, let's consider 72. The sum of two positive integers that are greater than one and multiply to 72 are 17 (9×8) and 27 (24×3). Since both 17 and 27 are possibilities for Samantha's sum, Patrick's number can't be 72.

To summarize, the only possible product that Patrick can have that requires Samantha's number to be 17 is 52. When she looks at 17, she is sure that Patrick can't figure out her number. This fact, and the fact that Patrick is looking at 52, leads Patrick to conclude that Samantha's number must be 17. If Samantha has 17 and Patrick has 52, the two hidden numbers on the table must be 13 and 4.

Debriefing 15.9 Working persistently and diligently on this problem has a tremendous range of side benefits. The students will notice a number of patterns when trying to work with the numbers. The students will get practice thinking what others are thinking and renormalizing after each new bit of information is learned. If a student starts this one and doggedly struggles with it for three weeks before finally figuring it out without any help, it will be a life-changing event. After an accomplishment like that, the student will not shy away from challenges, he or she will have more confidence and more determination and even walk a little taller.

Problem 15.10 A group of 20 people (feel free to insert the number of students in your class here, as the solution is the same whether there are 10 or 100 students) are given the following challenge: they must devise a procedure to determine when every member of the group has entered the "light" room.

Here's how it works. After a meeting to discuss strategy, all 20 people will be isolated in 20 rooms. The 21st room contains two indicator lights. One is green and the other is red. Each light is controlled with a simple on–off button. The host will select one person at random from one of the twenty rooms to take to the "light" room. There the person *must* push exactly one of the two buttons. For example, if the red light is off and the green light is on, the person has the choice to push the red light button to toggle it on or the green light button to toggle it off. That's it. If both are on, the person must turn only one of the lights off by pushing one button. This person is then placed back in his or her private room. Then another person is chosen randomly. And so on. Once the game starts, the only time any members of the group receive any information is when they enter the light room and can see the state of the two lights. For example, one person might have visited the room five times and have the following information:

Visit	Red	Green
1st	On	On
2nd	On	On
3rd	Off	Off
4th	On	Off
5th	Off	Off

What strategy will allow the team to determine when everyone has been in the room at least once? That is, at some point, a team member will walk into the room, see the state of the lights, and be able to say, "*Now I know by the state of the lights, that all 20 people have been in here.*"

The original state of the switches is unknown to the members of the group of 20.

Strategies Utilized Understand the problem. Take inventory. Simplify. Perform a gedanken. Build a model.

Discussion 15.10 This is yet another problem that seems impossible at first glance. There just doesn't seem to be enough information available in the state of the lights. This is a problem that requires a complete understanding, a thorough inventory, and the ability to utilize System 2 thinking.

> **Teacher Tip**
> We had the electronics shop make us a substantial metal box with two push button switches – each of which produces a satisfying CLICK when pressed.

(continued)

15 Grand Challenges   315

> One toggles a red light on and off and the other toggles a green light on and off. It is a remarkable aid to solving the problem, because it allows the student to visualize the situation. When a student is able to walk up to the box and depress a switch to toggle the state of a light, they are able frame the problem more quickly. If you have some student interested in electronics, it would make a great project.

Discussion 15.10 (cont) The inventory that the students usually gather is as follows:
1. There are four possible states of the lights and the original state of the lights is unknown.
2. Since the original state of the lights is unknown, when people go into the room for the first time, they have no idea how many people have been in the room previously.
3. Since students are chosen randomly to enter the light room, they may enter twice in a row and they may be not chosen 50 times in a row.
4. Every time a student enters the light room, he or she must toggle exactly one switch.

There are several questions that students usually ask about leaving something in the room to let the other members know that they have been there, for example, spitting on the floor, scratching something, dropping something, leaving fingerprints, etc. Simply remind them that the only information any group gets after the challenge starts is the state of the lights they see when they enter the room.

> **Teacher Tip**
> Perhaps a good way to get the students to focus only on the information contained in the state of the lights is to get them to imagine a long list of the states of the lights like the example given and imagine what information could be exchanged in this manner if they had an agreed-upon strategy for changing the state of the lights when they entered the light room.

Discussion 15.10 (cont) The students should try the "Simplify!" technique here. Instead of 20 people, start with two, then three, etc. Another simplification is to assume that both lights are off at the start.

They can work in groups or have a class discussion. After some time, they should realize that the only way the problem can be solved is if there is a "captain" and the captain is the only one that is counting how many people have entered the room.

Once this breakthrough is made, the path to the solution is easier to see. When both lights start in the off position, here are the instructions the captain gives to the other 19 members. Switch the green light from "off" to "on" at the first opportunity to do so and thereafter toggle the red light. If you go into the room for the first time and find the green light on, toggle the red light. Every time the captain gets called into the light room, if the green light is on he adds one to his tally and resets it by switching the green light off. If the captain finds the green light off, he just toggles the red light without increasing his count. So, the state of the green light carries the information, whereas the red light is just a dummy switch. Once the captain counts to 19, he knows everyone has been in the room at least once.

Now we have to tackle the situation in which the original state of the lights is unknown. The problem with the previous instructions is that the captain doesn't know whether to start his tally if the green light is on when he enters the room for the first time; did someone switch it on or was it on originally?

To account for this, the captain has to modify his instructions only slightly. He says, "Switch the green light from "off" to "on" *twice*, before switching to toggling the red light." Now, if he counts up to 39, he can be assured that everyone entered the room at least once.

Debriefing 15.10 This is a very challenging problem. It is natural to want to provide hints to the students to help them along. However, the goal is not to get the students to find the answer. The goal is to get them use their System 2. The students are progressing the most when they are exploring the problem space aggressively looking for a way to make some progress on the problem. The usual metric for the progress of the students is whether or not they solved the problem. However, we would recommend that the indicator of the students' progress is how hard they are thinking and how much their problem-solving skills are developing. If the students can't make the breakthrough to solve this one, that's OK. It is much better to leave the problem unsolved than to tell them the answer. They don't have to solve it before moving on to another problem. We have gotten e-mails from students after the course was over informing us that they have finally solved the problem. In one case, a gentleman had graduated and been in the workforce for two years when he wrote to tell us of his *Aha!* moment.

Problem 15.11 Johnny Newman invites the world's top twenty-seven logicians to attend a conference to address the problems of humanity. During the preconference meet and greet, at which wine and cheese are served, Johnny puts a colored dot on the forehead of every participant. Each can see the color of the dots on all the other participants, but not his or her own. After the meet and greet, Johnny gathers everyone in a room and announces, "I'm going to ding this bell once every minute. When I ding the bell, if you have figured out – using logic alone – the color of the dot on your forehead, you must leave the room." At this point, one participant raises her hand and asks, "*So, everyone here will be able logically to deduce the color of his or her own dot?*" The organizer replies, "*Yes, this is the case. In less than ten minutes, I will be the only one in the room. My assistant is standing just outside the door to confirm that each participant that leaves has figured out his or her dot color.*"

15 Grand Challenges   317

When Johnny dings the bell for the first time, four participants get up and leave. The first person tells the assistant, "my color is chartreuse," and he is right! At the second bell, all the participants with red dots get up and leave the room. At the third bell, everyone looked around, but no one got up to leave. At the fourth bell, the participant who asked the question, her brother, and some others leave the room – and she and her brother had different color dots. The room is not yet empty. How many people are left in the room at this point?

Strategies Utilized Understand the problem. Take inventory. Simplify. Perform a gedanken.

> **Teacher Tip**
> Whenever we present this problem, we use an actual bell that has a nice, crisp "ding." This helps the students to think like a participant would rather than an "outsider." When we presented this problem at the Gedanken Institute for Problem Solving, the young students immersed themselves into the situation even more by getting some colored stickers that are used to tag file folders and stuck a number of them on the foreheads their classmates. This modeling allowed the students to visualize and frame the problem. The bell and forehead dots not only help the students make progress on the problem, they make it a lot of fun as well. Therefore, it might be helpful to have colored stickers available – but don't give them to the students before they ask for them.

Discussion 15.11 Once when we gave this problem, there were a lot of students guessing answers, but one was quietly thinking, obviously perplexed. When asked what she was thinking about, she replied, "*I'm trying to figure out how in the heck the first guy knew he was chartreuse!*" This is the key question.

> **Student Pitfall**
> The students are likely to drift away from the problem as stated simply because they cannot believe that the first person to leave could *logically* deduce he must be chartreuse. So, they will say that he saw it in a mirror, he saw the color in a reflection off a wine glass or he rubbed his forehead with his finger and some of the color came off on his finger. Or even, the dot color matches their shirt, or the color of the school they are from, or a lapel flower they were given upon their arrival. Try to keep the students thinking logically and assure them that all those who leave the room are logically certain of their colors; it is not a guess or a theory.

Discussion 15.11 (cont) To make progress on this one, the students can perform gedankens to explore the sample space and frame the problem. Some good questions include:
What if they all had the same color?
What if everyone had a different color?
How many colors were there?
How many people left on the second bell?

They can also simplify the problem by starting with two people, each with a dot on her forehead. Once this simplification is made, the path to the solution is relatively straightforward. When considering only two people, there are only two possibilities, their colors are the same or they are different. If they are different, say, black and blue, there is no way either of them can ever leave the room – and this was a condition of the problem. Dr. Newman assigned the dots so that everyone would be able to figure out his or her dot color and leave the room. This is the key part of the inventory of the problem that is often overlooked by the students. So, with two people, they must have the same color because it is the only way they will be able to leave.

The same is true of three people. They all must have the same color and they all must leave the room on the first bell.

With four people however, there are two possibilities. All four could have the same color or there could be two pairs of colors. The students should be able to put themselves in this situation (literally, if convenient). If a student saw three purple dots on the other three people, the student must have a purple dot because there is no other way he could possibly deduce his color otherwise. If the student saw two purple dots and a salmon dot, he would know he is salmon because there is no way the person with the salmon dot would ever be able to logically deduce that he was salmon unless he saw another salmon dot.

This observation was put succinctly by a student when he remarked, "*The person that knew he was chartreuse must have somehow been prompted to think of chartreuse.*"

Now we can return to the original problem. On the first bell, four people left. The first person to leave knew he was chartreuse. The only way he could know he was chartreuse is if he saw exactly one other chartreuse. After looking at everyone else's dots he reasoned, "*How is the only person I can see that has a chartreuse dot ever going to conclude that she must be chartreuse?*" The answer is that he must be chartreuse as well. The prompt for her to know that her dot is chartreuse is his chartreuse dot.

So, the four people that leave the room on the first bell are two pairs of dot colors that only appeared on two people.

> **Teacher Tip**
> Ask the students, "*When did the first guy to leave realize that he must be chartreuse? Was it before the first bell, during the meet-n-greet?*" The answer is that he knew he must be chartreuse as soon as Dr. Newman announced that everyone will be able to leave the room.

Discussion 15.11 (cont) What about the second bell? Again, the students can put themselves in the situation. Let's say that the student sees only two red dots, only two chartreuse dots, only two salmon dots, and at least three of all the remaining dot colors. At the first bell, the two chartreuses and the two salmons get up to leave, but the two reds just sit there. Why didn't those reds leave? It must be because they were looking at another red dot! Therefore the student in question and the two other reds now know they have red dots and leave on the second bell. So, there must have been exactly three people with red dots if they left on the second bell.

The fact that no one leaves on the third bell means that there were no groups of four of one color.

Let's say that you are in the room at this point, and you see four yellow dots, five green dots, and ten white dots. You wonder why the four yellow dots did not leave on the third bell and conclude that the other four people with yellow dots must be thinking the same thing and now they all know that they are yellow. Each of the four people with green dots is reasoning similarly because they see five yellow dots, four green dots, and ten white dots.

Therefore on the fifth bell the greens and yellows leave. The people with white dots were looking at five yellow, nine white, and five green and didn't know what color they are. When the greens and yellows leave on the fourth bell, all the whites know they must have white dots and they leave on the fifth bell. So, a distribution that would work is:

2 people have chartreuse dots
2 people have salmon dots
3 people have red dots
5 people have green dots
5 people have yellow dots
10 people have white dots

It is not immediately obvious that the 10 remaining people with white dots leave on the fifth bell – the students might think that they have to wait until the ninth bell.

Note that it would be impossible for a third group of five to leave the room on the fourth bell because that would leave only five people in the room which means they should have all left by the fourth bell irrespective of their dot color distribution.

Debriefing 15.11 This problem clearly demonstrates the usefulness of developing problem-solving skills. When we present this problem, the students are nonplussed at first. They see no possible way that a person can logically deduce that he *must* be chartreuse. However, an experienced problem-solver will not be stymied. A thorough inventory taking, a simplification of the problem, a few gedankens, and some persistence should eventually lead to the answer.

Problem 15.12 Jack is using the garden hose to fill up a watering can so his kids can water their garden. The can seems to be filling slowly, so Jack turns the water off and attaches a speed nozzle to the end of the hose to increase the speed at which the water exits the hose. How does this affect the time it takes to fill the watering can?

Strategies Utilized Recognize a pattern. Perform a gedanken.

> **Student Pitfall**
> It is very likely that the students will feel unqualified or untrained to answer this question. They may feel that this is a physics problem or a problem in fluid dynamics and unless they "know the formula" they will lack the confidence they need to attack the problem.

Discussion 15.13 This problem[1] can be solved by recognizing the similarity between reducing the size of the opening at the end of the hose and controlling a typical household faucet.

The only thing a high-speed nozzle does is to reduce the size of the opening. It is similar to putting your thumb over the end of the hose to increase the speed at which the water exits the hose, which increases the range of the water.

Now let's consider a typical household faucet. Turning the handle more increases the flow of water. Consider what happens when the handle is turned. Most students will realize that the turning the handle more opens a valve more. The flow of water is directly proportional to the size of the valve opening. Simply put, if you want more water to flow, increase the size of the hole through which it is flowing.

Since the nozzle decreases the size of the hole, it must decrease the volume flow rate as well.

> **Teacher Tip**
> If there is a garden hose somewhere in the school, this is a great experiment to perform. All that is needed is a high-speed nozzle and a stopwatch or just a clock with a second hand.

---

[1] This problem is taken from the book Meyer EF (2011) Naked physics – thinking problems for everyday people. Gedanken Publishing. ISBN 0-9654178-0-8.

15 Grand Challenges

Debriefing 15.12 Problem 10.3 asked what happens to the size of the hole in a metal washer when it is heated. Just as no formal training involving thermal expansion coefficients was necessary to get the answer, no knowledge of fluid dynamics is necessary to figure out what happens to the flow rate when a high-speed nozzle is attached to the end of a garden hose.

When students successfully tackle problems such as this one, they are more likely to take on challenges that seem to require specialized knowledge or training. Often, these problems will yield to a sustained logical attack involving simplification and a series of gedankens.

Problem 15.13 It is Christmas Eve at the Z house and the four Zs – Zachary, Zoey, Zeo, and Zayna – all want to open their presents from their Grandma in the evening rather than wait until the morning. Their dad agrees to give them a chance by presenting the following challenge. "*All the Zs have to go into the basement for ten minutes,*" he begins, "*I'm going to put each of the four presents in one of the four kitchen cabinets. I will invite the Zs up from the basement one at a time to check any TWO cabinets for their presents. There is no communication allowed among any of the Zs after they leave the basement. If all the Zs find their own presents in two tries, then all four Zs can open them tonight. If not, then everyone will have to wait until the morning.*" While in the basement, the Zs get together to try to devise a cabinet-checking strategy that will maximize the probability that all of them find their presents. What is the optimal strategy, and what is the probability that they all find their presents utilizing this strategy?

Strategies Utilized Understand the problem. Perform a gedanken. Simplify. Take Inventory. Draw a diagram. Build a model.

> **Teacher Tip**
> This is a great problem to perform with the students. That is, get four different boxes with easily removable lids to represent the four presents and four index cards with the names of four students on them. Send the four students out in the hall to discuss strategy and have them come back in one at a time to try to find their name by opening two presents. Alternatively, you could split the entire class into groups of four and have them discuss strategy simultaneously. After a sufficient planning period, prepare the presents and have the groups try the exercise. If you have prepared the students well and made the classroom atmosphere comfortable, it can be a lot of fun. After each group makes an attempt, engage the class to discuss the probability that their particular strategy is successful.

Discussion 15.13 To understand the problem, the students might have to read it through a couple of times. They might draw a diagram to better understand the problem, perhaps sketching four rectangles to represent each of the four cabinets, labeling them, say, A, B, C, and D. They might assign labels to the presents as well, perhaps P1, P2, P3, and P4. In many cases, when the problem is likely to take some time, it is a good investment to simplify the terminology as much as possible. So, let's use Z1, Z2, Z3, and Z4 for the children and P1, P2, P3, and P4 for their presents, respectively, and A, B, C, and D for the four cabinets.

It is a good idea to further frame the problem, by establishing a baseline or a starting point. This is the probability of success if the Zs do not devise any strategy and each simply guesses twice. If each child guesses two cabinets he or she has a 50/50 chance of finding his or her present. The probability that all four of the children are successful is $(1/2)^4$, which is 1/16.

Another framing calculation is to determine the total number of ways the four presents can be arranged in the four cabinets. If the students can't calculate this by multiplying, they certainly should be able to enumerate all 24 possibilities.

This problem will stump those without problem-solving experience because it is hard to see all the way to the answer, and without this vision, they will not be able to move forwards. However, not seeing the answer does not stop experienced problem-solvers. They just do something to see what they can learn.

> **Student Pitfall**
> Students need to appreciate the potential benefit of simply trying something. Sometimes problem-solving is similar to taking a machete and hacking your way through the jungle without seeing where you are going. Sometimes this leads directly to the answer, but most of the time it does not and you have to go back to the beginning. However, the trek down the wrong path almost always is progress towards the solution because either something is learned or experience and intuition are gained.

Discussion 15.13 (cont) Here is a reasonable starting strategy that can be evaluated:
Z1 tries to find P1 in cabinet A first then in cabinet B.
Z2 tries to find P2 in cabinet B first then in cabinet C.
Z3 tries to find P3 in cabinet C first then in cabinet D.
Z4 tries to find P4 in cabinet D first then in cabinet A.

Hopefully, the students make a table either before or immediately after they try to analyze this strategy. It is possible to get the answer without the table, but a table allows a visualization of the problem. Here is a table that shows all 24 possible distributions of the four presents for the purpose of determining how often this strategy is successful.

P1 in	A	A	A	A	A	A	B	B	B	B	B	B	C	C	C	C	C	C	D	D	D	D	D	D
P2 in	B	B	C	D	C	D	A	A	C	C	D	D	A	A	B	B	D	D	A	A	B	B	C	C
P3 in	C	D	B	B	D	C	C	D	A	D	C	A	D	B	A	D	A	B	B	C	A	C	A	B
P4 in	D	C	D	C	B	B	D	C	D	A	A	C	B	D	D	A	B	A	C	B	C	A	B	A
Z1	Y	Y	Y	Y	Y	Y	Y	Y	Y	Y	Y	Y												
Z2	Y	Y	Y		Y				Y	Y					Y	Y					Y	Y	Y	Y
Z3	Y	Y			Y	Y	Y	Y		Y	Y		Y			Y				Y		Y		
Z4	Y		Y				Y		Y	Y	Y			Y	Y	Y		Y				Y		A

The top four rows give the 24 distributions of the four presents in the four cabinets. The bottom four rows indicate when each child found his or her present with a Y for yes. This table makes a number of things clear. First, each row representing one of the four children has 12 Ys. This must be true irrespective of the cabinet-checking strategy.

The table also reveals that this strategy is successful in two of the 24 possible distributions – when the presents P1, P2, P3, and P4 are in cabinets A, B, C, and D and when they are in B, C, D, and A, respectively. So, this strategy has a 1/12 chance of success, which is already better than the 1/16 if the children guessed randomly.

It is reasonable to try something else at this point. What about this strategy?

Z1 tries to find P1 in cabinet A first then in cabinet B.

Z2 tries to find P2 in cabinet B first then in cabinet A.

Z3 tries to find P3 in cabinet C first then in cabinet D.

Z4 tries to find P4 in cabinet D first then in cabinet C.

Here is the table for that strategy:

P1 in	A	A	A	A	A	A	B	B	B	B	B	B	C	C	C	C	C	C	D	D	D	D	D	D
P2 in	B	B	C	D	C	D	A	A	C	C	D	D	A	A	B	B	D	D	A	A	B	B	C	C
P3 in	C	D	B	B	D	C	C	D	A	D	C	A	D	B	A	D	A	B	B	C	A	C	A	B
P4 in	D	C	D	C	B	B	D	C	D	A	A	C	B	D	D	A	B	A	C	B	C	A	B	A
Z1	Y	Y	Y	Y	Y	Y	Y	Y	Y	Y	Y	Y												
Z2	Y	Y					Y	Y					Y	Y	Y	Y			Y	Y	Y	Y		
Z3	Y	Y			Y	Y	Y	Y		Y	Y		Y			Y				Y		Y		
Z4	Y	Y	Y	Y			Y	Y	Y			Y		Y	Y				Y		Y			

This strategy works in four of the 24 cases, so we improved the previous solution even further. Here the presents P1, P2, P3, and P4 can be in cabinets A, B, C, D; A, B, D, C; B, A, C, D; and B, A, D, C, respectively. So, this strategy has a 1/6 chance of success.

At this point the students might take inventory yet again. They should realize that there will always be 12 Ys in each row and the goal is to implement a strategy that will maximize the number of times that four Ys appear in the same column. In other words, there is nothing that can be done to increase the number of Ys in each row, but there are strategies that produce more columns in which four Ys appear.

To go further, the students need to have a breakthrough (*Aha!* moment). They have to base their second cabinet choice on the owner of the present that they found in the first cabinet they checked. To start, a correlation or mapping between the Zs and the cabinet must be made. The natural way to do this is to assign Z1 to cabinet A, Z2 to cabinet B, Z3 to cabinet C, and Z4 to cabinet D. With these assignments, we can devise the following strategy:

Z1 tries to find P1 in cabinet A first and then looks in the cabinet that corresponds to the owner of present that was found in cabinet A.

Z2 tries to find P2 in cabinet B first and then looks in the cabinet that corresponds to the owner of present that was found in cabinet B.

Z3 tries to find P3 in cabinet C first and then looks in the cabinet that corresponds to the owner of present that was found in cabinet C.

Z4 tries to find P4 in cabinet D first and then looks in the cabinet that corresponds to the owner of present that was found in cabinet D.

So, if Z1 finds Z3's present in cabinet A, Z1 will next check in cabinet C. Similarly, if Z4 finds Z2's present in cabinet D, she will check cabinet B with her second attempt to find her present.

This "cyclic" strategy produces the following success table:

P1 in	A	A	A	A	A	A	B	B	B	B	B	B	C	C	C	C	C	C	D	D	D	D	D	D
P2 in	B	B	C	D	C	D	A	A	C	C	D	D	A	A	B	B	D	D	A	A	B	B	C	C
P3 in	C	D	B	B	D	C	C	D	A	D	C	A	D	B	A	D	A	B	B	C	A	C	A	B
P4 in	D	C	D	C	B	B	D	C	D	A	A	C	B	D	D	A	B	A	C	B	C	A	B	A
Z1	Y	Y	Y	Y	Y	Y	Y	Y					Y		Y							Y		Y
Z2	Y	Y	Y				Y	Y	Y								Y	Y	Y			Y	Y	Y
Z3	Y	Y	Y				Y	Y	Y			Y			Y		Y			Y	Y		Y	Y
Z4	Y	Y	Y				Y	Y	Y	Y				Y	Y			Y					Y	Y

Note that there are still 12 Ys in each row, meaning that each Z is still 50/50 to find their individual present. Here, however, the Ys are stacked up as much as they can possibly be. This strategy produces success in 10 of the 24 possible distributions, giving it a success rate of 5/12 or nearly 42 %. And this is the maximum.[2]

Note further that in 6 of the 24 instances, none of the four Zs find the right present. So, if this strategy is employed, none of the four Zs would find the right present 1/4 of the time. If they guessed randomly, the chance of this happening is only 1/16.

Debriefing 15.13 This problem demonstrates the importance of trying something to establish a baseline to get an idea of the architecture of the solution space. It also demonstrates that taking inventory more than once when solving a problem can be

---

[2] The proof of this fact is outside the scope of this book – it involves cycle length and group theory. If anyone wants to try, here is a link to start: http://mathworld.wolfram.com/PermutationCycle.html

15 Grand Challenges

useful and that drawing a diagram can be very helpful. Finally, it encourages the student not to give up when it doesn't seem like there is any strategy that will work.

Problem 15.14 Your Aunt Matilda is a professor of statistics and a card counter at blackjack. She recently had a good night at the casino and offers you an opportunity to select either a $100 casino chip or a $1 casino chip. Of course, you would prefer the $100 chip. She presents two bags – one blue and one red – that each contains five casino chips. One of the bags contains three $100 chips and two $1 chips and the other contains four $100 chips and one $1 chip, but you don't know which is which.

You get to reach into the bag of your choice and randomly select one chip to keep. However, before you select your chip, Aunt Matilda allows you to "sample" a bag by removing two chips from it without replacement. After this sampling, the bag you sampled from will have only three chips remaining and the other will still have five chips. It is at this point that you can take one chip from either bag that is yours to keep.
(a) If you sample two chips from the red bag and *both* are $1 chips, which bag do you select your chip from and what is your chance of selecting a $100 chip from that bag?
(b) If you sample two chips from the red bag and one is a $1 chip and the other is a $100 chip, which bag do you select your chip from and what is your chance of selecting a $100 chip from that bag?
(c) If you sample two chips from the red bag and *both* are $100 chips, which bag do you select your chip from and what is your chance of selecting a $100 chip?

Strategies Utilized Understand the problem. Take inventory. Simplify. Enumerate the possibilities. Perform a gedanken. Draw a diagram.

> **Student Pitfall**
> This is a challenging problem that requires sustained, careful thought. In a typical educational system, students usually get stimulus – response types of problems rather than problems that require a lot of thought before the calculations can begin. Therefore the students might give up or ask for help because they don't immediately see a direct path to the solution. However, thinking about the best way to attack a problem is a crucial problem-solving skill that needs to be developed.

Discussion 15.14 An inventory of the relevant facts includes the following:
1. There are two bags – one red and one blue – each of which has five casino chips.
2. One bag contains three $100 chips and two $1 chips and the other contains four $100 chips and a $1 chip and you don't know which is which.
3. Two chips are sampled from the red bag.

4. The goal is to maximize the probability of getting a $100 chip.

Starting with part (a), the students should be able to realize that once two $1 chips are selected from the red bag, it must have been the bag that contained two $1 chips and three $100 chips. Therefore, the red bag now contains three $100 chips. It is now 100 % certain that a chip selected from the red bag will be a $100 chip and 80 % (four out of five) that a $100 chip will be selected from the blue bag.

For part (b), the answer is much less obvious because after choosing two chips, we still don't know which bag was the one that originally contained four $100 chips. In fact, taking a vote among the class to see how many would choose from red bag and how many would choose from blue bag is likely to result in a somewhat even distribution. Stepping back from the problem to take a vote also gives the students some mental "breathing room" and is a good way to clear any frustration from not making progress on this part of the problem. That said, you know your students a lot better than the authors of this book. You should follow your instincts regarding the level of coaching, the degree of independence, and the optimal teaching style – keeping the goal of the course firmly in mind.

> **Teacher Tip**
> Whenever a problem does not have a clear answer, the teacher can make the students a lot more curious about it by asking the students to provide an educated guess. There are numerous ways to do this. In many cases, dividing the students up into groups of three and asking them to reach a consensus "gut-feeling" is likely to develop the students' thinking process and many times students are more likely to contribute in smaller groups rather than speak out in front of the entire class.

Discussion 15.14 (cont) The calculation involves a two-step process. The first step is to calculate the probability that each bag originally contained either three or four $100 chips, and the second step is to calculate the probability of selecting a $100 chip for each of those two possibilities.

That is, to calculate the probability of selecting a $100 chip from the red bag, which now contains only three chips, we have to calculate the probability that it is the bag that originally contained four $100 chips and one $1 chip and then multiply that by the probability of selecting a $100 chip under that assumption, *and* we have to calculate the probability that it is the bag that originally contained three $100 chips and two $1 chips and then multiply that by the probability of selecting a $100 chip under that assumption.

To start, let's determine the likelihood of selecting a $1 chip and a $100 chip from a bag that contains four $100 chips and one $1 chip.

The chance of picking the $1 chip first and a $100 chip next is

$$P = \frac{1}{5} \times \frac{4}{4} = \frac{1}{5}$$

The chance of picking the $100 chip first and a $1 chip next is

$$P = \frac{4}{5} \times \frac{1}{4} = \frac{1}{5}$$

So, the chance of getting one of each chip denomination from the 4-1 bag is 2/5. If the students are not sure of this calculation, perform a gedanken for, say, 5,000 draws of two chips from a bag that contains four $100 chips and one $1 chip.

Now let's calculate the probability of getting one of each denomination chip from a bag that contains three $100 chips and two $1 chips.

The chance of picking the $1 first and a $100 next is

$$P = \frac{2}{5} \times \frac{3}{4} = \frac{3}{10}$$

The chance of picking the $100 first and a $1 next is

$$P = \frac{3}{5} \times \frac{2}{4} = \frac{3}{10}$$

So, the chance of getting one of each chip denomination from the 3-2 bag is 6/10, which reduces to 3/5. Again, if the students are not sure of this calculation, perform a gedanken for, say, 5,000 draws of two chips from a bag that contains three $100 chips and two $1 chips. They should be able to calculate the number of times that two $100 chips are drawn, the number of times that one of each is drawn, and the number of times that both $1 chips are drawn.

Now, we know that we started with one bag containing three $100 chips and two $1 chips and the other containing four $100 chips and one $1 chip. Further we know that the probability of selecting one $100 chip and one $1 from a bag that contains three $100 chips and two $1 chips is 3/5 and the chance of selecting one $100 chip and one $1 from a bag that contains four $100 chips and one $1 chip is 2/5.

So, once one of each denomination chip is selected from the red bag, it is 3/5 that the red bag is the one that originally contained only three $100 chips and 2/5 that the blue bag is the one that originally contained only three $100 chips.

Student Pitfall

In general, there is a tendency to assume that simply because there are two possibilities, the likelihood of either occurring must be 50 %. In this problem, a student might reason "Well, the bag either contained four $100 chips or three $100 chips so the probability of each is 50-50." This is not the case after a $1 and a $100 chip are removed from the red bag – this is a dependent problem again. This can be made clear by referring the student back to part (a) of this problem or asking them to take inventory at each step. Once two $1

(continued)

chips are removed from the red bag, it is 100 % that the red bag was the one that contained two $1 chips.

Discussion 15.14 (cont) Now we can calculate the probability of choosing a $100 chip from the red bag, which now contains only three chips:

$$P = \left(\frac{3}{5} \times \frac{2}{3}\right) + \left(\frac{2}{5} \times \frac{3}{3}\right) = \frac{12}{15} = \frac{4}{5}$$

To become a good probabilistic thinker, it is important for the student to understand what each of these four fractions represents. Starting from the left, the three-fifths is the probability that the red bag is the one that originally contained three $100 chips and two $1 chips. The two-thirds is the probability of choosing a $100 chip if it was the one that originally contained three $100 chips and two $1 chips. The two-fifths is the probability that the red bag originally contained four $100 chips and one $1 chip, and the three-thirds is the probability of selecting a $100 chip from that bag if it was the bag that originally contained four $100 chips and one $1 chip.

Now let's perform a similar calculation to determine the likelihood of selecting a $100 chip from the blue bag, remembering that it still has five chips in it:

$$P = \left(\frac{3}{5} \times \frac{4}{5}\right) + \left(\frac{2}{5} \times \frac{3}{5}\right) = \frac{18}{25}$$

Again, it is important for the student to understand where each of the fractions comes from. Starting from the left, the three-fifths is the likelihood that the blue bag contains four $100 chips and one $1 chip. The four-fifths is the probability of selecting a $100 chip if it contains four $100 chips and one $1 chip. The two-fifths is the probability that the blue bag contains three $100 chips and two $1 chips, and the three-fifths is the probability of selecting a $100 chip if it contains three $100 chips and two $1 chips.

Since 4/5 is greater than 18/25, it is better to choose the chip to keep out of the red bag, which now contains only three chips, than to select a chip from the blue bag, which still contains five chips.

So, the answer to part (b) is as follows: when the two chips selected from the red bag are one of each denomination, it is better to select your chip to keep from the red bag. The probability of selecting a $100 chip from the red bag is 20/25, and the probability of selecting a $100 chip from the blue bag is only 18/25.

Whenever a problem is this complicated, it is always a good idea to perform a gedanken in an attempt to replicate the answer. So, let's consider what would happen if we pulled two chips out of the red bag for 100,000 separate trials. It is given that the red bag contains either four $100 chips and one $1 chip or three $100 chips and two $1 chips, and these two possibilities are equally likely. So, in the

15 Grand Challenges

100,000 trials, the red bag contains four $100 chips and one $1 chip in 50,000 of them, and in the other 50,000 trials, it will contain three $100 chips and two $1 chips.

When two chips are removed from a 4-1 bag for 50,000 trials, they will be one of each denomination in about 20,000 instances, and when two chips are removed from a 3-2 bag for 50,000 trials, they will be one of each in about 30,000 instances. Since it is given that the chips are one of each denomination in part (b), only these 50,000 trials will be considered because these are the ones in which one $1 chip and one $100 chip were removed. At this point the following table might be very helpful.

Situation	# Occurrences
Red bag is 4-1 bag and two $100 chips are removed	30,000
Red bag is 4-1 bag and one of each denomination is removed	20,000
Red bag is 4-1 bag and two $1 chips are removed	0
Red bag is 3-2 bag and two $100 chips are removed	15,000
Red bag is 3-2 bag and one of each denomination is removed	30,000
Red bag is 3-2 bag and two $1 chips are removed	5,000

This table shows that there were 50,000 trials in which the red bag was the 4-1 bag and 50,000 trials in which the red bag was the 3-2 bag. For part (b), we are only focusing on the 50,000 trials in which one of each chip denomination was removed.

Out of the 30,000 trials in which the red bag was the 3-2 bag, a $100 chip will be selected 20,000 times because two out of the three chips remaining in bag one are $100 chips. Of the 20,000 trials in which the red bag was the 4-1 bag, a $100 chip will be selected in every one of the 20,000 trials because all three chips remaining in the bag are $100 chips. So, of the 50,000 trials in which one of each denomination chip was removed, a $100 chip was drawn from red bag in 40,000 of them. This indicates that the probability of drawing a $100 chip from red bag after one of each denomination was removed is 4/5. So the gedanken is in agreement with the calculation.

Now let's consider switching to the blue bag in an attempt to select a $100 chip after removing one of each denomination from the red bag. As before, we only consider the 50,000 trials in which one of each denomination chip was removed from the red bag.

In the 30,000 trials in which the red bag was the 3-2 bag, a $100 chip will be selected from the blue bag in 24,000 of them because 4 of the 5 chips in blue bag are $100 chips. In the 20,000 trials in which the red bag was the 4-1 bag, a $100 chip will be selected from the blue bag in 12,000 of them because 3 of the 5 chips in the blue bag are $100 chips. So, out of the 50,000 trials in which one of each denomination was removed from the red bag, a $100 chip was selected from blue bag in 36,000 of them. This is a probability of 36,000/50,000, which reduces to 18/25. This is in agreement with the answer calculated using fractional probabilities. As before the calculation reveals that it is better to stick with the red bag than to switch to the blue bag.

With the knowledge gained by solving part (b), it will be easier to determine which bag to select from after removing two $100 chips from the red bag, which is part (c) of this problem.

The probability of removing two $100 chips from a bag that contains four $100 chips and one $1 chip is

$$P = \frac{4}{5} \times \frac{3}{4} = \frac{3}{5}$$

The probability of removing two $100 chips from a bag that contains three $100 chips and two $1 chips is

$$P = \frac{3}{5} \times \frac{2}{4} = \frac{3}{10}$$

So, when two $100 chips are removed, it is twice as likely that they were from the bag that originally contained four $100 chips and one $1 chip. This means that the probability that the two chips were removed from the 4-1 bag is 2/3, and the probability that they were from the 3-2 bag is 1/3.

Now we can calculate the probability of selecting a $100 chip from the red bag:

$$P = \left(\frac{2}{3} \times \frac{2}{3}\right) + \left(\frac{1}{3} \times \frac{1}{3}\right) = \frac{5}{9}$$

As before, it is important to understand what the fractions in the above equation represent. Again, starting from the left, the first two-thirds is the probability that the red bag originally contained four $100 chips and one $1 chip. The second two-thirds is the probability of removing a $100 chip from this bag if it originally contained four $100 chips and one $1 chip. The one-third is the probability that the red bag originally contained three $100 chips and two $1 chips, and the second one-third is the probability that a $100 chip is removed from this bag if it originally contained three $100 chips and two $1 chips.

Now let's calculate the probability of selecting a $100 chip from the blue bag after two $100 chips were removed from the red bag:

$$P = \left(\frac{2}{3} \times \frac{3}{5}\right) + \left(\frac{1}{3} \times \frac{4}{5}\right) = \frac{10}{15} = \frac{2}{3}$$

So, in this case, it is better to switch to the blue bag, even though it is more likely that the red bag was the one that originally contained four $100 chips.

We can again perform a gedanken and refer to the table constructed previously to confirm this answer. From the table we can see that out of 100,000 trials, two $100 chips will be removed from the red bag in 45,000 of them. In 30,000 of them the red bag was the 4-1 bag and in 15,000 of them the red bag was the 3-2 bag. Of the 30,000 trials in which the red bag was the 4-1 bag, a $100 chip will be selected in 20,000 of them because two of the three chips remaining in the bag are $100 chips. Of the 15,000 trials in which the red bag was the 3-2 bag, a $100 chip will be selected in 5,000 of them because one of the three chips remaining in the bag is a $100 chip. So, of the 45,000 trials, a $100 chip will be selected from the red bag in 25,000 of them, for a probability of 5/9. This is in agreement with the calculation using fractional probabilities.

For completeness, let's look at the probability of selecting a $100 chip from the blue bag when two $100 chips are removed from the red bag. Again, we are only considering 45,000 of the 100,000 trials.

Of the 30,000 trials in which the red bag was the 4-1 bag, a $100 chip will be selected from the blue bag in 18,000 of them because two of the five chips in the blue bag are $100 chips. Of the 15,000 trials in which the red bag was the 3-2 bag, a $100 chip will be selected from the blue bag in 12,000 of them because four of the five chips in the blue bag are $100 chips. So, of the 45,000 trials, a $100 chip will be selected from the blue bag in 30,000 of them, for a probability of 2/3. This is in agreement with the calculation using fractional probabilities.

It is a good idea to look at the table summarizing these results.

	Probability of getting $100 chip from red bag	Probability of getting $100 chip from blue bag	Best choice
(a) Two $1 chips were removed from red bag	1	4/5	Red
(b) One $100 chip and one $1 chip were removed from red bag	4/5	18/25	Red
(a) Two $100 chips were removed from red bag	5/9	2/3	Blue

And here is a table of results for 100,000 trials of sampling two chips from the red bag:

		Red bag		Blue bag		
Result	# of occ.	$1 chip removed	$100 chip removed	$1 chip removed	$100 chip removed	Best choice
(a) Two $1 chips were removed from red bag	5,000	0	5,000	1,000	4,000	Red
(b) One $100 chip and one $1 chip were removed from red bag	50,000	10,000	40,000	14,000	36,000	Red
(a) Two $100 chips were removed from red bag	45,000	20,000	25,000	15,000	30,000	Blue

Debriefing 15.14 This is a multilayer problem that requires a depth of focus that is not often required of students. The ability to marshal one's mental facilities on a single problem and maintain this focus without distraction is a key problem-solving skill. The best way to develop both mental strength and mental stamina is by working on challenging problems that require them.

Problem 15.15 You are the director of manufacturing for a major pharmaceutical company. There is strong evidence that one of the 1,000 bottles of cancer medicine in a recent batch has been tainted. The medicine is very expensive to manufacture, so you would rather not scrap the entire batch.

There is a test that will determine if a sample is tainted and it needs only a minute fraction of the medicine in a single bottle. However, the samples must be shipped to another facility to be tested and the test is very expensive to perform. Obviously, you would like to perform the minimum number of tests and make the minimum number of shipments needed to isolate the tainted sample.

What if time was crucial? If you had time for only one shipment, what is the minimum number of samples that need to be sent?

Strategies Utilized Understand the problem. Take inventory. Simplify. Perform a gedanken.

Discussion 15.15 As is often the case in the real world, there is more than one objective to be optimized. A simple example might involve hiring a new employee. You would like to hire the best possible candidate, but you do not want to pay a salary that is too high. How many candidates do you interview, how much do you spend on travel for the candidates, and how many people are going to interview the candidates?

In this problem there are two things to consider: the number of tests performed and the number of shipments. Ask the students to start by trying to find the minimum number of shipments, because this will allow a nice framing of the problem.

The first breakthrough the students usually have is the realization that they can test multiple bottles simultaneously. If one vial has a small sample from 100 different bottles, a single test will either clear all those bottles or reveal that the tainted one is among the 100.

A good first step in this direction is to follow a binary search technique and send one vial that has a small amount from bottles 1 to 500. If it is tainted, then send one vial that has a small amount from bottles 1 to 250. If it is not tainted, then send one vial that has a small amount from bottles 501 to 750. By continually cutting the possibilities in half, the tainted sample can be uncovered in just ten shipments.

So, this testing strategy will require ten shipments and ten tests.

Another possibility is the following: send ten vials, the first containing a small sample from bottles 1 to 100, the second vial containing a small sample from bottles 101 to 200, etc. If the seventh vial is tainted, the second shipment will contain another ten vials, the first with samples from 601 to 610, the second from 611 to 620, etc. If the second of these is tainted, then ten vials will be sent, each with an individual sample of bottles 611–620. At this point someone might suggest sending 12 vials instead of ten on the third shipment, one with samples from bottles 611 to

15 Grand Challenges

615 and another with samples from bottles 616 to 620. When these two are tested first, the process will be more efficient.

With this strategy, there are only three shipments and a maximum of 27 tests.

Note that, in this case, the maximum number of tests need not be performed. For example, if the second of the ten vials is tainted, there is no reason to test the remaining eight because only one vial is tainted.

After much discussion, you might volunteer that theoretically, the problem should be able to be solved in just one shipment of ten vials. The result of each test is a yes–no answer, and ten yes–no answers should be able to isolate one number in 1,000.

The inefficient part of the processes discussed so far is that the results are not independent of each other. That is, when each of ten vials contains 100 samples from 100 bottles, the result is likely to be that the sample is not tainted and if one is tainted, the others must not be. The key is to send vials that have independent results and each result chops possibilities in half.

This is a problem that very seldom is solved during regular hours in a class, but it is presented here because the solution is so fascinating that the students should appreciate it.

> **Teacher Tip**
> It would be nice if the students had some familiarity with binary numbers. If the students do not have any experience with base two, you can present a quick lesson on base two after they have struggled with this one for some time. Struggling with a problem is not a bad thing. Students will appreciate the beauty of the solution much more if they try for some time without coming close to the elegant solution of one shipment of ten vials.

Discussion 15.15 (cont) Here's how it is done. First, the 1,000 medicine bottles are renumbered from one to one thousand in base *two*. All of the base-two bottle numbers are extended to 10 digits simply by adding zeros to the left. For example, bottle number one is 0000000001 in base 2; bottle number two is numbered 0000000010. The third bottle is 0000000011, bottle #678 is 1010100110, and bottle number one thousand is 1111101000.

Next, ten empty sample vials are numbered one through ten in base ten, as shown below.

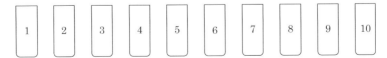

In this sampling scheme, the ten vials make a one-to-one correspondence with the ten digits of the bottle numbers. So, medicine bottle number 1010100110 (678 in base ten) is associated with the ten sample vials as follows:

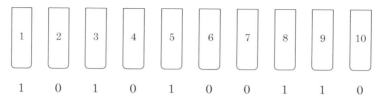

Drops of each of the 1,000 bottles are put into the vials according to the following scheme: if there is a one in the base-two bottle number that is associated with the vial, a sample from that bottle is placed in the corresponding vial. If there is a zero, no sample of that bottle is placed in the corresponding vial. So, samples of bottle number 678 are placed in vials, 1, 3, 5, 8, and 9. Further, only vial #10 contains a sample from bottle number one, which is numbered 0000000001. Similarly, samples of bottle number 1000100100 are placed in the vials numbered 1, 5, and 8. Samples of bottle number 1101011011 are placed in vials 1, 2, 4, 6, 7, 9, and 10. After the preparation, the ten vials are sent away for testing.

At this point it is a good idea to test students' understanding of the solution by challenging them with the following questions or others like them:
- The lab reports that vials 3, 4, 7, and 9 were tainted. Which bottle of medicine was tainted?
- The lab reports that vials 1, 3, 4, 8, and 10 were tainted. Which bottle of medicine was tainted?
- The lab reports that vials 1, 2, 6, and 7 were tainted. Which bottle of medicine was tainted?

So, this strategy results in ten tests and only one shipment.

Debriefing 15.15 While it is very challenging to solve this problem completely, students can make progress by figuring out how to combine the vials to make the process more efficient. As bonus, they get to learn about base two. The questions that ask to identify the tainted batch from specific test results are good candidates for tests or exams.

Reference

1. Meyer EF (2011) Naked physics – thinking problems for everyday people. Gedanken Publishing, Berea, Ohio. ISBN 0-9654178-0-8

Summary

> *We're less concerned about grades and transcripts and more interested in how you think. We're likely to ask you some role-related questions that provide insight into how you solve problems. Show us how you would tackle the problem presented – don't get hung up on nailing the "right" answer.*
> *– from Google's hiring page[1]:*

We hope that this book has provided the material and guidance needed to present a course on Puzzle-based Learning and that you have witnessed your students developing and becoming more confident as your course has continued. There is no doubt that the world needs more independent thinkers with the confidence to tackle challenging problems, and if your students have made progress down this path, then you have performed a great service for them and the human race as a whole!

We also hope that this text illustrates two important aspects of problem-solving: *be patient and persistent*. Remember that *every* problem-solver gets stumped now and then. We must have *patience* to investigate and understand the problem. Perhaps we might guess at the solution, but even if we are wrong, this guess might lead to an interesting discovery. However, we must *understand* the problem before going on to do anything else. Sometimes the answers seem so clear that they just *have to be right*, but they are not and, of course, we should not then rush to the solution.

Furthermore, experience is usually an asset, but we have to use it carefully. Do not allow yourself to go about solving problems using the same method every time. Be aware of your biases and prejudices, and do not let them get the better of you! Take inventory. Formulate a model. Think about different ways to represent (and model) the information that we have. Think "outside the box"– at least outside the structure implied by the problem. Simplify: concentrate on the essentials. We often face problems with so many details that we do not know where to start. Once we have identified the objective, we will have a much better chance of filtering out the "noise" in a problem and focusing in on the things that really matter.

---

[1] At the time of writing available at http://www.google.com/about/careers/lifeatgoogle/hiringprocess/

When we find a solution, we should remember that it does not have to be the end of the process; sometimes it is just a new beginning! Is the solution unique? Are there other possibilities? It might be that the solution found is no good and we have to try something else. We need to keep trying and not give up! *Persistence* is extremely important.

We also hope that students of this text will pick up additional skills or characteristics, such as scientific curiosity, creative thinking, and the understanding of basic terminology. But one of the most important additional skills – needed at least in science, business, and engineering – is the precision of language. After all, natural language is often used to formulate problems and describe solutions to these problems! So, by working through the many puzzles presented in this text, the students are trained (between the lines, so to speak) to understand the language of mathematics: What does it mean by x is a member of set X? What does it mean by one statement implies the other? What does it mean to prove something? What is a contradiction? What is probability? What does it mean "to optimize"? What does "for all" or "there exists" mean? How do we negate a sentence that starts with "For all x there exists y such that..."? We believe that after completing a puzzle-based learning course, students will develop an appreciation for the precision of formulating problems and their solutions.

We also hope that you and your students have had a lot of fun working on the puzzles. From our experience, the course evolves and matures each time it is taught. Lesson plans are developed, adjusted, and even dropped altogether. Materials are developed, puzzles are purchased, cabinets get filled, and the program expands. If your experience mirrors ours, you may observe the following: (a) puzzle-based learning is one of the most fun courses to teach and (b) puzzle-based learning is one of the most fun and memorable courses for students to take. (Many freshmen who have taken our course have commented four years later at their graduation that their Puzzle-based Learning course ranked in the top set of courses they took as an undergraduate.)

As the course matures, your school might host problem-solving nights, start a problem-solving club, or hold a problem-solving contest. Your class might post a Problem of the Month in the school newspaper. All of these indicate that you have changed the way that your students think, because they are now seeking to solve puzzles and problems as they *want to*, and not because they *have to* (i.e., to get a mark).

Above all, we hope that the readers feel that they have learned something *useful*, that they want to share with their colleagues and their students, and that this will make their future life as educators more rewarding and more productive.

Coming full circle, it is helpful to keep in mind that puzzles are just a means to an end. Puzzle-based learning forms a foundation for problem-based and project-based learning that your students will encounter in their future courses and career. As students work through your curriculum, we encourage you to explore the ideas of Puzzle-based Learning and their potential application in subsequent discipline-based courses you may teach. If you have taken one idea from this book and used it for teaching, or shared it with someone else and seen their delight at this new approach or knowledge, then we are happy.

List of Puzzles

All problems/puzzles are listed in the same order they appear in the text with a brief description/keywords. Some of them are named as "puzzles" (these are in the Introduction and in the first two chapters); some of them are icebreakers (we presented these in Chap. 3); some of them do not have any number assigned; and most of them are called "Problems," and they are identified by two numbers connected with a hyphen – e.g., Problem 7.4 means that this is the 4th problem presented in Chap. 7.

The purpose of the following list is to assist the reader in locating a particular problem – often we just remember a few keywords, e.g., we recall a good puzzle on some jars and chocolates, but we do not remember what this puzzle was about and we do not remember its placement. So the following list may assist forgetful readers in searching for the puzzle they have in mind.

Introduction
Puzzle – a river crossing puzzle (farmer, wolf, goat, cabbage, and boat). See also a discussion in Chap. 5 (between Problem 5.1 and Problem 5.2).

Chapter 1
Puzzle – dropping eggs from a 100-story building. See also Problem 11.6.
Puzzle – buying a shirt at a discount and applying sales tax.
Puzzle – throwing a biased coin.
Puzzle – the weight of the $10 gold coin.
Puzzle – a farmer selling 100 kg of mushrooms; 99 % moisture.
Puzzle – a metal washer with a hole in the middle. See also Problem 10.3.
Puzzle – backpackers sharing rice for dinner.

Chapter 2
Puzzle 1 – weighing out grain from 1 to 40 pounds in only one weighing using a two-pan balance.
Puzzle 2 – unreliable clock gaining exactly 12 minutes every hour; determining the correct time.
Puzzle 3 – five suspects for a stolen pie.
Puzzle 4 – 10 countries broken into chunks of letters.

Chapter 3
Icebreaker 1 – three-letters body parts.
Icebreaker 2 – checking one of two boxes.
Icebreaker 3 – writing down the lowest unique positive integer.
Icebreaker 4 – putting a sweatshirt inside out; handcuffs.
Icebreaker 5 – handcuffing two students.
Icebreaker 6 – *Bridg-it* game.
Icebreaker 7 – connecting a graph.
Icebreaker 8 – joining objects in a rectangular space.

Chapter 5
Problem 5.1 – *St. Ives* puzzle; a man with seven wives.
Problem – a river crossing puzzle (farmer, wolf, goat, cabbage, and boat). See also footnote 12 in the Introduction.
Problem – three jars; chocolate peanuts and chocolates; labeling the jars.
Problem 5.2 – calculating the number of different ways to fold four stamps.
Problem 5.3 – factory workers; ten-minute coffee breaks; calculating the probability of overlapping breaks.
Problem 5.4 – cutting a square and getting a rectangle with a larger area.
Problem 5.5 – the famous *Monty Hall* problem; three doors, one price.
Problem 5.6 – three humans and three zombies; crossing a river.
Problem 5.7 – a red car passes a blue car.
Problem 5.8 – producing bracelets; black and white beads; calculating the number of different six-beads bracelets.
Problem 5.9 – the surface of a soccer ball; calculating the ratio of number of pentagons to number of hexagons.
Problem 5.10 – calculating the speed of a car.
Problem 5.11 – cancer; tests; probability of having a cancer when the test is positive.
Problem 5.12 – survey among students; calculating the percentage of students who have never gone scuba diving.
Problem 5.13 – counting the numbers of small cubes that have some characteristics.

Chapter 6
Problem 6.1 – is a married person looking at an unmarried person?
Problem 6.2 – colony of algae; surface of a pond.
Problem 6.3 – idle Ivan; crossing a bridge; doubling money.
Problem 6.4 – two-player game; a pile of pebbles; *Nim*.
Problem 6.5 – cupcake sale; calculating the initial number of cupcakes.
Problem 6.6 – rectangular farm; three workers; crossing paths made in snow.
Problem 6.7 – analyzing samples of blood; 5 cc vial and 7 cc vial.

List of Puzzles

Chapter 7
Problem 7.1 – sequences of 4-digit numbers; number assignment.
Problem 7.2 – missing letter in a sequence.
Problem 7.3 – the *M-heart-8 sequence*; missing last symbol.
Problem 7.4 – *Nim* game; single pile of 100 pebbles. Teacher Tip section of this problem lists several other *Nim* games that consist of more than one pile of pebbles.
Problem 7.5 – a tournament with 512 tennis players; calculating the total number of games.
Problem 7.6 – calculating the number of segments generated by chords on a circle.

Chapter 8
Problem 8.1 – finding the colors of three hats placed on three men.
Problem 8.2 – two bears, black and white; determining some probabilities.
Problem 8.3 – three bags with black/white marbles; calculating the probability of having a black marble in a bag after removal of one marble.
Problem 8.4 – a game of *Chuck-a-Luck*; betting $1 on a number from 1 to 6.
Problem 8.5 – guessing the age of three sons; product of their ages is 36.
Problem 8.6 – two tribes on the island; liars and truth-tellers; who is telling the truth, who is lying? Reference to Problem 13.4 and Problem 13.5.

Chapter 9
Problem 9.1 – man and a picture; a father and his son.
Problem 9.2 – job interview; two piles of cards; separation of the deck of cards into two piles.
Problem 9.3 – riding a bike; average speed.
Problem 9.4 – 2-meter long, one-lane vine; 100 ants.
Problem 9.5 – pirate puzzle; 10 pirates divide 100 gold pieces.
Problem 9.6 – a village of one-eyed aliens; brown and blue eyes; new information; future developments in the village. See also Problem 15.7.

Chapter 10
Problem 10.1 – Samantha, Allison, and their age.
Problem 10.2 – a roller coaster and its speed.
Problem 10.3 – a metal disk with a hole in the middle. See also a puzzle from the beginning of Chap. 1. Reference to Problem 15.12.
Problem 10.4 – three bags with black/white marbles.
Problem 10.5 – three students; logic game; five hats (red and white).
Problem 10.6 – writing an integer number that is 4/5 of all written numbers (in a group).

Chapter 11
Problem 11.1 – a boy being late for school; calculating the probability of being late.
Problem 11.2 – crowd of people in the room; selection of a person; emergence of a pattern.

Problem 11.3 – estimating the value of π.

Problem 11.4 – four travelers crossing a bridge. See also Teacher Tip connected with this problem for versions with six and seven travelers.

Problem 11.5 – the shortest connection between cities on opposite sides of a river.

Problem 11.6 – breaking a rectangular chocolate bar into individual pieces. See also the first page of the Introduction.

Problem 11.7 – dropping eggs from a 36-story building. See also a puzzle listed at the beginning of Chap. 1.

Chapter 12

Problem – committee of ten people; calculating the probabilities of selecting the chairperson and the secretary.

Problem – rolling of a single six-sided die; two rolls of a single six-sided die.

Problem 12.1 – three boys and two girls in a line; calculating the probability of some particular arrangement. Reference to this problem is in the final paragraph of Sect. 11.1.

Problem 12.2 – four bags; two (white/black) marbles per bag; calculating the probability of a particular draw. Reference to Problem 8.3 and Problem 10.4. Reference to this problem is in the final paragraph of Sect. 11.1.

Problem 12.3 – calculating the probability of getting a "radar note." Reference to this problem is in the final paragraph of Sect. 11.1.

Problem 12.4 – handling a missed luggage; two misdirected suitcases. Reference to this problem is in the final paragraph of Sect. 11.1.

Problem 12.5 – calculating the probability of getting four deuces in five-card poker. Reference to this problem is in the final paragraph of Sect. 11.1.

Problem 12.6 – tossing pennies on parallel lines. Reference to this problem is in the final paragraph of Sect. 11.1.

Problem 12.7 – playing *Cash Wheel* game; optimal strategies for winning. Reference to this problem is in the final paragraph of Sect. 11.1.

Problem 12.8 – three contestants; trivia challenge; probabilities of winning the event. Reference to this problem is in the final paragraph of Sect. 11.1.

Problem 12.9 – speed dating; top choices; strategies and probabilities. Reference to this problem is in the final paragraph of Sect. 11.1.

Problem 12.10 – getting aces in players' hands of 13 cards. Reference to this problem is in the final paragraph of Sect. 11.1.

Problem 12.11 – airline flight with 100 seats; getting the assigned seat. Reference to this problem is in the final paragraph of Sect. 11.1. Reference to Problem 7.6.

Problem 12.12 – three monarch butterflies and two swallowtail butterflies; calculating the probability of swallowtail emerging as the third butterfly. Reference to this problem is in the final paragraph of Sect. 11.1.

Problem 12.13 – Yahtzee game; rolling five dice; getting straights on a single roll. Reference to this problem is in the final paragraph of Sect. 11.1.

Problem 12.14 – two bags; white and black marbles; choosing a white marble. Reference to this problem is in the final paragraph of Sect. 11.1. Reference to Problem 11.6.

List of Puzzles

Chapter 13
Problem 13.1 – a version of the Monty Hall problem with five doors. Reference to Problem 5.5.
Problem 13.2 – TV show; winning a car; estimating the chances to win; strategy.
Problem 13.3 – tossing a gun; finding a gun. Reference to the Monty Hall problem (Problem 5.5).
Problem 13.4 – liars and truth-tellers. Reference to Problem 8.6.
Problem 13.5 – liars and truth-tellers; days of the week.
Problem 13.6 – determining whether some statements about a single strand of beads are true or false.
Problem 13.7 – finding a set of integer numbers that sum to 50 and produce the largest product.
Problem 13.8 – finding a counterfeit among eight gold coins using a balance without weights.
Problem 13.9 – weighing grain from the range 1–40 with the minimum number of standard weights.
Problem 13.10 – finding the unique ten-digit autobiographical number.
Problem 13.11 – three college students; markers on faces; logical thought process.
Problem 13.12 – calculating the age of the monkeys.
Problem 13.13 – black and white hats; team-of-three intellectual competition; ten events.
Problem 13.14 – five hats of (possibly) different colors; planning a guessing strategy.

Chapter 14
Problem 14.1 – enumerate all the possible pentominoes. Reference to Problem 5.7.
Problem 14.2 – constructing a rectangle using five tetrominoes.
Problem 14.3 – *Pentomino Pasture Problem*; enclosing the largest area with a set of twelve pentominoes.
Problem 14.4 – arranging twelve pentominoes in groups of three.
Problem 14.5 – looking from a distance at the US Pentagon building.
Problem 14.6 – proving Pythagorean theorem.
Problem 14.7 – proving Pythagorean theorem.
Problem 14.8 – finding a radius of a circle inscribed in a square.
Problem 14.9 – estimating the distance to the horizon.
Problem 14.10 – finding the minimum length of the wire on a three-dimensional box.
Problem 14.11 – putting twenty pieces of cloth to form a square quilt.
Problem 14.12 – arranging square- and rectangle-shaped tabletops from two pieces.
Problem 14.13 – cutting a treasured rug to form a rectangle.
Problem 14.14 – calculating the area of a fish-shape figure.
Problem 14.15 – four corridors; animal shelter; placement of screens.
Problem 14.16 – cutting a piece of wood; making a square tabletop.

Chapter 15

Problem 15.1 – crafty crab; catching raccoon; finding a hole-checking protocol.

Problem 15.2 – weighing twelve cannonballs.

Problem 15.3 – guessing the color of a hat; seven red and two blue hats.

Problem 15.4 – value of the stock; limiting losses.

Problem 15.5 – guessing the color of a hat; five red and four white hats.

Problem 15.6 – cutting a rectangular block of wood into the minimum number of pieces that can be arranged to form a cube. Reference to Problem 14.12.

Problem 15.7 – one-eyed aliens; new information. Reference to Problem 9.6.

Problem 15.8 – three logicians; three hats with numbers; one number is the total of the other two.

Problem 15.9 – four cards; guessing two numbers; sum of two numbers; product of two numbers.

Problem 15.10 – entering the "light" room; two light indicators; finding the strategy.

Problem 15.11 – conference of logicians; color dots on foreheads.

Problem 15.12 – filling water can; speed nozzle. Reference to Problem 10.3.

Problem 15.13 – Zachary, Zoey, Zeo, and Zayna search for their Christmas presents.

Problem 15.14 – selecting $100 and $1 chips from two bags.

Problem 15.15 – testing 1,000 bottles for tainted medicine.

Index

A
Abbott, Derek, 27
Abbott, Robert, 43
ABC murders, 120
Aha, Insight, x, xi, 262
Alcuin, x
Alien, 145–147, 306, 307
Antoine de Saint-Exupery, 3
Assessment, xii, 5, 8, 10, 14–16, 19, 41, 47, 49, 52–54, 57–58, 60–62
Atman, Cynthia, 16
Autobiographical, 245, 246, 341
Autotelic, 4

B
Beads, 86, 87, 240–242, 261, 338, 341
Bridge, 6, 99, 170–177, 218, 221, 222, 242, 338, 340
Bridg-it, 28, 31, 32, 338
Buckminsterfullerene, 89

C
Cash wheel, 169, 209, 340
Chocolate Bar Puzzle, vii, 178–180, 232, 340
Christie, Agatha, 120
Chrysalides, 225
Chuck-a-Luck, 129, 339
Circuit board, 33
Claude Shannon, 29
Cluster, 15, 165
Cognitive apprenticeship
   methods, 38–39
   sequencing, 38–40
   sociology, 38, 40–42
Computational thinking (CT), 11
Confidence, xii, 39, 41, 47, 62, 80, 83, 202, 225, 234, 248, 313, 320, 335
Contract bridge, 218
Crab, 287–290, 342
Critical, 18, 44, 292
   thinking, viii, 4–6, 15, 116
CS Unplugged, ix, 168

D
Danesi, Marcel, 4
Davis, Bruce, 27
Deck of cards, 140–141, 192, 197, 202, 203, 218, 220, 339
Deus ex machina, 72
Dice, 46, 49, 129, 130, 169, 190–192, 227–228, 340
Domino, 260, 263
Dualism, 138

E
Effective assessment, 41, 54, 57–62
Egg-drop, 183
Einstein, Albert, 10, 69, 149
Einstellung Effect, 9
Eleusis, 43
Evaluation function, 170, 173

F
Fermat point, 283
Finite state automata, 71
Fisher, Alex, viii
Fogel, D.B, 184, 249

G
Gambler's fallacy, 10, 169, 295
Gardner, Martin, x, xi, xiii, 29, 33, 43, 187
Gates, Bill, vii
Gedanken, 66, 67, 149–158, 185, 195–197, 199, 201, 206, 207, 209, 214, 218,

219, 222, 224, 225, 227, 228, 230,
231, 233–236, 238, 239, 241–244, 246,
247, 250–253, 262, 264, 267, 268, 270,
275, 276, 280, 281, 283, 284, 290, 295,
296, 300, 301, 308–310, 314, 317, 318,
320, 321, 325, 327–330, 332
Gedanken Institute for Problem Solving, 27, 317
Gladwell, Malcolm, 9
Grading, 19, 24, 27, 57, 60
Graphs, 83, 300

H
Hexagon, 31, 87–89, 338
Holmes, Sherlock, 65, 95, 123

I
Icebreakers, 1, 21–35, 44, 156, 337, 338
Intuition, x, 9, 25, 76, 127, 161, 165, 171, 172, 175–176, 179, 187–189, 224, 272, 322

K
Kahneman, Daniel, 8, 10
Kerf, 93, 283, 303

L
Learning continuum, 1, 6

M
Marbles, 127–128, 153–155, 169, 194, 196, 197, 225, 226, 229–232, 239–240
Marking, 31, 38, 52, 60–62
Marking scheme, 57–61
Measurement, 51, 104, 259
Metal washer, 3, 152, 321, 337
Methods, 5, 11, 38–39, 43, 67, 72, 84, 87, 99, 123, 129, 130, 160, 167, 176, 180, 181, 184, 193, 218, 219, 221, 227, 228, 270, 277, 296, 335
Michalewicz, Zbigniew, xiii, 9, 116, 159, 184, 249
Microsoft, 4
Minimum wage, 149
Modeling a problem, 65
Monkey, 248, 249, 341

Monte Carlo, 160, 167–169, 296
Monty Hall problem, 40, 46, 79, 160, 233, 236, 237, 338, 341
Multiplication principle, 191, 199

N
Naked Physics, ix, 320
Nim, 100, 113–116, 338, 339

O
One-eyed aliens, 145, 306, 339, 342
Optimization, vii, xi, 66, 67, 123, 159–185, 232, 242, 243, 265, 282, 290
Oxford murders, 111, 113

P
Passenger, 199–201, 222–224
Pattern recognition, 66, 107–120, 214, 232, 236, 257, 265, 279
Peer teaching, xii, 62–63
Pentagon, 87, 88, 267, 268, 341
Pentominoes, 185, 261, 262, 264–266, 341
Pirates, 46, 142–145, 166, 339
Polya, Gyorgy, ix, 9, 108
Postage stamp, 74
Poundstone, William, 4
Project, Problem, Puzzle-based learning, 1, 6, 11, 14, 62, 336
Puzzle contest, 1, 17–20
Pythagorus, 185, 269–276, 282, 284, 341

Q
Quality measure, 170, 173, 175, 180

R
Real world, vii, viii, xi, 5, 7, 9, 11, 15, 34, 85, 93, 131, 159, 160, 166, 167, 187, 244, 332
Reverse auction, 27
Reverse raffle, 227
River, x, 70, 71, 80, 99, 175–177, 337, 338, 340
River crossing puzzle, 70, 337, 338
Rubik's cube, 93, 94, 260
Rug, 278, 341

S

Silicon Valley, 4
Simpsons, 111, 112
Simulation, 67, 74, 118, 159–184, 199, 201, 202, 205
Situated learning, 38
Soccer ball, 87–89, 111, 338
Steiner, 283
STEM, 5
Subway and underground systems, 84
Suitcase, 199, 200, 340
Swallowtail, 225–227, 340
System 1 thinking, 1, 8–10, 100, 150, 276
System 2 thinking, 1, 8–9, 21, 78, 100, 151, 201, 227, 260, 276, 287, 288, 314

T

Taking inventory, 67, 69–73, 209, 214, 222, 227, 252, 260, 288, 302, 303, 305, 307, 308, 310, 314, 320, 321, 324, 325, 332, 335

Taleb, Nassim, 188
Thinking
  critical, viii, 4–6, 15, 116
  system 1, 1, 8–10, 100, 150, 276
  system 2, 1, 8–9, 21, 78, 100, 151, 201, 227, 260, 276, 287, 288, 314
Trapezoid, 77, 78

W

Wolves from Sheep, viii

Y

Yahtzee, 227, 228, 340

Z

Zeng, Q., 27
Zombie, 80, 81, 338